# Springer Proceedings in Mathematics & Statistics

Volume 433

This book series features volumes composed of selected contributions from workshops and conferences in all areas of current research in mathematics and statistics, including data science, operations research and optimization. In addition to an overall evaluation of the interest, scientific quality, and timeliness of each proposal at the hands of the publisher, individual contributions are all refereed to the high quality standards of leading journals in the field. Thus, this series provides the research community with well-edited, authoritative reports on developments in the most exciting areas of mathematical and statistical research today.

Emmanuel Franck · Jürgen Fuhrmann ·
Victor Michel-Dansac · Laurent Navoret
Editors

# Finite Volumes for Complex Applications X—Volume 2, Hyperbolic and Related Problems

FVCA10, Strasbourg, France, October 30, 2023–November 03, 2023

Springer

*Editors*
Emmanuel Franck
Université de Strasbourg, CNRS, Inria
IRMA
Strasbourg, France

Jürgen Fuhrmann
Weierstrass Institute for Applied Analysis and Stochastics, Numerical Mathematics and Scientific Computing
Berlin, Germany

Victor Michel-Dansac
Université de Strasbourg, CNRS, Inria
IRMA
Strasbourg, France

Laurent Navoret
Université de Strasbourg, CNRS, Inria
IRMA
Strasbourg, France

ISSN 2194-1009 ISSN 2194-1017 (electronic)
Springer Proceedings in Mathematics & Statistics
ISBN 978-3-031-40859-5 ISBN 978-3-031-40860-1 (eBook)
https://doi.org/10.1007/978-3-031-40860-1

Mathematics Subject Classification: 65-06, 65Mxx, 65Nxx, 74Sxx, 76Mxx, 80Mxx, 85-08, 86-08, 92-08



This Springer imprint is published by the registered company Springer Nature Switzerland AG
The registered company address is: Gewerbestrasse 11, 6330 Cham, Switzerland

Paper in this product is recyclable.

# Organization

## Organizing committee

*Joubine Aghili*, Université de Strasbourg, IRMA UMR CNRS 7501, Inria, Strasbourg, France
*Clémentine Courtès*, Université de Strasbourg, IRMA UMR CNRS 7501, Inria, Strasbourg, France
*Emmanuel Franck*, Université de Strasbourg, CNRS, Inria, IRMA, Strasbourg, France
*Jürgen Fuhrmann*, Weierstrass Institute for Applied Analysis and Stochastics, Numerical Mathematics and Scientific Computing, Berlin, Deutschland
*Philippe Helluy*, Université de Strasbourg, IRMA UMR CNRS 7501, Inria, Strasbourg, France
*Victor Michel-Dansac*, Université de Strasbourg, CNRS, Inria, IRMA, Strasbourg, France
*Laurent Navoret*, Université de Strasbourg, IRMA UMR CNRS 7501, Inria, Strasbourg, France
*Andrea Thomann*, Université de Strasbourg, CNRS, Inria, IRMA, Strasbourg, France

## Scientific Committee

*Marianne Bessemoulin-Chatard*, LMJL, Université de Nantes, France
*Franck Boyer*, IMT, Université Toulouse 3—Paul Sabatier, France
*Konstantin Brenner*, Université Côte d'Azur, Inria, CNRS, LJAD, Nice, France
*Donna Calhoun*, Department of Mathematics, Boise State University, USA
*Claire Chainais-Hillairet*, Université de Lille, France
*Pietro Marco Congedo*, Inria, Centre de Mathématiques Appliquées, École Polytechnique, IPP, Palaiseau, France

# Preface

The Finite Volume method is a spatial discretization technique for partial differential equations based on the physical principle of local conservation. It has been successfully used in many applications, including fluid dynamics, magnetohydrodynamics, nuclear physics, or plasma physics. Motivated by their large applicability for real-world problems, finite volumes have been the purpose of an intensive research effort in the last decades, yielding significant progress in the design, the numerical analysis, or the practical implementation of the methods.

Research on finite volumes remains very active, as problems are becoming more and more complex. Among the current challenges addressed by the scientific community, let us mention for instance the design of robust (with respect to the mesh and/or physical parameters) numerical methods, of high-order methods, of methods preserving structural properties (positivity or dissipation of a prescribed quantity), or methods enhanced by machine learning approaches.

Previous conferences in this series have been held in Rouen (1996), Duisburg (1999), Porquerolles (2002), Marrakech (2005), Aussois (2008), Prague (2011), and Berlin (2014), Lille (2017) and Bergen (2020, online).

The present volumes contain the invited and contributed papers presented as posters or talks at the 10th International Symposium on Finite Volumes for Complex Applications held at the University of Strasbourg, from October 30 to November 3, 2023.

The first volume contains the invited contributions, as well as contributed papers focusing on finite volume schemes for elliptic and parabolic problems. Topics of the contributed papers include structure-preserving schemes, convergence proofs, and error estimates.

The second volume is focused on finite volume methods for hyperbolic and related problems, such as methods compatible with the low Mach number limit or able to exactly preserve steady solutions, the development and analysis of high-order methods, or the discretization of kinetic equations.

The volume editors thank the authors for their high-quality contributions, the members of the scientific committee for supporting the organization of the review process, and all reviewers for their thorough work on the evaluation of each of the contributions.

Finally, we warmly thank the "Cellule Congrès" of the University of Strasbourg, as well as the organizing committee for helping make this conference a great success.

Strasbourg, France — Emmanuel Franck
Berlin, Germany — Jürgen Fuhrmann
Strasbourg, France — Victor Michel-Dansac
Strasbourg, France — Laurent Navoret

# Contents

# Contributed Papers

# A Finite Volume Scheme with a Diffusion Control Parameter on Unstructured Hybrid Mesh: Application to Two-Dimensional Euler Equations

**Wassim Aboussi, Moussa Ziggaf, Imad Kissami, and Mohamed Boubekeur**

**Abstract** This paper presents a new approach to controlling the numerical diffusion in the finite volume characteristic (FVC) scheme. The approach is a generalization of an existing one-dimensional method and employs the backward method of characteristics to create interface states. The approach was assessed using two-dimensional Euler equations on unstructured hybrid meshes. The results demonstrate that the proposed approach is effective in controlling numerical diffusion and provides accurate solutions.

**Keywords** FVC Scheme · Unstructured hybrid meshes · 2D Euler equations · Diffusion control parameter

## 1 Introduction

In recent years, Computational Fluid Dynamics (CFD) has played a crucial role in various engineering and scientific applications, particularly in the modeling and simulating of complex flow phenomena. One of the essential components of CFD is the numerical scheme employed to solve the governing equations. The finite volume characteristics (FVC) scheme is a widely used method in this context, [1–5] owing to its ability to provide accurate and efficient results. However, there is always room for improvement in terms of adaptability and robustness, especially when dealing with

W. Aboussi (✉) · M. Ziggaf · M. Boubekeur
Sorbonne Paris Nord University, LAGA, CNRS, UMR, 7539 Villetaneuse, France
e-mail: aboussi@math.univ-paris13.fr

W. Aboussi
Faculty of Sciences Dhar El Mahraz, LAMA, Sidi Mohamed Ben Abdellah University, Fez, Morocco

M. Ziggaf
ENSAO, LMCS, Complexe Universitaire, B.P. 669, 60000 Oujda, Morocco

M. Ziggaf · I. Kissami
MSDA, Mohammed VI Polytechnic University, Lot 660, 43150 Ben Guerir, Morocco

 
E. Franck et al. (eds.), *Finite Volumes for Complex Applications X—Volume 2, Hyperbolic and Related Problems*, Springer Proceedings in Mathematics & Statistics 433,
https://doi.org/10.1007/978-3-031-40860-1_1

unstructured hybrid meshes. In this paper, we present a significant enhancement to the FVC scheme by introducing a novel formulation for the $\alpha$ parameter implemented in the manapy[1] code. Unlike the initial version of the scheme, where $\alpha$ was a freely chosen constant. The new formulation $\alpha_{ij}^{n}$ (see Sect. 3.2) depends on time and space, and is computed locally at each interface between cells $i$ and $j$, thereby providing improved robustness and adaptability.

This advancement paves the way for greater flexibility in handling complex computational scenarios with the FVC scheme, striking a balance between accuracy and computational efficiency.

In Sect. 2, we present the mathematical model employed. Following this, Sect. 3 introduces the finite volume characteristics method, incorporating the diffusion control parameter for unstructured hybrid meshes. Section 4 is dedicated to the application of the proposed technique to two-dimensional Euler equations. The paper is concluded in Sect. 5 with final conclusions and perspectives.

## 2 Governing Equations

We consider the two-dimensional Euler equations for modeling the dynamics of non-viscous gases. In the form presented hereafter, the variables retained to describe the conservation laws are the density $\rho$, the velocity vector $\mathbf{u} := (u, v)$, the total energy $E$, and the pressure $p$

$$\begin{aligned} &\partial_t \rho + \nabla \cdot (\rho \mathbf{u}) = 0, \\ &\partial_t (\rho \mathbf{u}) + \nabla \cdot (\rho \mathbf{u} \otimes \mathbf{u}) = -\nabla p, \\ &\partial_t (\rho E) + \nabla \cdot (\mathbf{u} \rho E) = -\nabla \cdot (\mathbf{u} p) \cdot \end{aligned} \tag{1}$$

For ideal gases, the energy of the system is related to these unknowns with the following state law

$$\rho E = \frac{1}{2}\rho |\mathbf{u}|^2 + \frac{p}{(\gamma - 1)}, \tag{2}$$

with $\gamma$ is the ratio of specific heats, it is a constant that depends on the particular gas, e.g. $\gamma = 1.4$ for air. To simplify the presentation of the mathematical model, we rewrite the system (1) in a conservative vector form

$$\partial_t \mathbf{W} + \nabla \cdot \mathbf{F}(\mathbf{W}) = 0, \tag{3}$$

where

$$\mathbf{W} = \begin{pmatrix} \rho \\ \rho u \\ \rho v \\ \rho E \end{pmatrix}, \qquad \mathbf{F}(\mathbf{W}) = \begin{pmatrix} \rho u & \rho v \\ \rho u^2 + p & \rho u v \\ \rho u v & \rho v^2 + p \\ u(\rho E + p) & v(\rho E + p) \end{pmatrix}, \tag{4}$$

[1] https://github.com/imadki/manapy.

with **W** is the vector of conservative variables and **F** is a non-linear flux function that can depend on space as well as on time.

# 3 A 2D Formulation for FVC Flux with $\alpha_{ij}^n$

In this section, we formulate the Finite Volume Characteristic scheme with a diffusion control parameter  to solve equation (1). The method uses an unstructured hybrid mesh to facilitate grid generation for complex configurations of the computational domain. As mentioned in previous works [1, 3, 4], the method requires two steps and can be interpreted as a predictor-corrector following the first idea presented in [6]. In the predictor step, the method of characteristics is used to determine the intermediate values to evaluate the numerical flux, while in the correction step, a fully conservative solution is obtained.

## 3.1 *The FVC Flux in Hybrid Unstructured Mesh*

Suppose that the computational domain $\Omega$ is divided into a finite number of control volumes $\Omega = (V_i)_{i\in I}$, with $I$ a finite set of indices. We will further assume that the mesh is admissible in the finite volumes discretization sense under cell-centered formulation [7, Sect. 10.1].

Integrating equation (3) over a control volume $V_i$, gives the following integral system

$$\frac{d}{dt}\int_{V_i} \mathbf{W} dV + \int_{\partial V_i} \mathbf{F}(\mathbf{W}) \cdot \mathbf{n}\, d\sigma = 0, \tag{5}$$

$\mathbf{n}$ is the normal vector to the edge $\partial V_i$ of the cell $V_i$ in the outward direction, $dV$ and $d\sigma$ are respectively the surface element and the length element. According to the framework of the finite volume method, the semi-discrete equation associated with (3) is defined as

$$\frac{d\mathbf{W}_i}{dt} = -\frac{1}{|V_i|}\sum_{j\in N(i)} |\gamma_{ij}|\Phi(\mathbf{W}_{ij}, \mathbf{n}_{ij}), \tag{6}$$

where

$$\mathbf{W}_i = \frac{1}{|V_i|}\int_{V_i} \mathbf{W} dV, \quad \text{and} \quad \Phi(\mathbf{W}_{ij}, \mathbf{n}_{ij}) \simeq \frac{1}{|\gamma_{ij}|}\int_{\gamma_{ij}} \mathbf{F}(\mathbf{W}) \cdot \mathbf{n}_{ij} d\sigma, \tag{7}$$

$\mathbf{W}_i$ represents the average quantity stored at the cell center of cell $V_i$. $\mathbf{W}_{ij}$ and $\Phi(\mathbf{W}_{ij}, \mathbf{n}_{ij})$ refer to the intermediate state and the numerical flux, respectively, computed at the interface $\gamma_{ij}$ between cells $V_i$ and $V_j$.

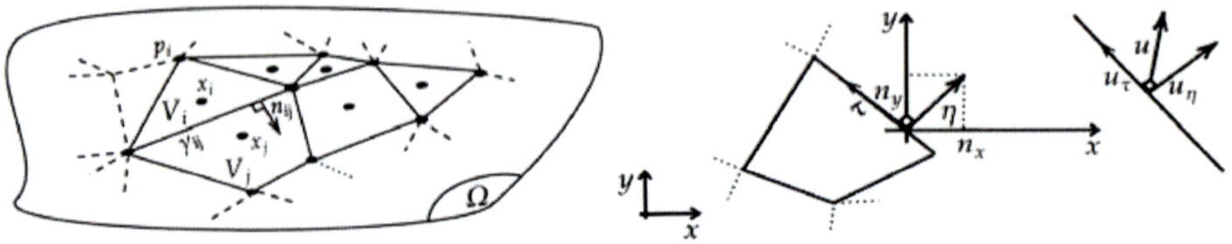


**Fig. 1** Generic control cells (left) and projected velocity frame (right)

**Notations:**

- $p_i$: vertex of $V_i$,
- $x_i$: centroid of the cell $V_i$,
- $\gamma_{ij}$: boundary face between the cells $V_i$ and $V_j$,
- $|\gamma_{ij}|$: length of $\gamma_{ij}$,
- $|V_i|$: volume of the cell $V_i$,
- $\partial V_i$: boundary of the cell $V_i$,
- $\mathcal{P}_i$: the perimeter of the cell $V_i$,
- $\mathbf{n}_{ij}$: unit normal to $\gamma_{ij}$, outward to $V_i$ such as, $\mathbf{n}_{ji} = -\mathbf{n}_{ij}$,

and $N(i)$ is the set of neighboring cells of the cell $V_i$. The spatial discretization of equation (6) is complete when a numerical construction of the flux $\Phi\left(\mathbf{W}_{ij}, \mathbf{n}_{ij}\right)$ is chosen. In general, this construction requires the solution of Riemann problems at the interfaces. From an algorithmic point of view, this procedure is very demanding. It may limit the application of the method when the solutions to the Riemann problem are complex or even impossible. This is why we have opted for constructing intermediate states $\mathbf{W}_{ij}$ using the method of characteristics, and the numerical flux is then calculated using the physical flux as $\Phi\left(\mathbf{W}_{ij}, \mathbf{n}_{ij}\right) := \mathbf{F}\left(\mathbf{W}_{ij}\right) \cdot \mathbf{n}_{ij}$. The basic idea of this method is the imposition of a non-regular spatial grid at the subsequent temporal instance and retracing the flow paths to the preceding temporal instance.

To construct the intermediate states $\mathbf{W}_{ij}$, we first rewrite the system in an advection form. For this, we use a projected velocity model whose velocity components are projected onto the frame $\mathcal{R} = (\mathcal{V}_i; \boldsymbol{\tau}, \boldsymbol{\eta})$ (Fig. 1 right). The advection model we found is an equation whose solution we can calculate in an almost exact way. Let $\boldsymbol{\eta} := (n_x, n_y)^T$ be the unit outward normal to the edge of the cell $V_i$ and $\boldsymbol{\tau} := (\tau_x, \tau_y)^T$ is the tangential vector, knowing that, $(\tau_x, \tau_y) = (-n_y, n_x)$. The projected velocity model associated with the Euler equations (3) is reformulated as (8), see [3, Sect. 3.2.1].

$$\frac{\partial \mathbf{U}}{\partial t}(t, X) + u_\eta(t, X)\frac{\partial \mathbf{U}}{\partial \eta}(t, X) = \mathbf{S}(\mathbf{U}(t, X)), \tag{8}$$

where

$$\mathbf{U} = \begin{pmatrix} \rho \\ \rho u_\eta \\ \rho u_\tau \\ \rho E \end{pmatrix}, \qquad \mathbf{S}(\mathbf{U}) = \begin{pmatrix} -\rho \partial_\eta(u_\eta) \\ -\rho u_\eta \partial_\eta(u_\eta) - \partial_\eta p \\ -\rho u_\tau \partial_\eta(u_\eta) \\ -\rho E \partial_\eta(u_\eta) - \partial_\eta(p u_\eta) \end{pmatrix}. \tag{9}$$

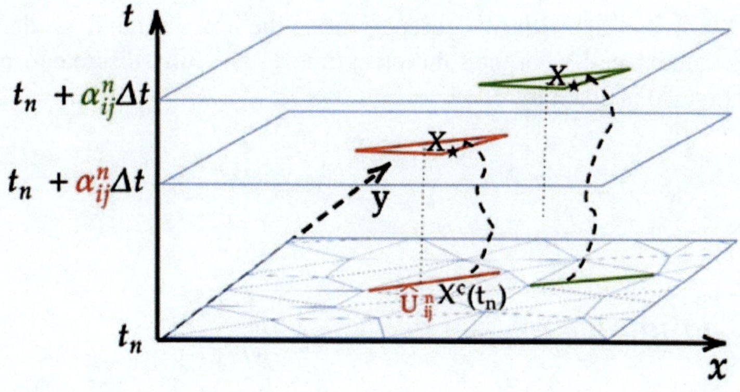

**Fig. 2** Sketch of the method of characteristics: A particle at $X_\star$ gridpoint is traced back in time to $X^c(t_n)$ where the intermediate solution $\hat{\mathbf{U}}_{ij}^n$ is interpolated

$\mathbf{U}$ is the projected conservative unknown, $u_\eta$ is the normal speed, and $\mathbf{S}(\mathbf{U})$ is the right-hand side that contains other terms of the system. The method of characteristics that is used traces the backward position at time $t_n$ of a particle that will reach the gridpoint point $X_\star$ (arrival gridpoint) of a fixed mesh at time $t_n + \alpha_{ij}^n \Delta t$ (see Fig. 2). The characteristic curves associated with (8) are the solutions of the following ODE.

$$\begin{cases} \dfrac{dX^c(t)}{dt} = u_\eta(t, X^c(t))\mathbf{n}, \quad t \in [t_n, t_n + \alpha_{ij}^n \Delta t], \\ X^c(t_n + \alpha_{ij}^n \Delta t) = X_\star. \end{cases} \tag{10}$$

A numerical integration method can calculate the solution (characteristic curves) of (10). Generally, the second-order methods lead to a non-linear equation in $X_c(t_n)$. A root-finding algorithm is subsequently required to solve this equation.

Once the characteristic curves are identified, the advection equation (8) can be solved using

$$\mathbf{U}(t_n + \alpha_{ij}^n \Delta t, X_\star) = \mathbf{U}(t_n, X^c(t_n)) + \int_{t_n}^{t_n + \alpha_{ij}^n \Delta t} \mathbf{S}(\mathbf{U}(s, X^c(s)))\, ds. \tag{11}$$

The projected solution on the interface $\gamma_{ij}$ is computed using the following formula

$$\mathbf{U}_{ij}^n = \hat{\mathbf{U}}_{ij}^n + \mathbf{I}_\mathbf{F}(\hat{\mathbf{U}}_{ij}^n), \tag{12}$$

where $\mathbf{I}_\mathbf{F}$ is the approximation of the integral in (11) and $\hat{\mathbf{U}}_{ij}^n = \sum\limits_{k \in V(X^c)} \beta_k(X^c)\mathbf{U}_k^n$, with $V(X^c)$ is the neighbor set of all cells that share at less a vertex with the face $\gamma_{ij}$ and $\beta_k(X^c)$ is the interpolation weight. The normal derivative terms in $\mathbf{S}$ are evaluated using the diamond scheme, see [3, Sect. 3.2.3] for more details. The time discretization of the semi-discrete equation (6) is based on an explicit scheme of order one. Let $t_n = n\Delta t$ with $t_0 = 0$ representing the initial time. If we denote $\mathbf{W}_j^n$ as the mean value in cell $V_j$ of the solution at time $t_n$. The fully-discrete formulation of the equation (5) is given by

$$\mathbf{W}_i^{n+1} = \mathbf{W}_i^n - \frac{\Delta t}{|V_i|} \sum_{j \in N(i)} |\gamma_{ij}| \Phi(\mathbf{W}_{ij}^n, \mathbf{n}_{ij}). \tag{13}$$

### 3.2 Local Diffusion Control Parameter $\alpha_{ij}^n$

In this section, we propose a control diffusion parameter that follows the same approach as the one we proposed for one-dimensional hyperbolic systems [8]. The objective of this article is to provide a simple way to extend this approach to the two-dimensional hyperbolic system. The construction of the diffusion control parameter is based on the analysis presented in [1].

On each cell $V_i$, we define the local diffusion control parameter $\alpha_i^n$ as follow

$$\alpha_i^n = \bar{\alpha}_i^n + \left(\frac{1}{2} - \bar{\alpha}_i^n\right) \psi_i \tag{14}$$

where

$$\bar{\alpha}_i^n = \max_{k \in N(i)} \left(\frac{|\gamma_{ik}|}{2\Delta t S_{ik}}\right), \quad \text{and} \quad S_{ik} = \max_l \left(\max\left(|\lambda_i^l|, |\lambda_k^l|\right)\right) \tag{15}$$

here $\lambda_i^l$ is the $l^{th}$ eigenvalue of normal flux, and $\psi_i$ is the Barth-Jespersen limiter function [9], we note that other limiter functions can be used. In this context, we have implemented a ratio that constrains the calculation involving Riemann invariants. Additional information can be found in [8]. Then the local diffusion control parameter for the interfaces is given by $\alpha_{ij}^n = \max(\alpha_i^n, \alpha_j^n)$.

## 4 Numerical Results

We test the method on a Cartesian mesh and a hybrid mesh. For time discretization, we use a first-order Euler scheme, and the time step $\Delta t$ is given as follows

$$\Delta t = Cr \min_{i,j \in I} \left(\frac{\gamma_{ij}}{\sqrt{2\alpha_{ij}^n \Lambda_{ij}^n}}\right). \tag{16}$$

$Cr$ is the Courant number and $\Lambda_{ij}$ is the spectral radius of the normal flux.

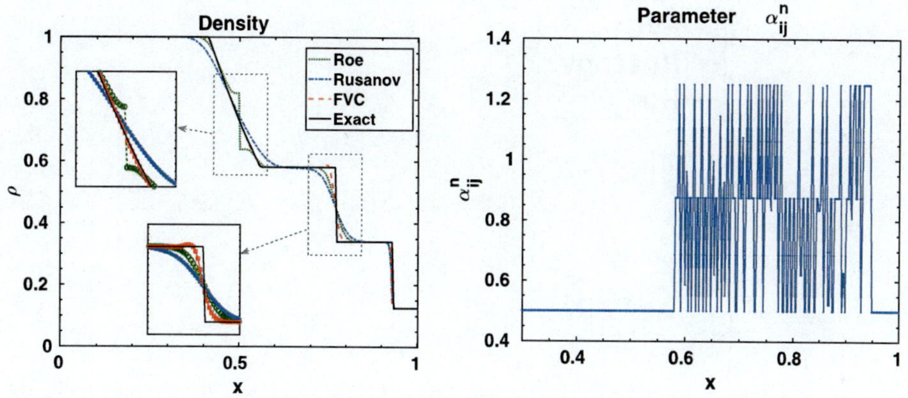


**Fig. 3** 2D Sod shock tube: cross-section of the density distribution (left) and cross-section of the control parameter $\alpha^n_{ij}$ variation (right) along the $x$-direction at time $t = 0.2s$

## 4.1 2D Sod Shock Tube Test

This section examines the one-dimensional shock tube problem in a two-dimensional setting $\Omega = [0, 0.06] \times [0, 1]$ and the initial condition is defined as

$$(\rho^0, u^0, v^0, p^0) = \begin{cases} (1, 0.75, 0, 1) & \text{if } x \leq 0.5, \\ (0.125, 0, 0, 0.1) & \text{otherwise}. \end{cases} \tag{17}$$

The numerical results presented in this section were calculated with a Cartesian mesh composed of 600×40 cells and $Cr = 0.8$.

In Fig. 3 (left), we present a cross-section in the x-direction of the numerical results computed using our method, Roe and Rusanov. In Fig. 3 (right), we present $\alpha^n_{ij}$ at t = 0.2s.

As the 1D shock tube test [8], we note that the shock, the contact, and the rarefaction are better captured using the FVC scheme. Moreover, the FVC scheme (unlike the Roe scheme) does not generate an artificial shock around the sonic point.

In Fig. 4 we compare the $L^1$ error using different schemes. It confirms that FVC method is more accurate than Rusanov scheme, and Roe scheme with Harten entropy correction [10].

## 4.2 2D Explosion Test

Here, we solve a two-dimensional explosion problem on the computational domain $\Omega = [-1, 1] \times [-1, 1]$ and the following initial condition

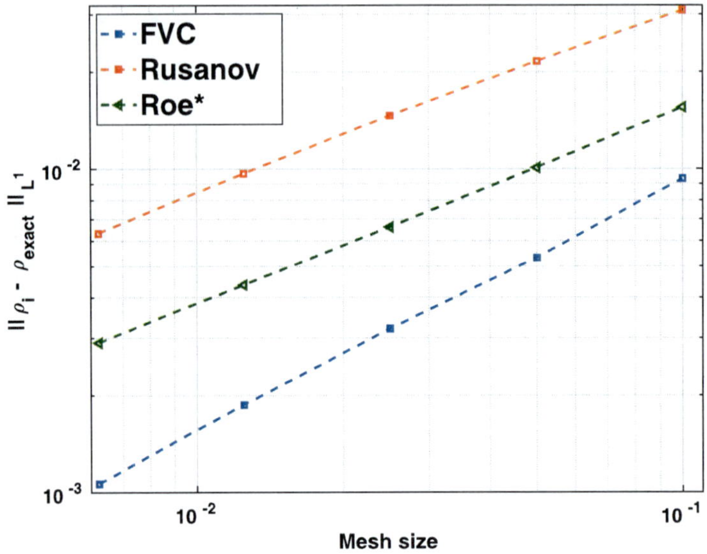


**Fig. 4** $L^1$ error (logarithmic scales) for 2D sod shock problem at time t = 0.2s

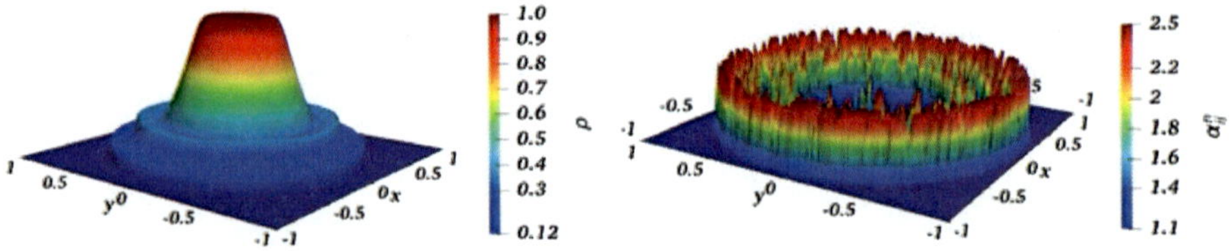

**Fig. 5** 2D explosion: the bird's eye view of density distribution (left) and bird's eye view of the local control diffusion parameter $\alpha_{ij}^n$ variation (right) at t = 0.2s

$$(\rho^0, u^0, v^0, p^0) = \begin{cases} (1, 0, 0, 1) & \text{if } \sqrt{x^2 + y^2} \leq 0.5, \\ (0.125, 0, 0, 0.1) & \text{otherwise}\cdot \end{cases} \tag{18}$$

The numerical results presented in this section were calculated with a hybrid mesh composed of 22 236 non-uniform cells and $Cr = 0.8$. We use the numerical solution obtained by Roe scheme on $10^6$ cells as a reference solution. Figure 5 shows the bird's eye views of the density (left) and the local diffusion parameter (right) at t = 0.2s. To better understand these results, we plot the cross-section in Fig. 6, which shows us that the method is more accurate than Roe and Rusanov schemes (left). In addition, the right figure shows us that $\alpha_{ij}^n$ adapts itself to get a less diffused solution. The FVC scheme exhibits a minor oscillation at the end of the rarefaction wave; this oscillation vanishes as the number of cells is increased.

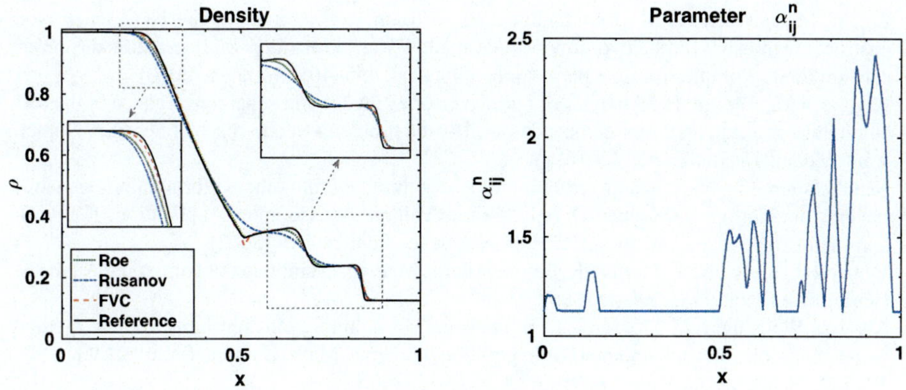


**Fig. 6** 2D explosion: cross-section of the density distribution (left) and cross-section of the control parameter $\alpha_{ij}^n$ variation along the radial direction(right) at t = 0.2s

# 5 Conclusions

In conclusion, this paper presents a significant enhancement to the FVC scheme by introducing the novel formulation of the $\alpha$ parameter, which was previously a free choice constant in the initial version of the scheme [1]. With the $\alpha_{ij}^n$ formulation, we have successfully made the FVC scheme more robust and adaptable, particularly for unstructured hybrid meshes. The numerical results of this study demonstrate remarkable shock resolution and high accuracy in smooth regions, while effectively eliminating nonphysical oscillations in proximity to shock zones. Furthermore, the absence of entropy defects highlights the robustness and reliability of the developed scheme. Although the numerical calculations have been focused on the Euler equations, the versatility of the present scheme allows for direct extension to address various fluid flow problems in both 2D and 3D dimensions. Future investigations will delve into these and other related issues.

**Acknowledgements** The authors thank Professor F. Benkhaldoun and Professor I. Elmahi for fruitful discussions, orientations, and helpful comments.

# References

1. Benkhaldoun, F., Seaïd, M.: A simple finite volume method for the shallow water equations. J. Comput. Appl. Math. **234**(1), 58–72 (2010)
2. Benkhaldoun, F., Sari, S., Seaïd, M.: Projection finite volume method for shallow water flows. Math. Comput. Simul. **118**, 87–101 (2015)
3. M. Ziggaf, I. Kissami, M. Boubekeur, F. Benkhaldoun, A well-balanced fvc scheme for 2d shallow water flows on unstructured triangular meshes. Adv. Appl. Math. Mech. (2013). https://doi.org/10.4208/aamm.OA-2022-0113

4. Ziggaf, M., Boubekeur, M., Benkhaldoun, F., El Mahi, I., Kissami, I.: The FVC scheme on unstructured meshes for the two-dimensional shallow water equations. In: International Conference on Finite Volumes for Complex Applications, pp. 455–465. Springer (2020)
5. Baccouch M., Temimi H.: Analysis of optimal error estimates and superconvergence of the discontinuous galerkin method for convection-diffusion problems in one space dimension. Numer. Anal. Model. **13**(3):403–434 (2016)
6. Benkhaldoun, F.: Analysis and validation of a new finite volume scheme for nonhomogeneous systems. In: Herbin, R., Kroner, D. (eds.), Finite Volumes for Complex Applications IV: Problems and Perspectives, pp. 269–276. Hermes Science Publications (2002)
7. Eymard, R., Gallouët, T., Herbin, R.: Finite Volume Methods: Handbook of Numerical Analysis, Ciarlet, P.G., Lions, J.L. (eds.) (2000)
8. Aboussi, W., Ziggaf, M., Kissami, I., Boubekeur, M.: A highly efficient finite volume method with a diffusion control parameter for hyperbolic problems. Math. Comput. Simul. **213**(1):177–193 (2023). https://doi.org/10.1016/j.matcom.2023.05.023
9. Barth, T., Frederickson, P.: Higher order solution of the Euler equations on unstructured grids using quadratic reconstruction. In: 28th Aerospace Sciences Meeting, p. 13 (1990)
10. Harten, A.: High resolution schemes for hyperbolic conservation laws. J. Comput. Phys. **135**(2), 260–278 (1997)

# Flow of Newtonian Fluids in a Pressurized Pipe

**Laila Baroukh and Emmanuel Audusse**

**Abstract** The Preissmann slot model is a hyperbolic model of Saint-Venant type that allows for modeling the transitions between free surface flows and pressure flows with a wave celerity corresponding to each type of flow. Despite its simplicity and a wide range of applications, numerical solutions of the Preissmann slot model suffer from spurious oscillations, particularly when flow conditions change from free surface flow to pressurized flow. To overcome this problem, we propose a modification of the finite volume characteristics (FVC) scheme by defining the diffusion term $\alpha$ with a self-adapting technique using limiter-type functions.

**Keywords** Preissmann slot · Free surface flow · Pressure flow · Wave celerity · Saint-Venant equation · Finite volume characteristics (FVC) scheme

## 1 Introduction

The Saint-Venant equations are commonly used to model free-surface fluid flows but they do not apply to modeling flows in closed pipes. The main obstacle is related to the abrupt transition from one flow regime to another, resulting in two distinct formulations of the flow. More precisely, the flow can transition from a free-surface regime, where the wave velocity is finite, to a pressurized regime, where the velocity is infinite and the Saint-Venant equations are no longer valid.

To unify the entire system, Preissmann [4] proposed adding a fictitious slot to model the transitions between free-surface and pressurized flows. This method transforms the entire system into a hyperbolic system of Saint-Venant type where infinite velocities are replaced by arbitrarily large ones, but this can lead to the appearance of numerical oscillations, see [3]. In that work, the authors propose ways to suppress these oscillations, mainly based on the possibility to enlarge the stencil of the numer-

L. Baroukh (✉) · E. Audusse
Laboratoire d'analyse, géométrie et application, Université Sorbonne Paris Nord, 99 Av. Jean Baptiste Clément, 93430 Villetaneuse, France
e-mail: baroukh@math.univ-paris13.fr



E. Franck et al. (eds.), *Finite Volumes for Complex Applications X—Volume 2, Hyperbolic and Related Problems*, Springer Proceedings in Mathematics & Statistics 433,
https://doi.org/10.1007/978-3-031-40860-1_2

ical scheme. We propose here another technique that allows us to keep a two-point stencil.

Our case study focuses on a hydraulic issue involving the filling of a bore in a pipe. This physical process marks the transition between a partially filled free-surface flow and a fully pressurized one. Our contribution is to propose a variant of the well-known finite volume characteristics scheme (FVC) introduced by Benkhaldoun and Seaïd [1] that consists in adding a self-adapting diffusion parameter that is sensitive to the flow transition and thus allows us to propose a scheme which is oscillation free near this transition.

The article is structured as follows: Sect. 2 is dedicated to a brief description of the Preissmann slot model. Section 3 presents the main steps of the finite volume characteristics (FVC) scheme and introduces the self-adapting diffusion term. Section 4 is devoted to the numerical simulations applied to the problem of filling a rectangular conduit. Finally, Sect. 5 completes this paper with some conclusions.

## 2 Governing Equations and Preissmann Slot Model

The Preissmann slot method [4] consists in placing an extremely narrow slot $T_s$ with infinite height on top of the conduct, effectively creating an open channel with a composite cross-section. The depth of water within this slot indicates the piezometric head of the pressurized flow within the original conduct, as illustrated in Fig. 1.

As a consequence, we consider the following Saint-Venant type model:

$$\begin{cases} \dfrac{\partial A}{\partial t} + \dfrac{\partial Q}{\partial x} = 0, \\ \dfrac{\partial Q}{\partial t} + \dfrac{\partial}{\partial x}\left(\dfrac{Q^2}{A} + gI_1\right) = gA(S_0 - S_f) + gI_2, \end{cases} \tag{1}$$

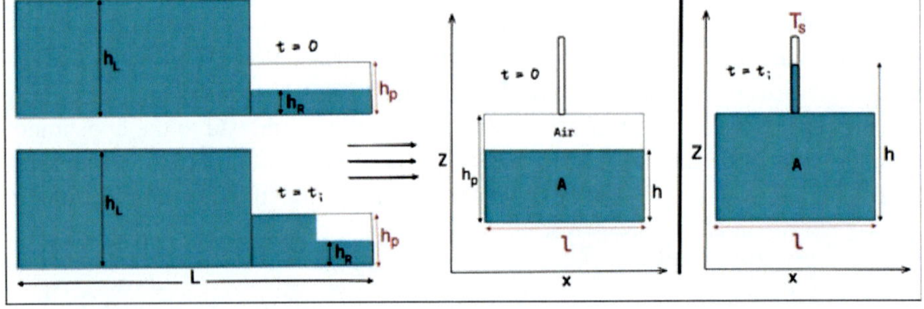


**Fig. 1** Filling of a closed pipe

with

$$I_1(h) = \int_{-h_b}^{h} (h - Z) l_(x, Z)\, dZ, \qquad I_2(h) = \int_{-h_b}^{h} (h - Z) \frac{\partial l(x, Z)}{\partial x}\, dZ,$$

where the unknowns are $A$ the wetted area in $m^2$ and $Q$ the volume flow rate $Q = Au$ in $\mathrm{m}^3.\mathrm{s}^{-1}$ and parameters are $g$ the acceleration of gravity in $m.s^{-2}$, $S_0$ is the bed slope in $m^{-1}$, $l$ is the width of the pipe in $m$, $T_s$ is the width of the Preissmann slot in $m$, $h_b$ is the bottom elevation, $S_f$ is the friction slope, which can be computed using, for example, the Manning-Strickler formula. The integral $I_1$ stands for the conservative pressure term whereas the integral $I_2$ takes into account the space variation of the rectangular section of the pipe. Here and to simplify the study, we assume that the bottom is flat, the section of the channel is constant and we do not take into account any friction term. Therefore, in the following $S_0 = S_f = I_2 = 0$. The pressure term $I_1$ is defined by

$$I_1(A) = \begin{cases} \dfrac{A^2}{2l}, & \text{si } A \leq A_f \\ A_f \left( \dfrac{(A - A_f)}{T_s} + \dfrac{h_p}{2} \right) + \dfrac{(A - A_f)^2}{2T_s}, & \text{si } A > A_f \end{cases}$$

where $l$ denotes the width of the pipe, $h_p$ is the height of the pipe, $A_f = lh_p$, and $T_s = gA_f/a^2$ with $a$ is the wave celerity.

The system can be expressed in the following vector form:

$$\frac{\partial U}{\partial t} + \frac{\partial F(U)}{\partial x} = 0$$

where $U = (A, Q)^T$ is the vector of conservative variables and $F(U) = (Q, Q^2/A + gI_1)$ is the physical flux. The Jacobian matrix of the physical flux $J_F$ is given by

✦ If $A \leq A_f$, i.e. $h \leq h_p$:

$$J_F = \begin{pmatrix} 0 & 1 \\ gh - u^2 & 2u \end{pmatrix}$$

✦ If $A > A_f$, i.e. $h > h_p$:

$$J_F = \begin{pmatrix} 0 & 1 \\ g\dfrac{A}{T_s} - u^2 & 2u \end{pmatrix}$$

The eigenvalues of this Jacobian write $\lambda_{1,2} = u \pm c$ where $c = \sqrt{gh}$ when $A \leq A_f$ and $c = \sqrt{gA/T_s}$ when $A > A_f$. In practice, the width of the narrow slot $T_s$ is often defined through the arbitrary large wave celerity $a$, and the velocity of the wave in the pressurized part of the flow, i.e. when $A > A_f$, is then given by $c = a\sqrt{A/A_f} > a$.

As we can see, the flux is continuous but not derivable and the eigenvalues of the Jacobian matrix change abruptly and show a discontinuity for $A = A_s$.

# 3 Numerical Method

## 3.1 Finite Volume Discretization

We discretize the problem using a classical colocated finite volume method. We divide the computational domain into small control volumes $C_i = [x_{i-\frac{1}{2}}, x_{i+\frac{1}{2}}]$ and the time interval with small time steps of size $\Delta t$ to be defined later through a CFL condition. We denote by $\Delta x$ the size of the control volumes. We then propose the following numerical scheme

$$U_i^{n+1} = U_i^n - \frac{\Delta t}{\Delta x}\left[F^n_{i+\frac{1}{2}} - F^n_{i-\frac{1}{2}}\right] \tag{2}$$

where $F^n_{i+\frac{1}{2}}$ is an explicit numerical approximation of the physical flux at interface $x_{i+1/2}$ and through the time interval $[t^n, t^{n+1}]$.

## 3.2 FVC Scheme: The Finite Volume Characteristics Method

This scheme was introduced in [1]. It is constructed into three steps: **computation of the departure point**, **predictor stage**, and **corrector stage**.

In order to do so, we first write the Preissmann model (1) into an advection form:

$$\begin{cases} \dfrac{\partial A}{\partial t} + u\dfrac{\partial A}{\partial x} = -A\dfrac{\partial u}{\partial x} \\ \dfrac{\partial Q}{\partial t} + u\dfrac{\partial Q}{\partial x} = -Q\dfrac{\partial u}{\partial x} - g\dfrac{\partial I_1}{\partial x} \end{cases}$$

that reads:

$$\frac{\partial U}{\partial t} + u\frac{\partial U}{\partial x} = W, \qquad W = \begin{pmatrix} -A\dfrac{\partial u}{\partial x} \\ -Q\dfrac{\partial u}{\partial x} - g\dfrac{\partial I_1}{\partial x} \end{pmatrix} \tag{3}$$

We refer to [1] for the details of the algorithm and just briefly describe here the three steps.

- ✧ **Departure point**: The first step consists in finding the foot of the characteristic curves associated with the system (3) which reach the interface $x_{i+1/2}$ at time $\alpha\Delta t$, assuming the right-hand side is zero, i.e. $W = 0$. In that case, characteristic curves are straight lines and unknowns are constant along these curves. It follows that assuming the characteristic curves do not cross and using a quadrature rule, the foot of the characteristic curves can be computed by solving a nonlinear equation.
- ✧ **Predictor stage**: once the departure point is known, we can compute a first approximation of the unknowns at the interface at time $\alpha\Delta t$. In a second step, this value is corrected by taking into account the right-hand side $W$ using a finite difference approximation to obtain an interface approximation $U_{i+\frac{1}{2}}$.
- ✧ **Corrector stage**: in the third step, we simply use the finite volume formula (2) with $F^n_{i+\frac{1}{2}} = F\left(U^n_{i+\frac{1}{2}}\right)$.

## 3.3 *Time Stepping*

It remains to explain how we compute the time step and the value of the parameter $\alpha$. Let us assume that the parameter $\alpha$ is bounded by a given value $\alpha_M$. To ensure compliance with the stability analysis presented in [1], the time step $\Delta t$ is computed by:

$$\Delta t = CFL\frac{\Delta x}{\sqrt{2\alpha_M}\max\limits_i\left(\Lambda_i\right)}$$

where $\Lambda_i = \max(\lambda_{i,1}, \lambda_{i,2})$ and $CFL < 1$. It remains to compute the diffusion parameter $\alpha$. It can be chosen fixed on the whole domain or it can be computed in a self-adapting way at each interface by the formula proposed in [2]

$$\alpha^*_{i+\frac{1}{2}} = \tilde{\alpha}_{i+\frac{1}{2}} + \left(\frac{1}{2}\times C_1 - \tilde{\alpha}_{i+\frac{1}{2}}\right)\phi(r_{i+\frac{1}{2}})$$

where $\phi$ is a Minmod-type flux limiter function [5]

$$\phi(r_{i+\frac{1}{2}}) = \max\left(0, \min\left(1, r_{i+\frac{1}{2}}\right)\right), \quad r_{i+\frac{1}{2}} = \frac{h_i - h_{i-1}}{h_{i+1} - h_i}.$$

The parameter $C_1$ has to be smaller than one and is fixed here to 0.9 and the quantity $\tilde{\alpha}_{i+\frac{1}{2}}$ satisfies

$$\tilde{\alpha}_{i+\frac{1}{2}} = \frac{\Delta x}{2\,\Delta t\, L_{i+\frac{1}{2}}}, \quad L_{i+\frac{1}{2}} = \max_k\left(\max(|\lambda_{i,k}|, |\lambda_{i+1,k}|)\right), \ k = 1, 2$$

Eventually, the interface value $\alpha_{i+\frac{1}{2}}$ is computed by

$$\alpha_{i+\frac{1}{2}} = \min\left(\alpha_M, C_2 * \max\left(C_3, \alpha^*_{i+\frac{1}{2}}\right)/C_1\right)$$

Here, we chose $C_2 = 5$, $C_3 = 3$ and $\alpha_M = 50$.

# 4 Numerical Results

## 4.1 Problem

We investigate the scenario of filling a closed pipe, which is initially partially filled and connected to a reservoir at the upstream end. To assess the performance of our variant of the FVC scheme, we employ the benchmark model proposed in [3] which is a Riemann problem whose solution is

$$h(x) = \begin{cases} h_L = 3.167 & if\ x < 10.067t \\ h_R = 0.6 & if\ x > 10.067t \end{cases} \quad \text{and} \quad u(x) = \begin{cases} u_L = 4.0334 & if\ x < 10.067t \\ u_R = 0 & if\ x > 10.067t \end{cases}$$

## 4.2 Results

To implement the numerical schemes, we used 400 cells and a CFL number equal to 0.9. The arbitrary wave celerity in the pressurized regime is fixed to $a = 100\,\text{m.s}^{-1}$. The width and height of the pipe are respectively taken by $l = 1$ cm and $h_p = 1$ cm. For the FVC scheme, we use a variant with a fixed diffusion parameter with is chosen to be $\alpha_f = 16.5$ and another variant with the self-adapted technique described in the previous section and which is denoted by $\alpha_c$ in the figures.

We compare our FVC schemes with three other classical finite volume schemes, namely the Roe, Rusanov [7] and HLL [6] ones. We observe in Figs. 2 and 3 that the four schemes quite accurately reproduce the exact solution. However, oscillations are observed in the pressure curve for Roe, Rusanov, and HLL schemes on the transition from pressurized to free surface regime. These oscillations have been reported in the literature, see [3]. In contrast, the FVC scheme provides a solution that is oscillation free when the self-adapting technique is used. Error curves are plotted in Fig. 4. It shows that the solution computed by the FVC scheme with self-adapted parameter $\alpha$ is more accurate for all the range of meshes that are considered.

The Table 1, shows the computation times for each scheme for a mesh of 400 points and in $t = 4s$. This computations have been executed on a macbook pro late

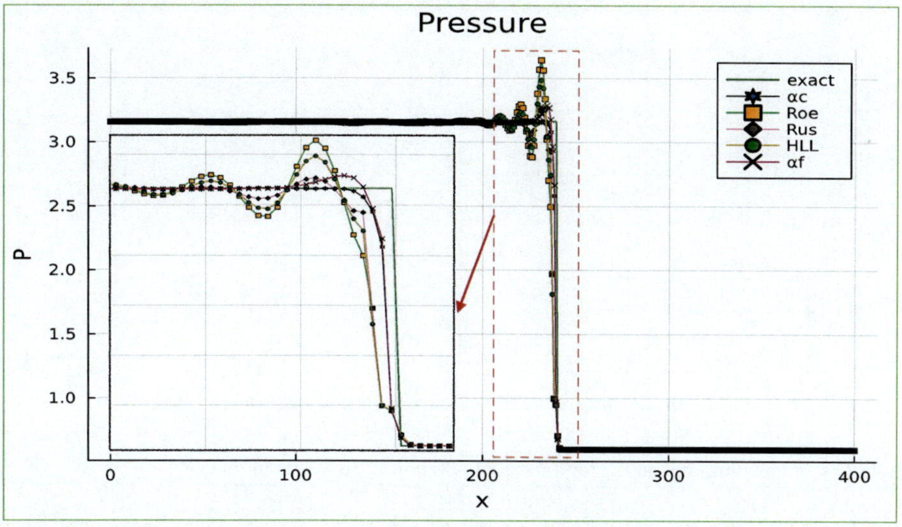


**Fig. 2** The behavior of pressure at time $t = 4$ s with 400 regular cells

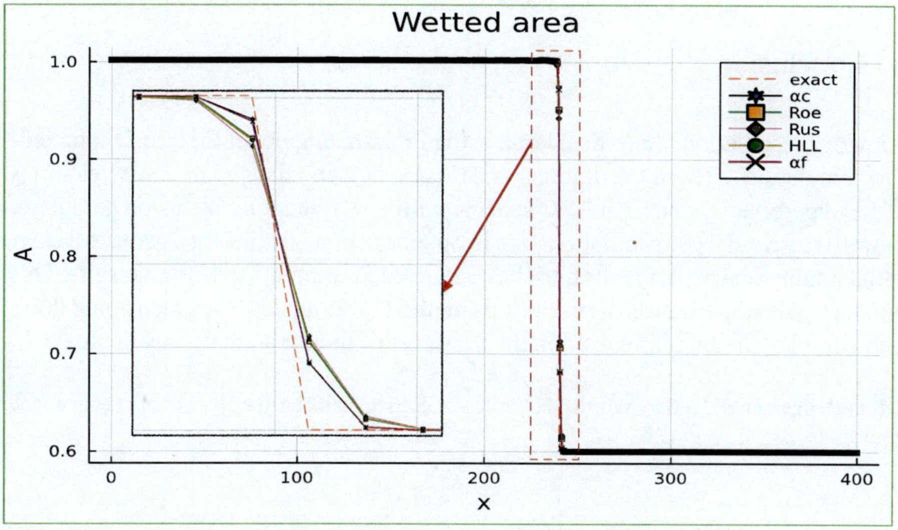


**Fig. 3** The behavior of wet section at time $t = 4$ s with 400 regular cells

2013 Intel Core i5 2.4 GHz double core, 8Go RAM. As shown in this table, our method with a fixed $\alpha_f$ or a controlled one $\alpha_c$ is faster than Roe scheme.

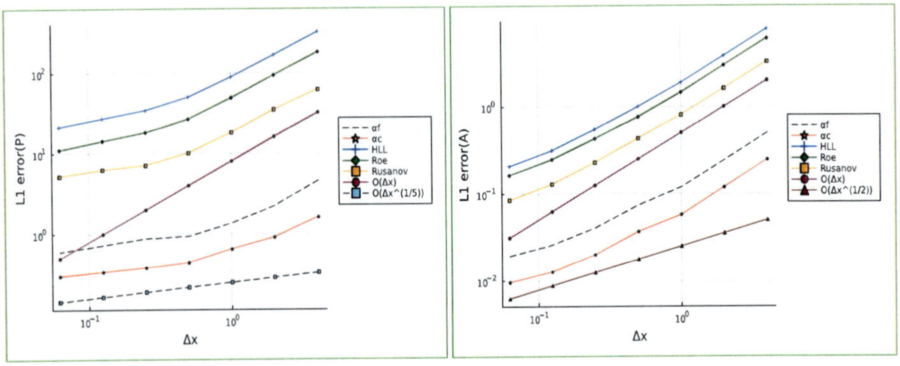


**Fig. 4** Order of convergence of pressure and wetted area in space at time $t = 4$ s

**Table 1** Computation time in $(s)$ for a mesh of 400 points, for a mixed flow

| Scheme | Roe | Rusanov | HLL | $\alpha_f$ | $\alpha_c$ |
|---|---|---|---|---|---|
| CPU in $(s)$ | 1.8177 | 0.0767 | 0.0584 | 0.6195 | 1.0055 |

## 5 Conclusion

Numerical oscillation is a critical issue in the Preissmann model for transient mixed-flow simulations. These oscillations arise from sudden changes in wave velocities of the hyperbolic system. An FVC scheme with a self-adapting diffusion parameter allows us to avoid these oscillations and to obtain accurate results for a given test case. Further numerical experiments must be conducted to complete these first results. This work is a first step towards the use of this method to simulate two-dimensional flows with complex rheology involved in the transport of industrial materials in pipes.

**Acknowledgements** L. Baroukh thanks Professor F. Benkhaldoun for all his help, advice, and motivation.

## References

1. Benkhaldoun, F., Seaïd, M.: A simple finite volume method for the shallow water equations. J. Comput. Appl. Math. **234**(1), 58–72 (2010). https://doi.org/10.1016/j.cam.2009.12.004
2. Aboussi, W., Ziggaf, M., Kissami, I., Boubekeur, M.: A highly efficient finite volume method with a diffusion control parameter for hyperboliq problems. Comput. Simul. (submitted after a first revision). https://doi.org/10.1016/j.matcom.2023.05.023
3. Malekpour, A., Karney, B.W.: Spurious numerical oscillations in the Preissmann slot method: origin and suppression. J. Hydraul. Eng. **142**(3), 04015060 (2016). https://doi.org/10.1061/(ASCE)HY.1943-7900.0001066
4. Preismann, A.: Propagation des intumescences dans les canaux et rivières. First Congress of the French Association for Computation. Grenoble, France (1961)

5. Roe, P.L.: Characteristic-based schemes for the Euler equations. Ann. Rev. Fluid Mech. **18**(1), 337–365 (1986). https://doi.org/10.1146/annurev.fl.18.010186.002005
6. Prebeg, M., Flåtten, T., M"uller, B.: Large time step HLL and HLLC schemes. ESAIM: Math. Model. Numer. Anal. **52**(4), 1239–1260 (2018). https://doi.org/10.1051/m2an/2018029
7. Clain, S., Rochette, D.: First-and second-order finite volume methods for the one-dimensional nonconservative Euler system. J. Comput. Phys. **228**(22), 8214–8248 (2009). https://doi.org/10.1016/j.jcp.2009.07.038

# Truly Multi-dimensional All-Speed Methods for the Euler Equations

**Wasilij Barsukow**

**Abstract** Several recent all-speed time-explicit numerical methods for the Euler equations on Cartesian grids are presented and their properties assessed experimentally on a complex application. These methods are truly multi-dimensional, i.e. the flux through an interface also depends on the values in cells adjacent to the endpoints of the edges (corners).

**Keywords** Low Mach number · Multi-dimensional methods

## 1 Introduction

The rescaled Euler equations read

$$\partial_t \rho + \nabla \cdot (\rho \mathbf{v}) = 0 \qquad \rho\colon \mathbb{R}_0^+ \times \mathbb{R}^d \to \mathbb{R}^+, \tag{1}$$

$$\partial_t (\rho \mathbf{v}) + \nabla \cdot (\rho \mathbf{v} \otimes \mathbf{v} + \frac{p}{\epsilon^2} \mathbb{1}) = 0 \qquad \mathbf{v}\colon \mathbb{R}_0^+ \times \mathbb{R}^d \to \mathbb{R}^d, \tag{2}$$

$$\partial_t e + \nabla \cdot ((e + p)\mathbf{v}) = 0 \qquad e, p\colon \mathbb{R}_0^+ \times \mathbb{R}^d \to \mathbb{R}^+, \tag{3}$$

$$\frac{p}{\gamma - 1} + \frac{1}{2} \epsilon^2 \rho |\mathbf{v}|^2 = e, \tag{4}$$

where $\epsilon \to 0$ corresponds to the limit of low Mach number. This is a singular limit: for well-prepared initial data, the solutions tend to those of the incompressible Euler equations as $\epsilon \to 0$.

Numerical diffusion that is used to stabilize time-explicit discretizations of (1)–(3) is often introduced by means of a Riemann solver, and thus inspired by supersonic phenomena. In subsonic and low Mach number flow, this diffusion is excessive and dominates the numerical solution unless the grid is refined strongly (see e.g. [11]).

W. Barsukow (✉)
Institut de Mathématiques de Bordeaux, CNRS UMR 5251, 351 Cours de la Libération, 33405 Talence, France
e-mail: wasilij.barsukow@math.u-bordeaux.fr



E. Franck et al. (eds.), *Finite Volumes for Complex Applications X—Volume 2, Hyperbolic and Related Problems*, Springer Proceedings in Mathematics & Statistics 433,
https://doi.org/10.1007/978-3-031-40860-1_3

The excessive diffusion is also worrying because it points to a fundamental deficiency of Godunov-type methods (see e.g. [4]).

The suggestions to overcome this problem traditionally amount to

1. removing certain terms from the diffusion matrix (or to multiply them by a factor $f \in \mathcal{O}(\epsilon)$) (**low Mach fixes**, e.g. [8]), or to
2. using central differences for at least the acoustic operator, which necessitates (partially) implicit time integration (**IMEX**, e.g. [5]).

Time explicit methods are easy to implement and are computationally inexpensive, but the CFL condition requires $\Delta t \in \mathcal{O}(\epsilon)$. Provided that the implicit methods are well-conditioned, as $\epsilon \to 0$, they will always outperform explicit methods in terms of computational efficiency. For good and well-implemented methods, the break-even point is typically at $\epsilon \sim 10^{-2}$.

This paper focuses on time-explicit methods for the following reason: If very low values of $\epsilon$ are to be used, and acoustic phenomena (sound waves, density stratifications, etc.) are of no interest, then it might be more practical to use an incompressible code. If the acoustic phenomena need to be resolved, out of considerations of accuracy, even for an implicit method the time step would be chosen close to that of an explicit method (in view of the above-mentioned break-even point, even choosing the time step to be 100 times that of an explicit method would barely justify using an implicit method).

Two strategies that reduce the numerical diffusion while retaining enough of it to allow explicit integration in time, and avoiding the introduction of ad-hoc factors have been developed recently:

3. In the methods from [2], the divergence $\nabla \cdot \mathbf{v} \in \mathcal{O}(\epsilon)$ plays the role of the factor $f \in \mathcal{O}(\epsilon)$, because it seems to arise naturally in just the right place.
4. [3] points out that, in fact, central differences can be stable upon *explicit* time integration for certain systems, e.g. the acoustic equations.

After some preliminary remarks concerning notation (Sect. 2) these two strategies shall be presented (Sects. 3 and 4) and compared to each other (Sect. 5). The numerical methods are considered on two-dimensional Cartesian grids, with cells $[x_{i-\frac{1}{2}}, x_{i+\frac{1}{2}}] \times [y_{j-\frac{1}{2}}, y_{j+\frac{1}{2}}]$.

## 2 Notation for Finite Difference Operators

In order to shorten finite difference formulae, and also in order to make them better readable, the following short hand notation is used:

$$[a]_{i+\frac{1}{2}} := a_{i+1} - a_i \quad \text{the jump/difference operator,} \tag{5}$$

$$\{a\}_{i+\frac{1}{2}} := a_{i+1} + a_i \quad \text{the summation operator.} \tag{6}$$

Both are centered at a cell interface, but one also has the cell-centered versions

$$[b]_{i\pm\frac{1}{2}} := b_{i+\frac{1}{2}} - b_{i-\frac{1}{2}}, \qquad [a]_{i\pm 1} := a_{i+1} - a_{i-1} \tag{7}$$

and similarly for the sums. This notation can be combined in one dimension, e.g. $[[a]]_{i\pm\frac{1}{2}} := [a]_{i+\frac{1}{2}} - [a]_{i-\frac{1}{2}} = a_{i+1} - 2a_i + a_{i-1}$ and $\{[a]\}_{i\pm\frac{1}{2}} = [a]_{i\pm 1}$ as well as in multiple dimensions (making sure not to interchange the order of the indices)

$$[a_i]_{j+\frac{1}{2}} \equiv [a]_{i,j+\frac{1}{2}} := a_{i,j+1} - a_{ij}, \tag{8}$$

$$[a]_{i+\frac{1}{2},j} \equiv [a_{\cdot,j}]_{i+\frac{1}{2}} := a_{i+1,j} - a_{ij}, \tag{9}$$

$$\{[a]_{i+\frac{1}{2}}\}_{j+\frac{1}{2}} = [a]_{i+\frac{1}{2},j+1} + [a]_{i+\frac{1}{2},j} \quad \text{etc.} \tag{10}$$

## 3 All-Speed Extension to Multiple Dimensions

The key observation of [2] is that the diffusive terms that are prohibiting low Mach compliance often are of the form $\partial_x(c_1\partial_x u)$ in the equation for the $x$-momentum $\rho u$ and $\partial_y(c_2\partial_y v)$ in the equation for the $y$-momentum $\rho v$, with $c_1, c_2$ being terms that depend on the method. While low Mach fixes would endow these two terms with prefactors $f \in \mathcal{O}(\epsilon)$, the strategy advocated in [2] is to extend them to $\partial_x(c_1(\partial_x u + \partial_y v))$ and $\partial_y(c_2(\partial_x u + \partial_y v)$, respectively. They then contain the divergence, which in the limit $\epsilon \to 0$ would become $\mathcal{O}(\epsilon)$, thus achieving the same as the prefactor. Due to the appearance of the mixed derivative $\partial_x\partial_y$ the resulting method necessarily is truly multi-dimensional. The advantage of this approach is that the one-dimensional method (with possibly all its good properties concerning stability, entropy, etc.) remains unmodified, that the method is all-speed, and that the low Mach compliance is achieved without ad-hoc factors.

As now the divergence would appear in both the central part and the diffusion, its respective discretizations need to be related, such that the cell-centered diffusion contains the difference of two edge-centered divergences and such that the central divergence is their average. Finally, due to the fact that there is a diffusion in $x$-direction and one in $y$-direction, the fundamental object is, in fact, a node-based divergence. A divergence centered on an edge is obtained as the average of the two divergences at the nodes. It has been shown in [1] that this is the unique procedure for symmetric discretizations.

The strategy of multi-dimensional all-speed extension, as explained above, has been applied to two methods in [2]. First, the Lagrange-Projection-type method from [7] was considered whose numerical flux in 1-d is

$$\hat{f}_{i+\frac{1}{2}} := \left(0, \frac{p^*_{i+\frac{1}{2}}}{\epsilon^2}, 0, p^*_{i+\frac{1}{2}}u^*_{i+\frac{1}{2}}\right) + u^*_{i+\frac{1}{2}}\left\{\frac{Q^n}{L}\right\}_{i+\frac{1}{2}} - |u^*_{i+\frac{1}{2}}|\left[\frac{Q^n}{L}\right]_{i+\frac{1}{2}}, \tag{11}$$

having defined (with $a > \max(\rho c)$)

$$u^*_{i+\frac{1}{2}} := \frac{\{u\}_{i+\frac{1}{2}}}{2} - \frac{1}{2a\epsilon}[p]_{i+\frac{1}{2}} \qquad p^*_{i+\frac{1}{2}} := \frac{\{p\}_{i+\frac{1}{2}}}{2} - \frac{a\epsilon}{2}[u]_{i+\frac{1}{2}}, \tag{12}$$

$$Q^n_i := \left(\rho^n_i, (\rho u)^n_i - \frac{\Delta t}{\Delta x}\frac{1}{\epsilon^2}[p^*]_{i\pm\frac{1}{2}}, (\rho v)^n_i, e^n_i - \frac{\Delta t}{\Delta x}[u^* p^*]_{i\pm\frac{1}{2}}\right), \tag{13}$$

$$L_i := 1 + \frac{\Delta t}{\Delta x}[u^*]_{i\pm\frac{1}{2}}. \tag{14}$$

Observe how the diffusion of the momentum equation takes its origin in $p^*$, and more precisely in the term $[u]_{i+\frac{1}{2}}$. It is this term that the authors of [7] propose to multiply by $\epsilon$ in order to achieve low Mach compliance. Here, instead, expressions (12) shall be replaced by

$$u^*_{i+\frac{1}{2},j} = \frac{\{\{\{u\}_{i+\frac{1}{2}}\}\}_{j\pm\frac{1}{2}}}{8} - \frac{1}{2a\epsilon}\frac{\{\{[p]_{i+\frac{1}{2}}\}\}_{j\pm\frac{1}{2}}}{4}, \tag{15}$$

$$p^*_{i+\frac{1}{2},j} = \frac{\{\{\{p\}_{i+\frac{1}{2}}\}\}_{j\pm\frac{1}{2}}}{8} - \frac{a\epsilon}{2}\left(\frac{\{\{[u]_{i+\frac{1}{2}}\}\}_{j\pm\frac{1}{2}}}{4} + \frac{\Delta x}{\Delta y}\frac{[\{v\}_{i+\frac{1}{2}}]_{j\pm 1}}{4}\right), \tag{16}$$

$$v^*_{i,j+\frac{1}{2}} = \frac{\{\{\{v\}\}_{i\pm\frac{1}{2}}\}_{j+\frac{1}{2}}}{8} - \frac{1}{2a\epsilon}\frac{[\{\{p\}\}_{i\pm\frac{1}{2}}]_{j+\frac{1}{2}}}{4}. \tag{17}$$

It is also natural to redefine

$$L_{ij} := 1 + \frac{\Delta t}{\Delta x}[u^*]_{i\pm\frac{1}{2},j} + \frac{\Delta t}{\Delta y}[v^*]_{i,j\pm\frac{1}{2}}, \tag{18}$$

$$(Q^n_{ij})_e := e^n_{ij} - \frac{\Delta t}{\Delta x}[u^* p^*]_{i\pm\frac{1}{2},j} - \frac{\Delta t}{\Delta y}[v^* p^*]_{i,j\pm\frac{1}{2}}, \tag{19}$$

where $(Q^n_{ij})_e$ is the energy-component of the vector $Q^n_{ij}$. This method shall be referred to as **Method A**. Now, an edge-based divergence is appearing in (16) (the average of two node-based divergences), while the divergence in (18) turns out to be the average of four such edge-based divergences. They thus all vanish simultaneously, which is one of the ingredients of the proof of low Mach compliance. For further details, see [2].

For the relaxation solver from [6], the other method investigated in [2], one finds for the intermediate states

$$u^*_{i+\frac{1}{2}} = \frac{\{u\}_{i+\frac{1}{2}}}{2} - \frac{1}{2a\epsilon}[p]_{i+\frac{1}{2}}, \qquad p^*_{i+\frac{1}{2}} = \frac{\{p\}_{i+\frac{1}{2}}}{2} - \frac{a\epsilon}{2}[u]_{i+\frac{1}{2}}, \tag{20}$$

$$\rho^*_{i+\frac{1}{2},\mathrm{L}} = \frac{\rho_i}{1 + \rho_i \epsilon \frac{[u]_{i+\frac{1}{2}}}{2a} - \rho_i \frac{[p]_{i+\frac{1}{2}}}{2a^2}}, \qquad \frac{e^*_{i+\frac{1}{2},\mathrm{L}}}{\rho^*_{i+\frac{1}{2},\mathrm{L}}} = \frac{e_i}{\rho_i} + \epsilon \frac{p_i u_i - p^*_{i+\frac{1}{2}} u^*_{i+\frac{1}{2}}}{a} \quad \text{etc.} \tag{21}$$

Equation (20) is the same as before, up to the definition of $a$ which now is more complicated (see [6] for more details). The flux associated with the intermediate state is (if $u^*_{i+\frac{1}{2}} > 0$)

$$f_{i+\frac{1}{2}} = \left( u^*_{i+\frac{1}{2}} \rho^*_{i+\frac{1}{2},\mathrm{L}}, \rho^*_{i+\frac{1}{2},\mathrm{L}} (u^*_{i+\frac{1}{2}})^2 + \frac{p^*_{i+\frac{1}{2}}}{\epsilon^2}, \rho^*_{i+\frac{1}{2},\mathrm{L}} u^*_{i+\frac{1}{2}} v_i, u^*_{i+\frac{1}{2}} (e^*_{i+\frac{1}{2},\mathrm{L}} + p^*_{i+\frac{1}{2}}) \right). \tag{22}$$

In [2], therefore, Eq. (20) are again replaced by (15)–(17), and the intermediate density is also naturally replaced by

$$\rho^*_{i+\frac{1}{2},j,\mathrm{L}} = \frac{\rho_{ij}}{1 + \frac{\rho_{ij}\epsilon}{2a} \left( \frac{\{\{[u]_{i+\frac{1}{2}}\}\}_{j\pm\frac{1}{2}}}{4} + \frac{\Delta x}{\Delta y} \frac{[\{v\}_{i+\frac{1}{2}}]_{j\pm 1}}{4} \right) - \frac{\rho_{ij}}{2a^2} \frac{\{\{[p]_{i+\frac{1}{2}}\}\}_{j\pm\frac{1}{2}}}{4}} \tag{23}$$

and similarly for $\rho^*_{i+\frac{1}{2},j,\mathrm{R}}$. This method shall be referred to as **Method B**. The proof of low Mach compliance is analogous to that of Method A. For more details, see [2].

## 4 Sequentially-Explicit Time Integration

Consider the following "off-diagonal" system (with $f, g\colon \mathbb{R} \to \mathbb{R}$ given)

$$\partial_t \begin{pmatrix} a \\ b \end{pmatrix} = \begin{pmatrix} f(b) \\ g(a) \end{pmatrix} \qquad a, b\colon \mathbb{R}^+_0 \to \mathbb{R}. \tag{24}$$

The sequential application of the forward Euler method already uses the value $a^{n+1}$ in the update of $b$:

$$\frac{a^{n+1} - a^n}{\Delta t} = f(b^n) \qquad \frac{b^{n+1} - b^n}{\Delta t} = g(a^{n+1}). \tag{25}$$

This method is not the forward Euler method. It is explicit, and thus also different from semi-implicit methods of e.g. [10] or [9], where similar ideas appear but the final systems remain implicit. The above method of time integration could be called "leap-frog", were it not for the many other, and different, methods that already carry this name, such as those in [15]. The name used here, and in [3] therefore is "sequential explicit".

One can show that methods using this integration in time have many good properties, such as non-dissipativity (the Fourier modes in the von Neumann stability analysis neither grow, nor decay), and there is a relation to symplectic or energy-conserving time integrators in cases when the PDE at hand possesses a Hamiltonian structure (e.g. those in [14] for the Maxwell equations).

Here, the sequential-explicit integrator shall be merely used for its ability to stabilize central differences. The Euler system can be thought of as being composed of acoustics and advection. It is a lucky coincidence that the acoustic sub-system is of the proper off-diagonal form. As it is only a sub-system of the full Euler equations, a relation to Hamiltonian systems etc. is neither relevant nor required. The advective sub-system, which is not off-diagonal, requires a conventional stabilization mechanism (upwinding). It has also been found in [3] that adding diffusion by way of a denominator as in Lagrange-Projection methods improves the stability of the overall method. The complete numerical flux (in $x$-direction, say) therefore is, schematically,

$$f^x_{i+\frac{1}{2},j} = \frac{\{\{ \boxed{\text{flux average}}_{i+\frac{1}{2}} \}\}_{j\pm\frac{1}{2}} - \boxed{\text{upwind difference}}_{i+\frac{1}{2},j}}{1 + \Delta t \boxed{\text{divergence}}_{i+\frac{1}{2},j}}.$$

In order to solve the acoustic sub-system with sequential-explicit time integration, and the advective one with forward Euler, it is sufficient to update the momentum equations first, and to use the new value of the momentum in the updates of $\rho$ and $e$. This way, the method has the same computational cost as the usual forward Euler method. It has been called Method G in [3].

It is important to make sure that the cell-based divergence, that arises as $\mathcal{O}(1)$ term in the energy equation is an average of the edge-based divergence appearing in the denominator, which is why the flux average is a multi-dimensional operator involving averaging in the perpendicular direction. The edge-based divergence, itself, is taken as the average of two node-based divergences, that have already appeared in Sect. 3. Under this condition one can then prove low Mach compliance (see [3]). This method shall be referred to here as **Method C**.

## 5 Numerical Results

We aim at comparing the performance of these three very different methods, common properties being merely the explicit nature and the fact that they are conservative methods on Cartesian grids. While method B derives from a Riemann solver, the other two use Riemann solvers at most for sub-systems, with method C essentially being Riemann-solver free. The numerical setup chosen for the comparison is a particular choice of a Kelvin-Helmholtz instability from [12, 13]. While the proposed methods might have individually been tested already on such setups, here they shall be compared to each other on the same setup for the first time, which also allows comparison to other low Mach number methods in future.

The setup is studied on a domain $[0, 2] \times [-\frac{1}{2}, \frac{1}{2}]$ with

$$H(y) := \begin{cases} -\sin(\pi \frac{y+\frac{1}{4}}{w}) & -\frac{1}{4} - \frac{w}{2} \le y < -\frac{1}{4} + \frac{w}{2}, \\ -1 & -\frac{1}{4} + \frac{w}{2} \le y < \frac{1}{4} - \frac{w}{2}, \\ \sin\left(\pi \frac{y-\frac{1}{4}}{w}\right) & \frac{1}{4} - \frac{w}{2} \le y < \frac{1}{4} + \frac{w}{2}, \\ 1 & \text{else}, \end{cases} \tag{26}$$

$$\rho = \gamma + rH(y), \qquad u = \mathcal{M}H(y), \qquad p = 1, \qquad v = \delta\mathcal{M}\sin(2\pi x). \tag{27}$$

In [12], $r = 0$. Here, we use $r = 10^{-3}$: In incompressible flow, the density acts as a passive scalar, and this way it is possible to visualize the interaction of the layers. Observe that $H \in C^1$. $\mathcal{M}$ is a free parameter that allows to modify the Mach number of the flow, and this requires a simulation time of $0.8/\mathcal{M}$. We choose $\delta = 0.1$ and the width of the shear layer as $w = \frac{1}{16}$. Periodic boundaries and a CFL number of 0.9 are used, and $\gamma = 1.4$.

One observes (Fig. 1, top) that indeed the results are independent of the Mach number, which is the essence of low Mach compliance. Despite the very different

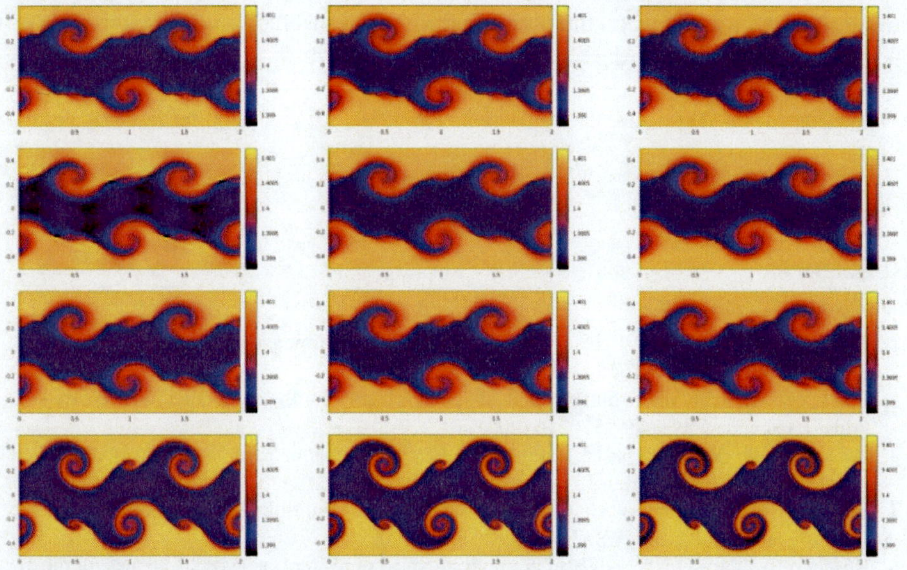

**Fig. 1** Density is color-coded (all figures using the same setup). *First row:* Grid 128 × 64, $\mathcal{M} = 10^{-2}, 10^{-3}, 10^{-4}$ (*left to right*), Method A. *Second row*: Same for Method B. *Third row*: Same for Method C. *Fourth row*: $\mathcal{M} = 10^{-2}$. Convergence study for Method A. *Left to right*: Grid 512 × 256, 1024 × 512, 2048 × 1024

strategies of derivation, the methods yield very similar results. One observes the appearance of artefacts (most pronounced for Method C), where a part of the shear layer seems not to become unstable in the proper way. They develop into vortices upon further grid refinement (Fig. 1, bottom), but the vortices keep decreasing in size and are purely numerical. They tend not to appear upon usage of a method of second or higher order of accuracy ([13]). On a grid of $128 \times 64$ ($256 \times 128$) one finds the total kinetic energy to have decayed by the following amounts by the end of the simulations: Method A 17.8% (13.8%), Method B 16.3% (12.5%), Method C 18.0% (13.7%). These values are indistinguishable for $\mathcal{M} = 10^{-2}$ and $\mathcal{M} = 10^{-3}$, which means that the decay is due to the advective diffusion. Methods A and C are designed to contain more of it, which might explain the slightly higher loss of kinetic energy.

## 6 Conclusions

This paper presents numerical results of three recently developed truly-multidimensional numerical methods for the Euler equations. They achieve the all-speed property without ad hoc fixes, by including the divergence operator in the numerical diffusion instead of derivatives of individual velocity components. Despite being first-order accurate they resolve details of a complex setup of a Kelvin-Helmholtz instability even on coarse grids and show asymptotically no Mach number dependence. They, however, display artefacts in the form of secondary vortices that are purely numerical. Future research will focus on the precise origin of the artefacts and on extensions of these methods to higher order.

**Acknowledgements** The author thanks Giovanni Leidi (HITS, Heidelberg) for inspiring discussions.

## References

1. Barsukow, W.: Stationarity preserving schemes for multi-dimensional linear systems. Math. Comput. **88**(318), 1621–1645 (2019)
2. Barsukow, W.: Truly multi-dimensional all-speed schemes for the euler equations on cartesian grids. J. Comput. Phys. **435**, 110216 (2021)
3. Barsukow, W.: All-speed numerical methods for the Euler equations via a sequential explicit time integration. Accepted in J. Sci. Comp. (2023)
4. Barsukow, W., Klingenberg, C.: Exact solution and a truly multidimensional Godunov scheme for the acoustic equations. ESAIM: M2AN **56**(1) (2022)
5. Boscheri, W., Pareschi, L.: High order pressure-based semi-implicit IMEX schemes for the 3D Navier-Stokes equations at all Mach numbers. J. Comput. Phys. **434**, 110206 (2021)
6. Chalons, C., Coquel, F., Godlewski, E., Raviart, P.-A., Seguin, N.: Godunov-type schemes for hyperbolic systems with parameter-dependent source: the case of euler system with friction. Math. Models Methods Appl. Sci. **20**(11), 2109–2166 (2010)

7. Chalons, C., Girardin, M., Kokh, S.: Large time step and asymptotic preserving numerical schemes for the gas dynamics equations with source terms. SIAM J. Sci. Comput. **35**(6), A2874–A2902 (2013)
8. Dellacherie, S.: Analysis of Godunov type schemes applied to the compressible Euler system at low Mach number. J. Comput. Phys. **229**(4), 978–1016 (2010)
9. Dimarco, G., Loubère, R., Vignal, M.-H.: Study of a new asymptotic preserving scheme for the Euler system in the low Mach number limit. SIAM J. Sci. Comput. **39**(5), A2099–A2128 (2017)
10. Guerra, J., Gustafsson, B.: A semi-implicit method for hyperbolic problems with different time-scales. SIAM J. Numer. Anal. **23**(4), 734–749 (1986)
11. Guillard, H., Murrone, A.: On the behavior of upwind schemes in the low Mach number limit: II. Godunov type schemes. Comput. Fluids **33**(4), 655–675 (2004)
12. Leidi, G., Birke, C., Andrassy, R., Higl, J., Edelmann, P.V.F., Wiest, G., Klingenberg, C., Röpke, F.K.: A finite-volume scheme for modeling compressible magnetohydrodynamic flows at low Mach numbers in stellar interiors. Astron. Astrophys. **668**, A143 (2022)
13. Leidi, G.: Private communication
14. Remaki, M.: A new finite volume scheme for solving Maxwell's system. COMPEL- Int. J. Comput. Math. Electr. Electron. Eng. (2000)
15. Thomas, J., Roe, P.: Development of non-dissipative numerical schemes for computational aeroacoustics. In: 11th Computational Fluid Dynamics Conference, p. 3382 (1993)

# Monotonicity for Genuinely Multi-step Methods: Results and Issues From a Simple Lattice Boltzmann Scheme

**Thomas Bellotti**

**Abstract** This work focuses on the convergence of a two velocities lattice Boltzmann scheme towards the weak entropic solution of a non-linear scalar conservation law. The result is proved using monotonicity properties of the corresponding multi-step Finite Difference scheme and techniques germane to Finite Volume schemes. Still, only the over-relaxation regime is straightforwardly handled and we emphasize the need for a monotonicity theory for genuinely multi-step schemes for PDEs. This analysis should take the role of the initialization routines into account.

**Keywords** Monotonicity · Multi-step · Lattice Boltzmann · Finite difference · Convergence

## 1 Introduction

Lattice Boltzmann schemes [1] are widely employed in Computational Fluid Dynamics and praised for their efficiency, ease of implementation and parallelization. However, numerous theoretical questions on these methods need to be addressed, especially concerning their consistency, stability, and eventually convergence. In particular, [2] have proved the convergence of a two velocities lattice Boltzmann scheme towards the unique entropic solution of a scalar hyperbolic conservation law in the under-relaxation regime. This is achieved by proving a maximum principle, total variation estimates and finally entropy inequalities for the discrete solution. Instead of working on the original lattice Boltzmann formulation, we have proposed [3] a way of recasting any scheme under the form of multi-step Finite Difference method. In the present contribution, we want to exploit this equivalent but different form of the scheme to prove the convergence of the original lattice Boltzmann scheme towards the entropic solution in the over-relaxation regime. Observe that this regime was not accessible to [2]. However, we will also emphasize the difficulties in encompass-

T. Bellotti (✉)
CMAP, CNRS, École polytechnique, Institut Polytechnique de Paris, 91120 Palaiseau, France
e-mail: thomas.bellotti@polytechnique.edu

 
E. Franck et al. (eds.), *Finite Volumes for Complex Applications X—Volume 2, Hyperbolic and Related Problems*, Springer Proceedings in Mathematics & Statistics 433,
https://doi.org/10.1007/978-3-031-40860-1_4

ing the result from [2], obtained for the under-relaxation regime, using our point of view, thus calling for a better understanding of monotonicity for genuinely multi-step schemes.

## 2 Target PDE, Numerical Scheme and Main Result

The problem we are interested in is the scalar conservation law under the form

$$\partial_t u(t,x) + \partial_x(\varphi(u))(t,x) = 0, \qquad t \in [0,T], \quad x \in \mathbb{R}, \tag{1}$$

$$u(0,x) = u^\circ(x), \qquad x \in \mathbb{R}, \tag{2}$$

where $\varphi \in C^1(\mathbb{R})$ and $u^\circ \in L^\infty(\mathbb{R})$. In the sequel, $u$ will denote the unique entropic weak solution according to the standard definition, see [4].

As far as time-space discretization are concerned, we work on an unbounded lattice $\Delta t\mathbb{N} \times \Delta x\mathbb{Z}$. The discrete points are $t^n := n\Delta t \in \Delta t\mathbb{N}$ for $n \in \mathbb{N}$ and $x_j := j\Delta x \in \Delta x\mathbb{Z}$ for $j \in \mathbb{Z}$. We use the so-called "acoustic scaling" which boils down to $\Delta x/\Delta t = \lambda$ kept fixed as $\Delta x \to 0$. As numerical scheme, we consider the original two velocities lattice Boltzmann scheme [2], which reads:

- **Initialization**. Initialize the conserved moment $\mathsf{u}$ using the averages of $u^\circ$ and take the non-conserved moment $\mathsf{v}$ at its equilibrium value:

$$\mathsf{u}_j^0 = \frac{1}{\Delta x}\int_{x_j}^{x_{j+1}} u^\circ(x)\mathrm{d}x, \qquad \mathsf{v}_j^0 = \varphi(\mathsf{u}_j^0), \qquad j \in \mathbb{Z}. \tag{3}$$

- **Scheme** for $n \in \mathbb{N}$, split into two phases:
  - **Collision phase**. This phase is local in space and reads

$$\mathsf{u}_j^{n,\star} = \mathsf{u}_j^n, \qquad \mathsf{v}_j^{n,\star} = (1-s)\mathsf{v}_j^n + s\varphi(\mathsf{u}_j^n), \qquad j \in \mathbb{Z}, \tag{4}$$

  where the relaxation parameter $s \in ]0,2]$. The post-collision distribution densities are retrieved point-by-point using $\mathsf{f}_j^{\pm,n,\star} = \frac{1}{2}\mathsf{u}_j^{n,\star} \pm \frac{1}{2\lambda}\mathsf{v}_j^{n,\star}$.
  - **Stream phase**. This linear phase is non-local and reads

$$\mathsf{f}_j^{\pm,n+1} = \mathsf{f}_{j\mp 1}^{\pm,n,\star}, \qquad j \in \mathbb{Z}. \tag{5}$$

  Then we recover $\mathsf{u}_j^{n+1} = \mathsf{f}_j^{+,n+1} + \mathsf{f}_j^{-,n+1}$ and $\mathsf{v}_j^{n+1} = \lambda(\mathsf{f}_j^{+,n+1} - \mathsf{f}_j^{-,n+1})$.

In [3, 5], we have seen that the conserved moment $\mathsf{u}$ computed by the lattice Boltzmann scheme above and the one computed using the corresponding Finite Difference scheme right underneath are the same.

- **Initialization**

$$\mathsf{u}_j^0 = \frac{1}{\Delta x}\int_{x_j}^{x_{j+1}} u^\circ(x)\mathrm{d}x, \qquad j \in \mathbb{Z}. \tag{6}$$

- **Initial scheme**, depending on the choice for $\mathsf{v}^0$, namely (3).

$$\mathsf{u}_j^1 = \frac{1}{2}(\mathsf{u}_{j-1}^0 + \mathsf{u}_{j+1}^0) + \frac{1}{2\lambda}\big(\varphi(\mathsf{u}_{j-1}^0) - \varphi(\mathsf{u}_{j+1}^0)\big), \qquad j \in \mathbb{Z}, \tag{7}$$

which is a Lax-Friedrichs scheme.
- **Bulk scheme** for $n \in \mathbb{N}^*$.

$$\mathsf{u}_j^{n+1} = \Big(1 - \frac{s}{2}\Big)(\mathsf{u}_{j-1}^n + \mathsf{u}_{j+1}^n) + \frac{s}{2\lambda}\big(\varphi(\mathsf{u}_{j-1}^n) - \varphi(\mathsf{u}_{j+1}^n)\big) + (s-1)\mathsf{u}_j^{n-1}, \quad j \in \mathbb{Z}. \tag{8}$$

The scheme (8) is combination of three well-known Finite Difference schemes

  - A scheme for the wave equation at velocities $\pm\lambda$ obtained when $s = 0$, since time and space steps are linked by $\Delta t = \Delta x/\lambda$.
  - The Lax-Friedrichs scheme—obtained when $s = 1$. This scheme is known for being TVD under the CFL condition.
  - The leap-frog scheme—obtained when $s = 2$. This scheme is the prototype of non-TVD scheme and generates oscillating solutions.

We can then identify two regimes, with names inherited from the lattice Boltzmann standpoint, with (4) seen as a relaxation towards the equilibrium $\varphi(\mathsf{v}_j^n)$.

- The over-relaxation, when $s \in [1, 2]$, where (8) is a convex combination of the Lax-Friedrichs (weight $2 - s$) and the leap-frog scheme (weight $s - 1$). To the best of our knowledge, this regime has never been studied in the non-linear framework. For the linear framework, refer to [6].
- The under-relaxation for $s \in ]0, 1]$, where the weight for the leap-frog scheme is negative, meaning that we are adding "way more" Lax-Friedrichs. Still, the scheme is a convex combination of a scheme for the wave equation (weight $1 - s$) and Lax-Friedrichs scheme (weight $s$). It is harder to study using the corresponding Finite Difference scheme as far as monotonicity is involved. This regime has been studied by [2] on the original lattice Boltzmann formulation (3), (4) and (5), exploiting the fact that the second equation in (4) features a convex combination of $\mathsf{v}_j^n$ and $\varphi(\mathsf{u}_j^n)$.

Observe that the initialization scheme (7) is different from the bulk scheme (8) for general $s$ because of the freedom in the initialization of the non-conserved moment $\mathsf{v}$. Taking its value at equilibrium yields (7) being the Lax-Friedrichs scheme, which coincides with (8) only when $s = 1$, that is, unsurprisingly, when the relaxation (4) is done on the equilibrium and the bulk scheme has one step, thus does not need to be complemented by an initialization scheme.

The aim of our work is to study the original lattice Boltzmann scheme (3), (4) and (5) using its corresponding Finite Difference scheme (6), (7) and (8), forgetting about where the latter comes from, knowing that it contains all information that we need. Since we have taken $u^{\circ} \in L^{\infty}(\mathbb{R})$, we shall denote $\underline{\mathsf{u}}^0 := \inf_{j\in\mathbb{Z}} \mathsf{u}_j^0$ and $\overline{\mathsf{u}}^0 := \sup_{j\in\mathbb{Z}} \mathsf{u}_j^0$ and we introduce the Courant number for the compact set $[\underline{\mathsf{u}}^0, \overline{\mathsf{u}}^0]$ given by $C := \left(\max_{\mathsf{u}\in[\underline{\mathsf{u}}^0,\overline{\mathsf{u}}^0]} |\varphi'(\mathsf{u})|\right)/\lambda$. The main result about convergence that shall be proved in Sect. 3 by means of several intermediate results is the following.

**Theorem 1** *Under the condition*

$$s \in [1, 2], \qquad \textit{and} \qquad |C| \leq \frac{2-s}{s}, \tag{9}$$

*setting*

$$\mathsf{u}_{\Delta t,\Delta x}(t, x) := \sum_{n\in\mathbb{N}}\sum_{j\in\mathbb{Z}} \mathsf{u}_j^n 1_{[t^n,t^{n+1}[}(t) 1_{[x_j,x_{j+1}[}(x), \qquad (t, x) \in \mathbb{R}_+ \times \mathbb{R}, \tag{10}$$

*where $\mathsf{u}^n$ is the solution of the corresponding Finite Difference scheme (6), (7) and (8), there exists a sub-sequence, also denoted $(\mathsf{u}_{\Delta t,\Delta x})_{\Delta t,\Delta x}$ for brevity, and a function $\overline{u}$ such that $\overline{u} \in L^{\infty}(\mathbb{R}_+ \times \mathbb{R}_+) \cap L^1([0, T] \times \mathbb{R}) \cap BV([0, T] \times \mathbb{R})$ for all $T > 0$, with $\mathsf{u}_{\Delta t,\Delta x} \to \overline{u}$ in $L^1_{loc}(\mathbb{R}_+ \times \mathbb{R})$ as $\Delta t, \Delta x \to 0$. Moreover, $\overline{u}$ is the unique weak entropic solution $u$ of (1) and (2).*

## 3 Proof

The proof of Theorem 1 is split into several steps. Using the monotonicity of (8), we show that this Finite Difference scheme admits an invariant compact set (maximum principle). This secures total variation estimates in space and time, allowing to show the boundedness of the numerical solution for suitable norms and thus employ results on relative compactness. The convergence of the extracted sub-sequence to the entropic weak solution is guaranteed by discrete entropy inequalities based on Krushkov entropies.

### *3.1 Monotonicity*

Following the usual notations for one-step explicit $(2k + 1)$-points schemes [4], (8) can be written as

$$\mathsf{u}_j^{n+1} = H(\mathsf{u}_{j-1}^n, \mathsf{u}_{j+1}^n, \mathsf{u}_j^{n-1}), \tag{11}$$

with $H(a, b, c) = ((1 - \frac{s}{2})a + \frac{s\varphi(a)}{2\lambda}) + ((1 - \frac{s}{2})b - \frac{s\varphi(b)}{2\lambda}) + (s - 1)c$.

**Proposition 1** (Monotone Finite Difference scheme) *The non-linear Finite Difference scheme (8) is monotone, namely $H$ in (11) is monotone non-decreasing with* *respect to each of its arguments when they belong to the compact* $[\underline{\mathsf{u}}^0, \overline{\mathsf{u}}^0]$*, if and only if* (9) *holds.*

Observe that this kind of monotonicity property holds uniquely in the over-relaxation regime $s \geq 1$ and treats every point in the time-space domain of dependence of $H$ in the same fashion.

## 3.2 Maximum Principle

From Proposition 1, we have:

**Proposition 2** (Maximum principle) *Under condition* (9) *and using the initialization* (7)*, the solution of the non-linear Finite Difference scheme* (8) *is such that*

$$\mathsf{u}_j^n \in [\underline{\mathsf{u}}^0, \overline{\mathsf{u}}^0], \qquad n \in \mathbb{N}, \qquad j \in \mathbb{Z}. \tag{12}$$

***Proof*** The proof proceeds by induction. Using the mean value theorem, the scheme is "linearized" and its coefficients are positive and sum to one, thus it is a convex combination of the datum at the previous time steps at different locations in space. To provide an example, for any $j \in \mathbb{Z}$, the mean value theorem guarantees that there exists $\overline{\mathsf{u}}_j^n \in ]\underline{\mathsf{u}}^0, \overline{\mathsf{u}}^0[$ such that

$$\varphi(\mathsf{u}_{j-1}^n) - \varphi(\mathsf{u}_{j+1}^n) = \varphi'(\overline{\mathsf{u}}_j^n)(\mathsf{u}_{j-1}^n - \mathsf{u}_{j+1}^n), \tag{13}$$

because $\mathsf{u}_{j-1}^n, \mathsf{u}_{j+1}^n \in [\underline{\mathsf{u}}^0, \overline{\mathsf{u}}^0]$ by induction assumption and $\varphi \in C^1(\mathbb{R})$. All these arguments together yield the claim.

## 3.3 Total Variation Estimates

Thanks to the invariant compact set, we can control total variations.

**Proposition 3** (Total variation estimates) *Under condition* (9) *and using the initialization* (7)*, the solution of the non-linear Finite Difference scheme* (8) *is such that we have the space total variation estimate*

$$\mathrm{TV}(\mathsf{u}^n) \leq \mathrm{TV}(\mathsf{u}^0), \qquad n \in \mathbb{N}, \tag{14}$$

*and the time total variation estimate*

$$\sum_{j \in \mathbb{Z}} |\mathsf{u}_j^{n+1} - \mathsf{u}_j^n| \leq 2\mathrm{TV}(\mathsf{u}^0), \qquad n \in \mathbb{N}. \tag{15}$$

***Proof*** The proof proceeds by induction using the mean value theorem (for the solution at different spatial points for (14), similarly to (13), and for the solution at different times for (15)), relying on the fact that the solution lies in an invariant compact set at any point in space and time, and using the triangle inequality.

## *3.4 Relative Compactness and Extraction*

Then, we can extract a converging sub-sequence.

**Theorem 2** *Under condition (9), considering* $\mathsf{u}_{\Delta t,\Delta x}$ *given by (10), where* $\mathsf{u}^n$ *is the solution of the corresponding Finite Difference scheme (6), (7) and (8), there exists a sub-sequence, also denoted* $(\mathsf{u}_{\Delta t,\Delta x})_{\Delta t,\Delta x}$ *for brevity, and a function* $\overline{u}$ *such that* $\overline{u} \in L^\infty(\mathbb{R}_+ \times \mathbb{R}_+) \cap L^1([0,T] \times \mathbb{R}) \cap BV([0,T] \times \mathbb{R})$ *for all* $T > 0$*, with* $\mathsf{u}_{\Delta t,\Delta x} \to \overline{u}$ *in* $L^1_{loc}(\mathbb{R}_+ \times \mathbb{R})$ *as* $\Delta t, \Delta x \to 0$*.*

***Proof*** Let $\Omega$ be any open bounded Lipschitz set such that $\Omega \subset [0,T] \times [-C, C]$, for some suitable $T > 0$ and $C > 0$. Also, consider $n_T \in \mathbb{N}$ such that $(n_T - 1)\Delta t \leq T < n_T \Delta t$ and $N_x \in \mathbb{N}$ such that $(N_x - 1)\Delta x \leq C < N_x \Delta x$. Using Proposition 3, we get $|\mathsf{u}_{\Delta t,\Delta x}|_{\mathrm{BV}(\Omega)} \leq (2\lambda T + T + \Delta t)\mathrm{TV}(\mathsf{u}^0)$, which remains bounded as $\Delta t \to 0$. Furthermore, since Proposition 2 holds:

$$||\mathsf{u}_{\Delta t,\Delta x}||_{L^1(\Omega)} = \int\int_\Omega |\mathsf{u}_{\Delta t,\Delta x}(t,x)|\mathrm{d}x\mathrm{d}t \leq 2TC||\mathsf{u}^0||_{\ell^\infty} < +\infty. \tag{16}$$

We conclude by using the compactness theorem given in [4]

From now on, we understand the extraction of sub-sequence claimed by Theorem 2.

## *3.5 Convergence to a Weak Solution*

Using elementary algebra, we recast the Finite Difference scheme (8) under the form

$$\begin{aligned} \frac{\mathsf{u}_j^n - \mathsf{u}_j^{n-1}}{\Delta t} + \frac{1}{2\Delta x}(\varphi(\mathsf{u}_{j+1}^n) - \varphi(\mathsf{u}_{j-1}^n)) = &- \frac{1}{s\Delta t}(\mathsf{u}_j^{n+1} - 2\mathsf{u}_j^n + \mathsf{u}_j^{n-1}) \\ &+ \frac{\lambda}{\Delta x}\Big(\frac{1}{s} - \frac{1}{2}\Big)(\mathsf{u}_{j-1}^n - 2\mathsf{u}_j^n + \mathsf{u}_{j+1}^n), \end{aligned} \tag{17}$$

for $n \in \mathbb{N}^*$ and for $j \in \mathbb{Z}$. The left hand side is a (falsely) implicit discretization of the differential operators in the target equation, whereas we expect that, once the equation is multiplied by a test function and summations by parts are performed, the terms on the right hand side shall be vanishing because they respectively discretize $\Delta t/s\partial_{tt}$ and $-\lambda\Delta x(1/s - 1/2)\partial_{xx}$ of the test function (which are all bounded since the test function is smooth) at leading order. This formulation is non-standard for explicit methods like the one we are considering.

**Theorem 3** *Under condition* (9) *and using initialization* (7)*, the limit* $\overline{u}$ *given by Theorem* 2 *is a weak solution of* (1) *and* (2).

***Proof*** The proof is achieved by multiplying (17) by a test sequence being the discretization of a smooth function, using summations by parts, and passing to the limit in $\Delta x$. In this way, we obtain the weak formulation of (1) and (2).

### 3.6 Convergence to the Weak Entropic Solution

Contrarily to [2], we devise discrete entropy inequalities by making use of Krushkov entropies. The latter come under the form $\eta(u) = |u - \kappa|$ and $q(u) = \text{sign}(u - \kappa)(\varphi(u) - \varphi(\kappa)) = \varphi(u \top \kappa) - \varphi(u \perp \kappa)$ for any $\kappa \in \mathbb{R}$, where $a \top b := \max(a, b)$ and $a \perp b := \min(a, b)$ for any $a, b \in \mathbb{R}$. Hence, for any $\kappa \in \mathbb{R}$, we have the discrete entropy inequality for (8), which reads

$$\begin{aligned}
&\frac{|\mathsf{u}_j^n - \kappa| - |\mathsf{u}_j^{n-1} - \kappa|}{\Delta t} \\
&\quad + \frac{1}{2\Delta x}((\varphi(\mathsf{u}_{j+1}^n \top \kappa) - \varphi(\mathsf{u}_{j+1}^n \perp \kappa)) - (\varphi(\mathsf{u}_{j-1}^n \top \kappa) - \varphi(\mathsf{u}_{j-1}^n \perp \kappa))) \\
&\quad \le -\frac{1}{s\Delta t}(|\mathsf{u}_j^{n+1} - \kappa| - 2|\mathsf{u}_j^n - \kappa| + |\mathsf{u}_j^{n-1} - \kappa|) \\
&\quad + \frac{\lambda}{\Delta x}\Big(\frac{1}{s} - \frac{1}{2}\Big)(|\mathsf{u}_{j-1}^n - \kappa| - 2|\mathsf{u}_j^n - \kappa| + |\mathsf{u}_{j+1}^n - \kappa|), \qquad (18)
\end{aligned}$$

and is valid under (9) thanks to the monotonicity of $H$.

**Theorem 4** *Under condition* (9) *and using initialization* (7)*, the limit* $\overline{u}$ *given by Theorem* 2 *is the weak entropic solution of* (1) *and* (2).

***Proof*** The proof is achieved by multiplying (18) by a test sequence being the discretization of a smooth function, using summations by parts, and passing to the limit in $\Delta x$. In this way, we obtain the weak entropy inequality for the Krushkov entropies.

## 4 Numerical Simulations

To numerically check that the scheme converges in the $L^1$ norm (cf. Theorem 1) and that the maximum principle is fulfilled (cf. Proposition 2) under (9), we consider the Burgers flux $\varphi(u) = u^2/2$ with final time $T = 1$ and initial datum

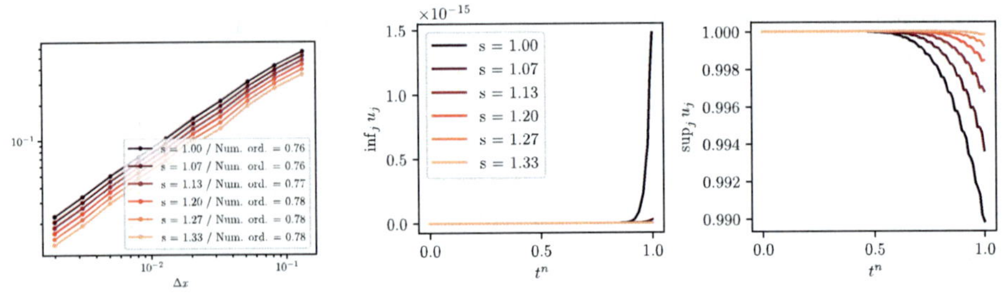


**Fig. 1** Burgers flux. Left: error at final time $T$ in the $L^1$ norm between numerical and exact solution with the numerical orders of convergence. Right: minimum and maximum of the numerical solution as function of the time

**Fig. 2** Advection flux. Error at final time $T$ in the $L^1$ norm between numerical and exact solution with the numerical orders of convergence

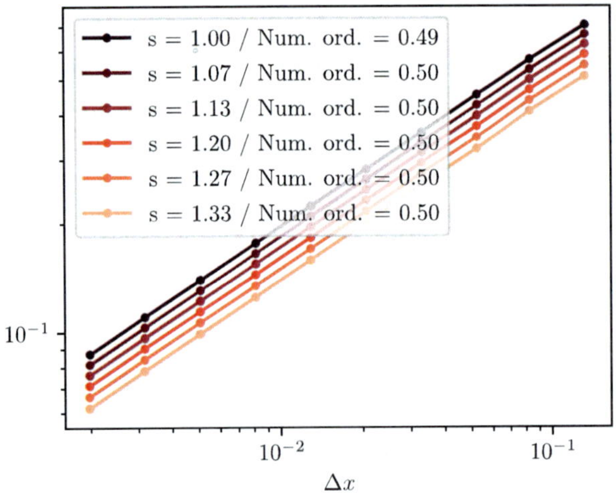


$u^{\circ}(x) = \mathbb{1}_{[0,1/2]}(|x|)$. We simulate using the original lattice Boltzmann scheme (3), (4) and (5) on the bounded domain $[-2, 2]$ with periodic boundary conditions and $\lambda = 2$. The Courant number is $C = 1/2$, thus (9) is met for $s \in [1, 4/3]$. The error and the minimum/maximum of the solution given in Fig. 1 confirm all the previous discussion. The convergence rate—measured numerically—is around 0.8. Repeating the test using with the advection equation $\varphi(u) = u$, cf. Fig. 2, the convergence rate is $1/2$, as theoretically expected [7].

## 5 Conclusions and Important Open Questions

We have shown convergence for a non-linear two velocities lattice Boltzmann scheme in the over-relaxation regime relying on its corresponding multi-step Finite Difference scheme and trivially generalizing monotonicity to multi-step methods. This boils down to treating the solution at times $t^{n-1}$ and $t^n$ yielding the new one at time

$t^{n+1}$ with the same "policy". The open question concerns how to obtain the same result—which we know to be true cf. [2]—for the under-relaxation regime, under the condition $s \in ]0, 1]$ and $|C| \leq 1$. Understanding this would allow to study the whole monotonicity regime of more general lattice Boltzmann schemes using the corresponding Finite Difference scheme [3]. The issue comes from the sign of the coefficient in front of $\mathsf{u}_j^{n-1}$ in (8), hence (11) is not monotone, at least when treating all this arguments in the same way. Let us indicate some results that we are still capable of obtaining and those which remain unattained.

- We can obtain the invariant compact set (12) by following the propagation of information along the characteristic directions $x \pm \lambda t$ indicated by (5) and heavily relying on the specific initialisation (7). By doing so, we have to climb back time until the initial datum, see [8]. This is not surprising since for multi-step time integration routines [9], authors have observed that "...it is crucial to consider the linear multi-step method in combination with suitable starting procedures ...More importantly, it turns out that the insistence on arbitrary starting vectors severely limits the class of methods for which monotonicity can be demonstrated."
- We can prove (14) but we have not been able to show (15) (which nevertheless holds, see [2]) under the assumption $u^\circ \in L^\infty(\mathbb{R})$.
- The last major question concerns the discrete entropy inequality to use. Some thoughts on this are as follows. Does the discrete entropy inequality need to be multi-step like (18) or not? For example, could we employ the one for the Lax-Friedrichs scheme ($s = 1$) even when $s < 1$? A path could be to use some comparison principle concerning dissipation, see [10]. The idea is that if one proves that a scheme is more dissipative than another one satisfying an entropy inequality (the Lax-Friedrichs scheme?), the former also satisfies this entropy inequality. The problem is that here schemes are multi-step. For the one-step Finite Volume schemes, it is easy to quantify the numerical viscosity. In our case, the situation is probably more involved.

**Acknowledgements** The author thanks his PhD advisors Benjamin Graille and Marc Massot, as well as Christophe Chalons, for the discussions on this topic.

## References

1. Higuera, F.J., Jiménez, J.: EPL **9**(7), 663 (1989)
2. Caetano, F., Dubois, F., Graille, B.: DCDS-S (2023). https://doi.org/10.3934/dcdss.2023072
3. Bellotti, T., Graille, B., Massot, M.: Numer. Math. **152**(1), 1–40 (2022)
4. Godlewski, E., Raviart, P.-A.: Ellipses (1991)
5. Bellotti, T.: arXiv preprint arXiv:2302.07558 (2023)
6. Dellacherie, S.: Acta Appl. Math. **131**, 69–140 (2014)
7. Tao, T., Teng, Z.H.: Math. Comput. **64**(210), 581–589 (1995)
8. Cheng, S.S., Lu, Y.-F.: Comput. Math. Appl. **38**(7–8), 65–79 (1999)
9. Hundsdorfer, W., Ruuth, S.J., Spiteri, R.J.: SINUM **41**(2), 605–623 (2003)
10. Tadmor, E.: Math. Comput. **43**(168), 369–381 (1984)

# A Low Mach Number Two-Speed Relaxation Scheme for Ideal MHD Equations

**Claudius Birke and Christian Klingenberg**

**Abstract** In this work we apply the two-speed relaxation technique to a relaxation system for the compressible ideal magnetohydrodynamic (MHD) equations. We show that the resulting approximate Riemann solver reduces the dissipation in the low Mach number regime and can thus generate accurate solutions in this regime even on coarse grids. These findings are supported by numerical results.

**Keywords** Finite volume methods · Relaxation methods · Approximate Riemann solver · Low Mach number regime

## 1 Introduction

In practical applications such as the simulation of gas flows in the interior of stars, flows with large scale differences can occur. One example is the low sonic Mach number regime, in which the sound speed is much higher than the speed of the fluid flow. Numerically, for standard finite volume methods two difficulties arise in the low Mach number regime: excessive dissipation and a restrictive CFL condition. Basically, there are two strategies in the literature to deal with these challenges:

1. (a) Introduce a low Mach number fix in the numerical flux function to reduce the dissipation and (b) use an implicit time integration to overcome the restrictive CFL condition [10, 11].
2. Perform a splitting into convective and acoustic parts and solve the acoustic parts implicitly [7, 8].

C. Birke: CB acknowledges the support by the German Research Foundation (DFG) under the project no. KL 566/22-1.

C. Birke (✉) · C. Klingenberg
Department of Mathematics, University of Würzburg, Würzburg, Germany
e-mail: claudius.birke@mathematik.uni-wuerzburg.de

 
E. Franck et al. (eds.), *Finite Volumes for Complex Applications X—Volume 2, Hyperbolic and Related Problems*, Springer Proceedings in Mathematics & Statistics 433,
https://doi.org/10.1007/978-3-031-40860-1_5

In this paper we focus only on part (a) of the first strategy, i.e. the design of the numerical flow function and for practical reasons combine it with an explicit time-stepping method. Typically, from implementing a low Mach number fix follows that the solver does not satisfy a discrete entropy inequality anymore. This is different for the two-speed relaxation technique. As already shown for isentropic Euler [3], full Euler [4] and full Euler equations with gravity [2], the resulting approximate Riemann solver reduces dissipation in the low Mach number regime and still satisfies a discrete entropy inequality. In the spirit of these papers, we apply the two-speed technique to an existing relaxation system for the ideal MHD equations [5, 6]. In a first step we show how the artificial dissipation is reduced in the low Mach number regime and substantiate this theoretical result in numerical tests. We assume that the approximate Riemann solver satisfies a discrete entropy inequality as in previous work, but do not show it here. This is left to future work.

## 2 MHD Equations

The compressible ideal MHD equations can be written as

$$
\begin{aligned}
\frac{\partial \rho}{\partial t} + \nabla \cdot (\rho \boldsymbol{U}) &= 0, \\
\frac{\partial (\rho \boldsymbol{U})}{\partial t} + \nabla \cdot [\rho \boldsymbol{U} \otimes \boldsymbol{U} + (p + \frac{1}{2}|\boldsymbol{B}|^2)\boldsymbol{I} - \boldsymbol{B} \otimes \boldsymbol{B}] &= \boldsymbol{0}, \\
\frac{\partial E}{\partial t} + \nabla \cdot [(E + p + \frac{1}{2}|\boldsymbol{B}|^2)\boldsymbol{U} - \boldsymbol{B}(\boldsymbol{B} \cdot \boldsymbol{U})] &= 0, \\
\frac{\partial \boldsymbol{B}}{\partial t} + \nabla \cdot (\boldsymbol{U} \otimes \boldsymbol{B} - \boldsymbol{B} \otimes \boldsymbol{U}) &= \boldsymbol{0},
\end{aligned}
\tag{1}
$$

where $\rho$ denotes the density, $\boldsymbol{U} = (u, u_\perp)$ the velocity field, $\boldsymbol{B} = (B_x, B_\perp)$ the magnetic field and $p$ the gas pressure. The total energy $E$ is defined as $E = \rho e + \frac{1}{2}\rho|\boldsymbol{U}|^2 + \frac{1}{2}|\boldsymbol{B}|^2$, where $e$ denotes the specific internal energy. The system is closed by an equation of state, which provides the numerical value of the gas pressure. Throughout this work we will consider an ideal gas law defined by $p(\rho, e) = (\gamma - 1)\rho e$. In addition to the set of equations in (1), the magnetic field satisfies the solenoidal constraint $\nabla \cdot \boldsymbol{B} = 0$. Solutions to the system (1) automatically satisfy this condition at all times if the initial field obeys the constraint.

In order to analyze the behaviour of solutions in the low Mach number regime it is of help to determine the dimensionless form of the MHD equations.

$$
\begin{aligned}
&\frac{\partial \hat{\rho}}{\partial \hat{t}} + \hat{\nabla} \cdot (\hat{\rho}\hat{\boldsymbol{U}}) = 0, \\
&\frac{\partial(\hat{\rho}\hat{\boldsymbol{U}})}{\partial \hat{t}} + \hat{\nabla} \cdot \left[ \hat{\rho}\hat{\boldsymbol{U}} \otimes \hat{\boldsymbol{U}} + \left( \frac{\hat{p}}{\hat{\mathcal{M}}_{\text{son}}^2} + \frac{1}{\hat{\mathcal{M}}_{\text{Alf}}^2} \frac{1}{2} |\hat{\boldsymbol{B}}|^2 \right) \boldsymbol{I} - \frac{\hat{\boldsymbol{B}} \otimes \hat{\boldsymbol{B}}}{\hat{\mathcal{M}}_{\text{Alf}}^2} \right] = \boldsymbol{0}, \\
&\frac{\partial(\hat{\rho}\hat{E})}{\partial \hat{t}} + \hat{\nabla} \cdot \left[ \left( \hat{\rho}\hat{E} + \hat{p} + \frac{\hat{\mathcal{M}}_{\text{son}}^2}{\hat{\mathcal{M}}_{\text{Alf}}^2} \frac{1}{2} |\hat{\boldsymbol{B}}|^2 \right) \hat{\boldsymbol{U}} - \hat{\boldsymbol{B}} \left( \hat{\boldsymbol{B}} \cdot \hat{\boldsymbol{U}} \right) \frac{\hat{\mathcal{M}}_{\text{son}}^2}{\hat{\mathcal{M}}_{\text{Alf}}^2} \right] = 0, \\
&\frac{\partial \hat{\boldsymbol{B}}}{\partial \hat{t}} + \hat{\nabla} \cdot (\hat{\boldsymbol{U}} \otimes \hat{\boldsymbol{B}} - \hat{\boldsymbol{B}} \otimes \hat{\boldsymbol{U}}) = \boldsymbol{0},
\end{aligned} \tag{2}
$$

The different variables have been rescaled by some reference quantity representative of the physical system of interest: $t = t_r\hat{t}$, $x = x_r\hat{x}$, $\rho = \rho_r\hat{\rho}$, $\boldsymbol{U} = U_r\hat{\boldsymbol{U}}$, $E = a_r^2\hat{E}$, $p = a_r^2\hat{p}$, $\boldsymbol{B} = B_r\hat{\boldsymbol{B}}$, $a_r^2 = p_r/\rho_r$. $\hat{\mathcal{M}}_{\text{son}} = |U_r|/a_r$ and $\hat{\mathcal{M}}_{\text{Alf}} = |U_r|/(|B_r|/\sqrt{\rho_r})$ represent the characteristic sonic and Alfvén Mach numbers of the flow. The system (2) has seven eigenvalues, which can be expressed using the normal vector $\boldsymbol{n}$ by

$$
\lambda_{1,7} = \hat{\boldsymbol{U}} \cdot \boldsymbol{n} \mp c_{\text{f}}, \ \lambda_{2,6} = \hat{\boldsymbol{U}} \cdot \boldsymbol{n} \mp c_{\text{A}}, \ \lambda_{3,5} = \hat{\boldsymbol{U}} \cdot \boldsymbol{n} \mp c_{\text{s}}, \ \lambda_4 = \hat{\boldsymbol{U}} \cdot \boldsymbol{n}. \tag{3}
$$

The fast and slow magnetosonic and Alfvén wave speeds are defined by

$$
c_{\text{f,s}} = \left[ \frac{1}{2} \left( a^2 + \frac{1}{\hat{\mathcal{M}}_{\text{Alf}}} \frac{|\hat{\boldsymbol{B}}|^2}{\hat{\rho}} \pm \sqrt{ \left( a^2 + \frac{1}{\hat{\mathcal{M}}_{\text{Alf}}} \frac{|\hat{\boldsymbol{B}}|^2}{\hat{\rho}} \right)^2 - 4a^2 c_{\text{A}}^2 } \right) \right]^{\frac{1}{2}}, \tag{4}
$$

$$
c_{\text{A}} = \frac{1}{\hat{\mathcal{M}}_{\text{Alf}}} \frac{|\hat{\boldsymbol{B}} \cdot \boldsymbol{n}|}{\sqrt{\hat{\rho}}}, \tag{5}
$$

with the sound speed $a = \frac{1}{\hat{\mathcal{M}}_{\text{son}}}\sqrt{\frac{\gamma\hat{p}}{\hat{\rho}}}$. Note that adding the root term in (4) results in the fast wave speed $c_{\text{f}}$ while subtracting it in the slow wave speed $c_{\text{s}}$. In this work we only consider low sonic Mach numbers, but not low Alfvén Mach numbers. Therefore, we set $\hat{\mathcal{M}}_{\text{Alf}} = 1$ in the rest of the paper.

## 3 Relaxation Model

Our goal is to derive a one-dimensional Riemann solver. Therefore, in the following we only consider one spatial dimension. In [5, 6] a relaxation system is described, which leads to a 5-wave approximate Riemann solver for the MHD equations. For problems with low Mach numbers, however, this solver introduces excessive dissipation, so that the solution is poorly resolved on coarse grids (see Sect. 5). To cure this defect, we reformulate this relaxation system using a two-speed relaxation technique [3, 4]. The resulting system writes

$$
\begin{aligned}
\partial_t \rho + \partial_x(\rho v) &= 0,\\
\partial_t(\rho u) + \partial_x(\rho u v + \pi) &= 0,\\
\partial_t(\rho u_\perp) + \partial_x(\rho u_\perp v + \pi_\perp) &= 0,\\
\partial_t E + \partial_x\left((E+\pi)\, v + \pi_\perp \cdot u_\perp\right) &= 0,\\
\partial_t B_x + v \partial B_x &= 0,\\
\partial_t B_\perp + \partial_x\left(B_\perp v - B_x u_\perp\right) + u_\perp \partial_x B_x &= 0,\\
\partial_t(\rho\pi) + \partial_x\left(\rho \pi v\right) + c_1 c_2 \partial_x v &= \rho \frac{p + |B_\perp|^2/2 - B_x^2/2 - \pi}{\varepsilon},\\
\partial_t(\rho\pi_\perp) + \partial_x\left(\rho \pi_\perp v\right) + c_a^2 \partial_x u_\perp &= \rho \frac{-B_x B_\perp - \pi_\perp}{\varepsilon},\\
\partial_t(\rho v) + \partial_x\left(\rho v^2\right) + \frac{c_1}{c_2}\partial_x \pi &= \rho \frac{u - v}{\varepsilon},\\
\partial_t c_1 + v \partial_x c_1 &= 0,\\
\partial_t c_2 + v \partial_x c_2 &= 0,\\
\partial_t c_a + v \partial_x c_a &= 0,
\end{aligned}
\tag{6}
$$

where the pressure variables $\pi$ and $\pi_\perp$ are defined in relaxation equilibrium by

$$
\pi = p + \frac{1}{2}|B_\perp|^2 - \frac{1}{2}B_x^2 \quad \text{and} \quad \pi_\perp = -B_x B_\perp. \tag{7}
$$

In this two-speed approach the relaxation speed for the normal direction is split into two different speeds and additionally a new relaxation equation for the velocity is introduced. At this point we would like to point out that the relaxation system (6) has a total number of three relaxation speeds ($c_1$, $c_2$ and $c_a$), but that we always refer to $c_1$ and $c_2$ when using the term *two-speed*.

As long as the relaxation speeds $c_1$, $c_2$ and $c_a$ satisfy the stability conditions

$$
\begin{aligned}
&\frac{1}{\rho} - \frac{B_x^2}{c_a^2} \geq 0, \quad c_1 c_2 - \rho^2 a^2 \geq 0,\\
&|B_\perp| \leq (c_1 c_2 - \rho^2 a^2)\left(\frac{1}{\rho} - \frac{B_x^2}{c_a^2}\right), \quad c_1 \geq c_2,
\end{aligned}
\tag{8}
$$

solutions of the relaxation system are viscous approximations of the solutions of the original MHD system. The conditions in (8) are the result of a Chapman-Enskog expansion for the relaxation system (6) that follows the same steps as the one for the one-speed system in [5]. The homogeneous system $(6)_{\varepsilon=\infty}$ is hyperbolic and admits 5 different eigenvalues

$$
\lambda_1 = v - \frac{c_1}{\rho}, \quad \lambda_2 = v - \frac{c_a}{\rho}, \quad \lambda_3 = v \quad \lambda_4 = v + \frac{c_a}{\rho} \quad \lambda_5 = v + \frac{c_1}{\rho}, \tag{9}
$$

where the eigenvalue $v$ has multiplicity 8. All characteristic fields are linearly degenerate and the Riemann invariants for the different waves are given as follows

$$
\begin{aligned}
v-\frac{c_1}{\rho}:&\quad v-\frac{c_1}{\rho},\ u-\frac{c_2}{\rho},\ \pi+\frac{c_1c_2}{\rho},\ \frac{B_\perp}{\rho},u_\perp,\ B_x,\ \pi_\perp,\ c_1,\ c_2,\ c_a,\\
&\quad \frac{2c_1c_2-c_2^2-2c_2\rho(v-u)+\rho\left(2\pi+|B_\perp|^2+B_x^2+2\rho e\right)}{2\rho^2},\\
v-\frac{c_a}{\rho}:&\quad \pi_\perp+c_a u_\perp,\ B_\perp-\frac{\rho u_\perp B_x}{c_a},\ e+\frac{c_a^2u_\perp^2+2c_a u_\perp(B_xB_\perp+\pi_\perp)-\rho B_x^2u_\perp^2}{2c_a^2},\\
&\quad \rho,\ u,\ B_x,\ \pi,\ v,\ c_1,\ c_2,\ c_a,\\
v:&\quad v,\ u_\perp,\ \pi,\ \pi_\perp,\\
v+\frac{c_a}{\rho}:&\quad \pi_\perp-c_a u_\perp,\ B_\perp+\frac{\rho u_\perp B_x}{c_a},\ e+\frac{c_a^2u_\perp^2-2c_a u_\perp(B_xB_\perp+\pi_\perp)-\rho B_x^2u_\perp^2}{2c_a^2},\\
&\quad \rho,\ u,\ B_x,\ \pi,\ v,\ c_1,\ c_2,\ c_a,\\
v+\frac{c_1}{\rho}:&\quad v+\frac{c_1}{\rho},\ u+\frac{c_2}{\rho},\ \pi+\frac{c_1c_2}{\rho},\ \frac{B_\perp}{\rho},\ u_\perp,\ B_x,\ \pi_\perp,\ c_1,\ c_2,\ c_a,\\
&\quad \frac{2c_1c_2-c_2^2+2c_2\rho(v-u)+\rho\left(2\pi+|B_\perp|^2+B_x^2+2\rho e\right)}{2\rho^2}.
\end{aligned}
$$

Let us now introduce the state vector

$$W=(\rho,\rho u,\rho u_\perp,E,B_x,B_\perp,\rho\pi,\rho\pi_\perp,\rho v,c_1,c_2,c_a)^T. \tag{10}$$

The solution to the Riemann problem associated with system $(6)_{\varepsilon=\infty}$ given by

$$W_0(x)=\begin{cases}W^L, & x<0,\\ W^R, & x>0,\end{cases} \tag{11}$$

consists of six constant states. Therefore the Riemann solver $W_{\mathcal{R}}(x/t;W^L,W^R)$ has the structure

$$W_{\mathcal{R}}(x/t;W^L,W^R)=\begin{cases}W^L, & \frac{x}{t}<\min(\lambda_1,\lambda_2),\\ W^{L*}, & \min(\lambda_1,\lambda_2)<\frac{x}{t}<\max(\lambda_1,\lambda_2),\\ W^{L**}, & \max(\lambda_1,\lambda_2)<\frac{x}{t}<\lambda_3,\\ W^{R**}, & \lambda_3<\frac{x}{t}<\min(\lambda_4,\lambda_5),\\ W^{R*}, & \min(\lambda_4,\lambda_5)<\frac{x}{t}<\max(\lambda_4,\lambda_5),\\ W^R, & \max(\lambda_4,\lambda_5)<\frac{x}{t}.\end{cases} \tag{12}$$

The intermediate states can be derived using the invariants from above. Since each Riemann invariant is constant accross its corresponding wave, the invariants pose a system of equations that can be solved for the intermediate states. Due to their large number, we will not list the intermediate states here. The definition of the relaxation speeds is given in Sect. 4.

## 4 Definition of the Relaxation Speeds and the Low Mach Number Property

Starting from the stability conditions in (8), the relaxation speeds for the two-speed solver can be derived with steps very similar to those in [6] for $k = 1, 2$:

$$\begin{aligned} c_k^L &= \rho^L a_{0,k}^L + \beta \rho^L \left( (v^L - v^R)_+ + \frac{(\pi^R - \pi^L)_+}{\rho^L a_q^L + \rho^R a_q^R} \right), \\ c_k^R &= \rho^R a_{0,k}^R + \beta \rho^R \left( (v^L - v^R)_+ + \frac{(\pi^L - \pi^R)_+}{\rho^L a_q^L + \rho^R a_q^R} \right), \\ c_a^L &= \sqrt{\frac{\rho^L}{x^L} \left( (B_x^L)^2 + |B_x^L B_\perp^L| \right)}, \quad c_a^R = \sqrt{\frac{\rho^R}{x^R} \left( (B_x^R)^2 + |B_x^R B_\perp^R| \right)}, \end{aligned} \tag{13}$$

with

$$a_{0,1}^2 = \frac{1}{\min\left(\phi^2, 1\right)} a^2 + \frac{|B_\perp|^2 + |B_x B_\perp|}{\rho x}, \tag{14}$$

$$a_{0,2}^2 = \min\left(\phi^2, 1\right) a^2 + \frac{|B_\perp|^2 + |B_x B_\perp|}{\rho x}, \tag{15}$$

$$a_q^2 = \min\left(\phi^2, 1\right) a^2 + \frac{|B_\perp|^2 + |B_x B_\perp|}{\rho}. \tag{16}$$

For more details on the definitions of the used quantities, we refer the reader to [6]. We want to point out that for $\phi = 1$, we get $c_1 = c_2$ and consequently the speeds coincide with the definitions in the one-speed solver. However, to get the desired effect of the two-speed approach, we set $\phi = \hat{\mathcal{M}}_{\text{son}}$. As a consequence, taking into account the results of Sect. 2, the speeds scale in this case by

$$c_1 = \mathcal{O}(1/\hat{\mathcal{M}}_{\text{son}}^2) \quad \text{and} \quad c_2 = \mathcal{O}(1). \tag{17}$$

The rescaling by the Mach number in (14)–(16) is essential to prevent excessive numerical dissipation in the low Mach number regime. This becomes clear when looking at the intermediate state of the pressure $\pi$ given by

$$\pi^* = \frac{c_2^R \pi^L + c_2^L \pi^R - c_2^L c_2^R (v^R - v^L)}{c_2^L + c_2^R}. \tag{18}$$

This state can be viewed as a central flux of the left and right pressure combined with a dissipation term including the velocity difference $v^R - v^L$. Thanks to the rescaling, $c_2$ is of order $\mathcal{O}(1)$, which keeps the amount of dissipation in the pressure independent of the Mach number. For the one-speed solver, on the other hand, the relaxation speed $c = c_1 = c_2$ scales with $\mathcal{O}(1/\hat{\mathcal{M}}_{son})$. Thus, in the low Mach number regime, the dissipation term becomes the dominant part of the momentum flux, causing the dissipation to increase with decreasing Mach number. The rescaling in the two-speed approach prevents this behaviour.

## 5 Numerical Results

The one-dimensional approximate Riemann solver (12) is incorporated into a second order two-dimensional unsplit finite volume method, that uses a standard linear reconstruction with minmod-limiter. By applying the Riemann solver separately in both spatial directions, the reduced dissipation in the low Mach number regime due to the two-speed approach is inherited by the overall scheme. In order to keep the divergence of the magnetic field to machine precision, the scheme is combined with a second order staggered constrained transport method [9]. For time discretisation, the explicit method of Heun is used under the CFL condition

$$\Delta t \leq \frac{\text{CFL}}{4} \frac{\min(\Delta x, \Delta y)}{\max_i(|\lambda_i|)} \quad \text{with} \quad \text{CFL} = 0.9. \tag{19}$$

For comparison, we also show results of the one-speed solver, in which the relaxation speeds are not rescaled with the Mach number and thus we set $c_1 = c_2$.

### 5.1 *Stationary Vortex*

We start with a stationary version of the Balsara vortex that can be set up with different maximum Mach numbers [1]. We compute the numerical solution on a $64 \times 64$ grid and evaluate the distribution of the local Mach number $\mathcal{M}_{loc} = |\boldsymbol{U}|/a$ relative to the maximum Mach number $\mathcal{M}_{max}$ in the initial data. The results for different $\mathcal{M}_{max}$ after one turn of the vortex ($t_f = 5.9055/\mathcal{M}_{max}$) are shown in Fig. 1. Clearly, the one-speed solver introduces too much artificial dissipation, so that the vortex is not resolved accurately for smaller Mach numbers. With the two-speed method, on the other hand, the dissipation is independent of the Mach number. This advantage is paid for by a smaller timestep in comparison to the one-speed method, since the rescaling of $c_1$ increases the fastest wave speed in the CFL condition.

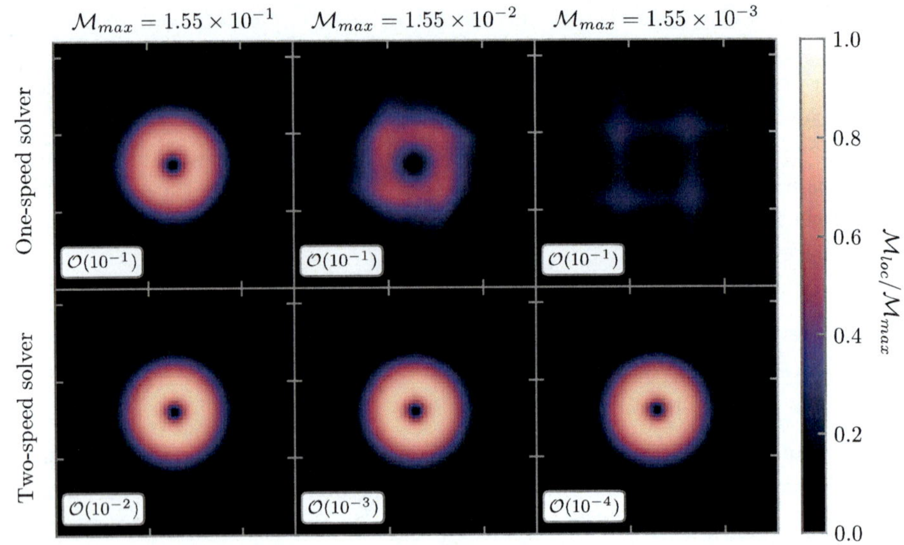


**Fig. 1** Distribution of the local sonic Mach number relative to $\mathcal{M}_{max}$ after one turn on a $64 \times 64$ grid. The value in the white box gives the order of the timestep

## 5.2 *Kelvin Helmholtz Instability*

A more sophisticated test is the magnetised version of the Kelvin Helmholtz instability [10]. This test is set up with different maximum Mach number of the initial horizontal flow denoted by $\mathcal{M}_x$. We compute the solution on a $128 \times 64$ grid until the final time $t_f = 0.8/\mathcal{M}_x$. Figure 2 shows the distribution of the local sonic Mach

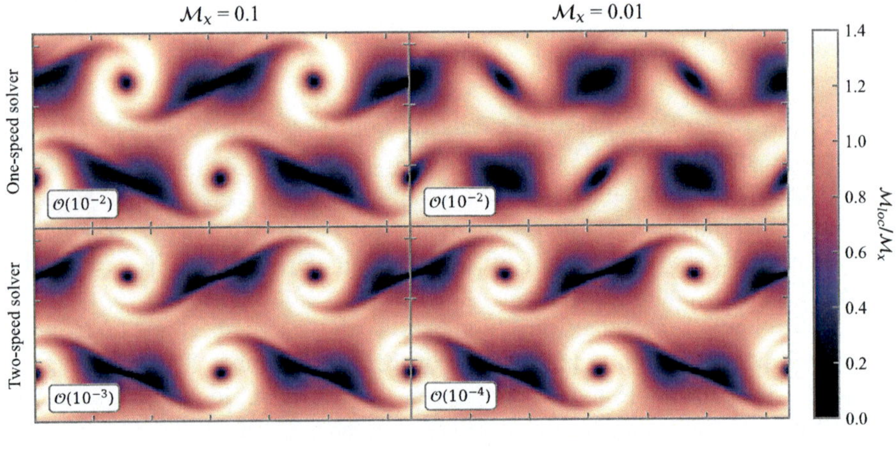


**Fig. 2** Distribution of the local sonic Mach number relative to $\mathcal{M}_x$ at time $t_f$ on a $128 \times 64$ grid. The value in the white box gives the order of the timestep

number $\mathcal{M}_{loc}$ relative to $\mathcal{M}_x$. The test again shows that the two-speed solver introduces significantly less numerical dissipation and can therefore resolve the details of the solution much better.

## 6 Summary

In this work, we have built an approximate Riemann solver using the two-speed technique, which gives accurate solutions even in the low Mach number regime. Typically, relaxation-based Riemann solvers satisfy a discrete entropy inequality when the subcharacteristic condition is satisfied. The goal of future work will be to prove such a result for the solver described here. Furthermore, it would be interesting to investigate whether the two-speed approach can also be applied or extended to the low Alfvén Mach number regime.

## References

1. Berberich, J.P., Chandrashekar, P., Klingenberg, C.: High order well-balanced finite volume methods for multi-dimensional systems of hyperbolic balance laws. Comput. Fluids **219**, 104858 (2021)
2. Birke, C., Chalons, C., Klingenberg, C.: A low Mach two-speed relaxation scheme for the compressible Euler equations with gravity (2023). arXiv:2112.02986v3
3. Bouchut, F., Chalons, C., Guisset, S.: An entropy satisfying two-speed relaxation system for the barotropic Euler equations: application to the numerical approximation of low Mach number flows. Numerische Mathematik **145**, 35–76 (2020)
4. Bouchut, F., Franck, E., Navoret, L.: A low cost semi-implicit low-Mach relaxation scheme for the full Euler equations. J. Sci. Comput. **83**(1), 24 (2020)
5. Bouchut, F., Klingenberg, C., Waagan, K.: A multiwave approximate Riemann solver for ideal MHD based on relaxation. I: theoretical framework. Numerische Mathematik **108**, 7–42 (2007)
6. Bouchut, F., Klingenberg, C., Waagan, K.: A multiwave approximate Riemann solver for ideal MHD based on relaxation II: numerical implementation with 3 and 5 waves. Numerische Mathematik **115**, 647–679 (2010)
7. Chen, W., Wu, K., Xiong, T.: High order asymptotic preserving finite difference WENO schemes with constrained transport for MHD equations in all Sonic Mach numbers (2022). arXiv:2211.16655
8. Dumbser, M., Balsara, D.S., Tavelli, M., Fambri, F.: A divergence-free semi-implicit finite volume scheme for ideal, viscous, and resistive magnetohydrodynamics. Int. J. Numer. Methods Fluids **89**(1–2), 16–42 (2019)
9. Gardiner, T.A., Stone, J.M.: An unsplit Godunov method for ideal MHD via constrained transport. J. Comput. Phys. **205**(2), 509–539 (2005)
10. Leidi, G., Birke, C., Andrassy, R., Higl, J., Edelmann, P.V.F., Wiest, G., Klingenberg, C., Röpke, F.K.: A finite-volume scheme for modeling compressible magnetohydrodynamic flows at low Mach numbers in stellar interiors. Astron. Astrophys. **668**(A143) (2022)
11. Minoshima, T., Miyoshi, T.: A low-dissipation HLLD approximate Riemann solver for a very wide range of Mach numbers. J. Comput. Phys. **446**, 110639 (2021)

# Domain of Dependence Stabilization for the Acoustic Wave Equation on 2D Cut-Cell Meshes

**Gunnar Birke, Christian Engwer, Sandra May, and Florian Streitbürger**

**Abstract** Cut-cell meshes are an attractive alternative to avoid common mesh generation problems. For hyperbolic problems they pose additional challenges, as elements can become arbitrarily small, leading to prohibitive time step restrictions for explicit time stepping methods. To alleviate this small cell problem we consider a particular stabilization method, the Domain of Dependence (DoD) method. So far, while possessing many favorable theoretical properties, in two dimensions the DoD method was essentially restricted to the transport equation. In this work we extend the DoD method to the acoustic wave equation in two dimensions and provide numerical results for validation.

**Keywords** Cut-cells · Small cell problem · Discontinuous Galerkin method · DoD stabilization · Wave equation

## 1 Introduction

The generation of traditional body fitted meshes can become very involved and time consuming, when working with complex geometries. One possible alternative is the use of cut-cell meshes. The drawback is that one has no control over the cut-cell shapes, in particular cut-cells can become arbitrarily small. Explicit time stepping methods, which are commonly used to simulate hyperbolic conservation laws, require a time step size that is chosen based on the smallest cell in the mesh. This becomes

G. Birke (✉) · C. Engwer
Applied Mathematics, Münster University, Münster, Germany
e-mail: gunnar.birke@uni-muenster.de

S. May
Department of Information Technology, Uppsala University, Uppsala, Sweden

F. Streitbürger
Fakultät für Mathematik, Dortmund University, Dortmund, Germany



E. Franck et al. (eds.), *Finite Volumes for Complex Applications X—Volume 2, Hyperbolic and Related Problems*, Springer Proceedings in Mathematics & Statistics 433,
https://doi.org/10.1007/978-3-031-40860-1_6

infeasible on a cut-cell mesh. Instead, one likes to choose the time step based on the size of the larger uncut cells. This is referred to as the small cell problem.

As discontinuous Galerkin (dG) methods are attractive to solve hyperbolic partial differential equations (PDEs), different stabilization approaches have been developed in recent years to handle the small cell problem on cut-cell meshes, see for example [9, 12]. We consider the Domain of Dependence (DoD) stabilization method which was introduced in [7] for the linear transport equation and extended, in one space dimension, to non-linear equations in [11]. In two dimensions, the method was originally restricted to certain flow/geometry combinations, e. g. a flow parallel to a domain boundary. In [4] the stabilization was generalized for $P^0$ discretizations of linear systems to handle triangular cut-cells with multiple inflow or outflow faces.

In this contribution we extend the lowest order DoD stabilization such that it can be applied to the acoustic wave equation. The main difference is that the wave equation cannot be written as a system of coupled transport equations, as the system is not globally diagonalizable. This adds significant complications in deriving the stabilization term. We focus on a $P^0$ dG discretization here, the development of appropriate higher-order stabilization terms for the acoustic wave equation is ongoing research. The construction of the DoD stabilization went hand in hand with an $L^2$-stability analysis for the semi-discretization in space, see [4, 11, 13], and the terms were designed such that we regain the spatial stability properties of the original dG scheme. This also holds true for the acoustic wave equation and the analysis was an important cornerstone in the development of the stabilization term. As the actual $L^2$-stability proof goes beyond the scope of this paper, we add appropriate remarks where necessary to understand the design of the actual formulation.

The outline of the paper is as follows: We first describe the problem setup and the underlying dG scheme. Then we introduce the DoD stabilization for the acoustic wave equation. We conclude with numerical results to validate our findings.

## 2 Problem Setup

Let $\Omega \subset \mathbb{R}^2$ be an open domain. We consider the acoustic wave equation for the solution $u = (p, v_1, v_2)$ given by

$$\begin{aligned} u_t + A_1 u_x + A_2 u_y &= 0 \quad \text{in } \Omega, \\ \tau u &= g \quad \text{on } \partial\Omega, \\ u(\cdot, 0) &= u_0 \quad \text{in } \Omega, \end{aligned}$$

with $p$ being the pressure, $v = (v_1, v_2)^t$ being the velocity and $\tau$ being a boundary operator imposing an inflow boundary condition for incoming waves. The system matrices are given as

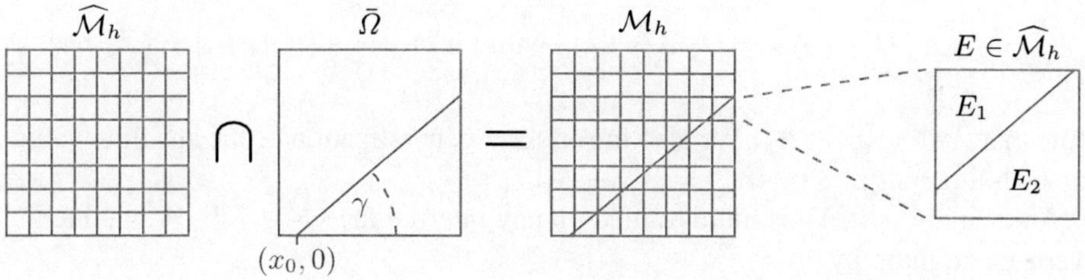


**Fig. 1** Construction of the mesh: Out of the structured grid $\widehat{\mathcal{M}_h}$ on the domain $\Omega$ the mesh $\mathcal{M}_h$ is constructed by introducing cut-cells $E_1, E_2 \subset E \in \widehat{\mathcal{M}_h}$ along the cut such that $\bar{E}_1 \cup \bar{E}_2 = \bar{E}$

$$A_1 = \begin{pmatrix} 0 & c & 0 \\ c & 0 & 0 \\ 0 & 0 & 0 \end{pmatrix} \text{ and } A_2 = \begin{pmatrix} 0 & 0 & c \\ 0 & 0 & 0 \\ c & 0 & 0 \end{pmatrix}.$$

Here $c > 0$ denotes the speed of sound. We want to stress that $A_1 A_2 \neq A_2 A_1$.

In our numerical tests we choose $\Omega = [0, 1]^2$ and discretize it by a structured grid $\widehat{\mathcal{M}_h}$. We then introduce an artificial cut, a straight line going through the square, starting at $(x_0, 0)$ and having an angle $\gamma$ relative to the $x$-axis. This creates an internal boundary with two subdomains which we will resolve by a cut-cell mesh $\mathcal{M}_h$. A sketch is contained in Fig. 1. The wave equation is then solved in the whole domain $\Omega$.

Our discrete function space is defined as

$$\mathcal{V}_h = \mathcal{V}_h^0 = \{v_h \in L^2(\Omega)^3 : (v_h)_i\big|_E \in \mathcal{P}^0(E)\, \forall\, 1 \leq i \leq 3,\ E \in \mathcal{M}_h\}.$$

We define the sets of internal and external faces as

$$\mathcal{F}_h^{\text{int}} = \{F = \partial E_1 \cap \partial E_2 : E_1, E_2 \in \mathcal{M}_h,\ E_1 \neq E_2,\ |F| > 0\},$$
$$\mathcal{F}_h^{\text{ext}} = \{F = \partial E \cap \partial \Omega : E \in \mathcal{M}_h,\ |F| > 0\}$$

and the set of neighbor cells $\mathcal{N}(E) := \{E' \in \mathcal{M}_h : |\bar{E}' \cap \bar{E}| > 0\}$.

So for any internal face $F \in \mathcal{F}_h^{\text{int}}$ there are always two unique elements $E_1, E_2 \in \mathcal{M}_h$ such that $\bar{E}_1 \cap \bar{E}_2 = F$. This face will often be denoted by $F_{E_1,E_2} = F_{E_2,E_1}$. We fix once and for all an orientation on $F = F_{E_1,E_2}$ by setting its outer normal vector $n$ to be $n = n_F = n_F(x) := n_{E_1}(x)$ for $x \in F$ where $n_{E_1}$ is the outer unit normal field on $\partial E_1$. The flux matrix in normal direction on the face $F$ will be denoted by

$$A_F := (n_F)_1 A_1 + (n_F)_2 A_2 = O_F \Lambda_F O_F^t$$

where $O_F \Lambda_F O_F^t$ is an eigenvalue decomposition of the matrix $A_F$ with $\Lambda_F$ being a diagonal matrix and $O_F$ being an orthonormal matrix. Based on this, we define matrices which encode the flux directions as

$$A_F^+ = O_F \Lambda_A^+ O_F^t, \quad A_F^- = O_F \Lambda_F^- O_F^t, \quad \text{with } (\Lambda_F^\pm)_{i,i} = \frac{1}{2}\big(|(\Lambda_F)_{i,i}| \pm (\Lambda_F)_{i,i}\big).$$

Note that $A_F = A_F^+ + A_F^-$. We also introduce a generalization of the absolute value for such flux matrices by $|A_F| = A_F^+ - A_F^-$.

An element $v_h \in \mathcal{V}_h$ is multi-valued on any internal face $F \in \mathcal{F}_h^{\text{int}}$. We define its average and jump by

$$\{\!\{v_h\}\!\} := \frac{1}{2}(v_h\big|_{E_1} + v_h\big|_{E_2}), \quad [\![v_h]\!] := v_h\big|_{E_1} - v_h\big|_{E_2}.$$

For exterior faces $F \in \mathcal{F}_h^{\text{ext}}$ we simply choose the unit outer normal and extend the definition of jump and average appropriately.

The (unstabilized) upwind semi-discretization in space is then given as: Find $u_h(t) \in \mathcal{V}_h$ such that

$$(\partial_t u_h(t), v_h)_{L^2(\Omega)} + a_h^{\text{upw}}(u_h(t), v_h) + l_h(v_h) = 0 \quad \forall\, v_h \in \mathcal{V}_h \tag{1}$$

with

$$\begin{aligned} a_h^{\text{upw}}(u_h, v_h) &= \sum_{F \in \mathcal{F}_h^{\text{int}}} \int_F \langle A_F \{\!\{u_h\}\!\}, [\![v_h]\!]\rangle + \frac{1}{2}\langle |A_F| [\![u_h]\!], [\![v_h]\!]\rangle ds \\ &\quad + \sum_{F \in \mathcal{F}_h^{\text{ext}}} \int_F \langle A_F^+ u_h, v_h\rangle ds, \\ l_h(v_h) &= - \sum_{F \in \mathcal{F}_h^{\text{ext}}} \int_F \langle A_F^- g, v_h\rangle ds. \end{aligned}$$

Here, $\langle \cdot, \cdot \rangle$ denotes the standard scalar product in $l^2$. While we use an upwind flux here, we want to note that our following construction for the stabilization terms also holds for the Lax-Friedrichs flux, where some terms cancel. We skipped this for brevity. Discretization in time is then accomplished by the explicit Euler scheme. If the time step choice does not reflect the size of smaller cut-cells, this causes stability issues, which is why we need stabilization terms.

## 3 Stabilization

The main idea of the DoD stabilization is to extend the numerical domain of dependence of the neighbor cells of a small cut-cell by introducing additional numerical fluxes on the cut-cell boundary depending on the extrapolated solutions of the neighbor elements.

Let $\mathcal{I} \subset \mathcal{M}_h$ be the set of cut-cells which are supposed to be stabilized (in practise this set will be chosen based on a volume fraction with respect to the background cells). For simplicity we assume that cells in $\mathcal{I}$ are not neighbors of each other. The $P^0$ stabilization fluxes are collected in a term

$$J_h^0(u_h, v_h) = \sum_{E \in \mathcal{I}} J_h^{0,E}(u_h, v_h)$$

and the stabilized scheme reads: Find $u_h(t) \in \mathcal{V}_h$ such that for any $v_h \in \mathcal{V}_h$

$$(\partial_t u_h(t), v_h)_{L^2(\Omega)} + a_h(u_h(t), v_h) + J_h^0(u_h(t), v_h) + l_h(v_h) = 0.$$

For any $E \in \mathcal{M}_h$ we introduce a map $\mathcal{L}_E^{\text{ext}} : \mathcal{V}_h(E) \to \mathcal{P}^0(\Omega)^3$, called an extension operator, such that $\mathcal{L}_E^{\text{ext}}(u_h)\big|_E = u_h\big|_E$. The cell stabilization terms $J_h^{0,E}(u_h, v_h)$ then contain extended fluxes

$$\eta_E \sum_{(E_1,E_2) \in \mathcal{N}(E)} \int_{F_{E,E_2}} \Big\langle (\omega_{F_{E,E_1}} A^+_{F_{E,E_2}})_{\text{sym}} (\mathcal{L}_{E_1}^{\text{ext}}(u_h) - u_h\big|_E), [\![v_h]\!] \Big\rangle ds,$$

with appropriate weighting matrices $\omega_{F_{E,E_1}} \in \mathbb{R}^{3\times 3}$ and a penalty parameter $\eta_E \in (0, 1]$. Here, we have denoted the symmetrization $\frac{1}{2}(A + A^t)$ for a matrix $A \in \mathbb{R}^{3\times 3}$ by $A_{\text{sym}}$. This introduces a direct mass transport from $E_1$ to $E_2$ for any pair $(E_1, E_2)$ of neighbors of a small cut-cell $E$. (For a triangular cut-cell, the sum contains 9 terms.) This is a result of the inherent nature of the wave equation, creating waves in all possible directions. The weighting matrices describe how much of an inflow coming from $E_1$ is transported to $E_2$. Figure 2 shows an illustration. Similarly to what was proposed in [4], we require for the weighting matrices that

$$\int_{F_3} \langle ((\omega_{F_1} A^+_{F_3})_{\text{sym}} (\mathcal{L}_{E_1}^{\text{ext}}(u_h) - u_h^{E_{\text{cut}}}) + (\omega_{F_2} A^+_{F_3})_{\text{sym}} (\mathcal{L}_{E_2}^{\text{ext}}(u_h) - u_h^{E_{\text{cut}}}) + (\omega_{F_3} A^+_{F_3})_{\text{sym}} (\mathcal{L}_{E_3}^{\text{ext}}(u_h) - u_h^{E_{\text{cut}}})\rangle [\![v]\!] ds$$

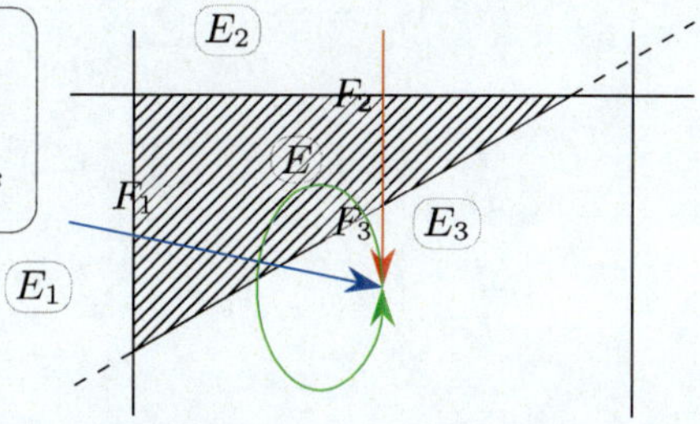


**Fig. 2** DoD stabilization for the acoustic equation on a triangular cut-cell. On the left are the stabilization fluxes on the face $F_3$ (without the parameter $\eta_E$), colored in correspondence with the arrows on the right indicating the different couplings. Note that there will be additional stabilization fluxes on $F_1$ and $F_2$. The green flux is somewhat of a curiosity and might just be a product of the way we construct our weighting matrices for the specific case of the acoustic equation. In a sense it acts similar to a reflecting wall

$$\sum_{E_2 \in \mathcal{N}(E)} \int_{F_{E,E_2}} \omega_{E,E_1} A^+_{F_{E,E_2}} ds = - \int_{F_{E,E_1}} A^-_{F_{E,E_1}} ds \qquad \forall E_1 \in \mathcal{N}(E), \tag{2a}$$

$$\sum_{E_1 \in \mathcal{N}(E)} \omega_{F_{E,E_1}} = \mathrm{Id}_{3\times 3}. \tag{2b}$$

The first equation describes how incoming flow is distributed among the cut-cell's neighbors while the second equation ensures that the overall amount of flow over a single face is preserved.

In [4] we were able to prove a discrete dissipation and $L^2$-stability result. One key step of the proof is an application of the binomial formula relying on the symmetry of the matrices $\omega_{F_{E,E_1}} A^+_{F_{E,E_2}}$, which holds given simultaneously diagonalizable (equivalently, commuting) system matrices. For the acoustic equation this matrix product will in general not be symmetric and the proof breaks down. A working fix is to instead take the symmetrization of the matrix $\omega_{F_{E,E_1}} A^+_{F_{E,E_2}}$.

Taking the symmetrization is not enough by itself though since the resulting flux matrices potentially possess negative eigenvalues leading to incorrect fluxes. This can be corrected by introducing for any pair $(E_1, E_2) \in \mathcal{N}(E)$ a term

$$-\kappa\eta_E \int_{F_{E,E_2}} \left\langle (\omega_{F_{E,E_1}} A^+_{F_{E,E_2}})^-_{\mathrm{sym}} (\mathcal{L}^{\mathrm{ext}}_{E_1}(u_h) - u_h\big|_{E_2}), \mathcal{L}^{\mathrm{ext}}_{E_1}(v_h) - v_h\big|_{E_2} \right\rangle ds.$$

For any $\kappa \geq 1$ this term ensures $L^2$-stability of the semi-discrete form. We observe that certain choices of $\kappa$ significantly improve the numerical approximation; the optimal choice still needs to be investigated. Note that this new term has an opposite sign, uses the negative part of the symmetrized form, and subtracts the solution from cell neighbor $E_2$ (instead of the cut-cell $E$).

Putting those components together leads to a cell stabilization term

$$\begin{aligned} J_h^{0,E}(u_h, v_h) = \eta_E & \sum_{(E_1,E_2) \in \mathcal{N}(E)} \\ & \int_{F_{E,E_2}} \left\langle (\omega_{F_{E,E_1}} A^+_{F_{E,E_2}})_{\mathrm{sym}} (\mathcal{L}^{\mathrm{ext}}_{E_1}(u_h) - u_h\big|_{E}), [\![ v_h ]\!] \right\rangle ds \\ & -\kappa \int_{F_{E,E_2}} \left\langle (\omega_{F_{E,E_1}} A^+_{F_{E,E_2}})^-_{\mathrm{sym}} (\mathcal{L}^{\mathrm{ext}}_{E_1}(u_h) - u_h\big|_{E_2}), \mathcal{L}^{\mathrm{ext}}_{E_1}(v_h) - v_h\big|_{E_2} \right\rangle ds. \end{aligned}$$

For this formulation together with the properties (2a) and (2b) there is a $L^2$-stability result for the semi-discretization in space. This was shown in [5] but due to limited space we cannot present the argument here.

The weighting matrices can be directly computed from (2a). Due to the system matrices being divergence-free, this equation is equivalent to

$$\omega_{F_{E,E_1}} \Big( \sum_{E_2 \in \mathcal{N}(E)} |F_{E,E_2}| A^-_{F_{E,E_2}} \Big) = |F_{E,E_1}| A^-_{F_{E,E_1}}.$$

The matrix $(\sum_{E_2 \in \mathcal{N}(E)} |F_{E,E_2}| A^-_{F_{E,E_2}})$ is invertible since on a triangular cut-cell we got three summands, each of rank one, and the three column spaces are linearly independent. Thus we have

$$\omega_{F_{E,E_1}} = |F_{E,E_1}| A^-_{F_{E,E_1}} \left( \sum_{E_2 \in \mathcal{N}(E)} |F_{E,E_2}| A^-_{F_{E,E_2}} \right)^{-1}.$$

## 4 Numerical Results

We present numerical results to support our findings. Our implementation uses the DUNE framework [1, 2], in particular the dune-udg module [3, 6] and the TPMC library [8]. We consider an analytic test case [10] with

$$u(x, y, t) = \frac{1}{c} \begin{pmatrix} -\cos(2\pi ct)(\sin(2\pi x) + \sin(2\pi y)) \\ \sin(2\pi ct)\cos(2\pi x) \\ \sin(2\pi ct)\cos(2\pi y) \end{pmatrix}$$

Initial condition $u_0$ and inflow boundary conditions $g$ are given by the exact solution. The speed of sound is chosen as $c = \frac{1}{2}$, the final time is $T = 0.3$ and the number $N$ of fundamental cells in one direction is chosen from the set {400, .., 1200}, starting at 400 with a stepsize of 23. The resulting volume fractions of cut-cells are in the interval $[7.24 \cdot 10^{-10}, 5.37 \cdot 10^{-5}]$. We classify a cell $E \in \mathcal{M}_h$ as a small cut-cell if $\frac{|E|}{(1/N)^2} \leq 0.4$. For the CFL-condition we choose $\Delta t = 0.3\frac{\Delta x}{c}$ where $\Delta x = \frac{1}{N}$. The factor 0.3 was added to handle cut-cells which are not classified as small. We made here a conservative choice and the factor could in principal be chosen sharper, our main goal was to avoid stabilization of cut-cells with an $O(1)$ volume fraction. For the cut angle we set $\gamma = 35°$ and choose $x_0 = 0.2001$ for the start of the cut. The penalty parameter $\eta_E$ is chosen as $(1 - \eta_E) = \frac{|E|}{\Delta t c \max_{F \subset \partial E} |F|}$.

We compute the $L^2$-error of the discrete solution at the final time $T$ with respect to the exact solution. We also compute the pointwise error at certain quadrature points, which gives an approximation to the error in the $L^\infty$-norm. We will display results for $\kappa \in \{1.0, 7.5\}$ to display the effect of this additional parameter.

The computed errors are plotted in Fig. 3. We observe the expected convergence order in the $L^2$-norm in all components, regardless of the choice of $\kappa$. For the $L^\infty$-norm the situation differs. The pressure component shows again optimal convergence behavior, independent of $\kappa$. However for the velocity components the convergence order is influenced by the choice of $\kappa$ and we obtain better absolute errors and a better convergence order (at the cost of some mild wiggles) for a larger $\kappa = 7.5$. The wiggles suggest that the parameter $\kappa$ should be chosen cell dependent, but how to choose it exactly requires further investigation.

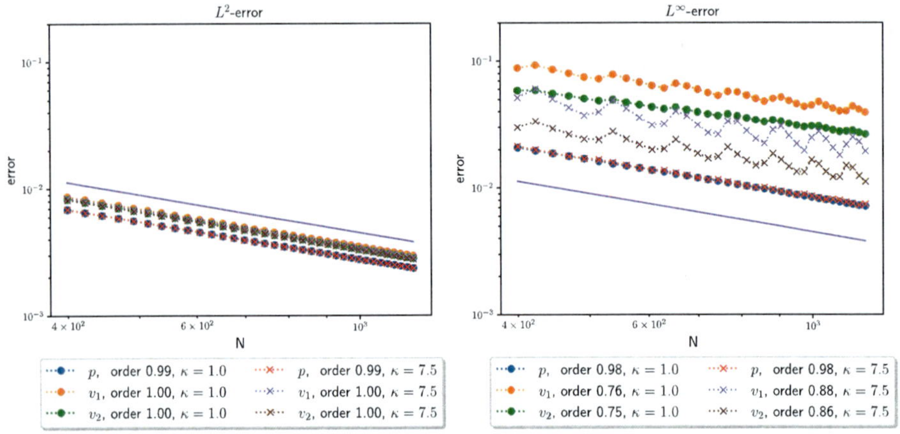


**Fig. 3** Error at $T = 0.3$ in the $L^2$ norm (left) and in the $L^\infty$ norm (right). The straight purple line is for reference and denotes convergence of order 1

## 5 Discussion and Outlook

We presented an extension of the DoD stabilization to the acoustic wave equation for $P^0$ trial and test functions on a structured grid with triangular cut-cells. We observed a decent convergence behavior in our numerical test setup. We note that the new parameter $\kappa$ included in the extension has a considerable influence on the measured error. Its precise effect and optimal choice will be investigated more in the future. An extension to higher-order approximations, as well as to non-linear systems, e. g. the Euler equations, is ongoing research.

**Acknowledgements** The authors acknowledge support by the Deutsche Forschungsgemeinschaft as project 439956613 under contract numbers EN 1042/5-1 and MA 7773/4-1/2, as well as under Germany's Excellence Strategy EXC 2044 390685587, Mathematics Münster: Dynamics–Geometry–Structure.

## References

1. Bastian, P., Blatt, M., Dedner, A., Dreier, N.-A., Engwer, C., Fritze, R., Gräser, C., Grüninger, C., Kempf, D., Klöfkorn, R., Ohlberger, M., Sander, O.: The Dune framework: basic concepts and recent developments. Comput. Math. Appl. **81**, 75–112 (2021)
2. Bastian, P., Blatt, M., Dedner, A., Engwer, C., Klöfkorn, R., Kornhuber, R., Ohlberger, M., Sander, O.: A generic grid interface for parallel and adaptive scientific computing. Part II: implementation and tests in DUNE. Computing **82**, 121–138 (2008)
3. Bastian, P., Engwer, C.: An unfitted finite element method using discontinuous Galerkin. Int. J. Numer. Methods Eng. **79**(12), 1557–1576 (2009)
4. Birke, G., Engwer, C., May, S., Streitbürger, F.: DoD stabilization of linear hyperbolic PDEs on general cut-cell meshes. Proc. Appl. Math. Mech. (2022)
5. Birke, G.: Masterthesis, University of Münster (2023). (online link not available yet)

6. Engwer, C., Heimann, F.: Dune-UDG: A Cut-Cell Framework for Unfitted Discontinuous Galerkin Methods, Advances in DUNE, pp. 89–100. Springer, Berlin, Heidelberg (2012)
7. Engwer, C., May, S., Nüßing, A., Streitbürger, F.: A stabilized DG cut cell method for discretizing the linear transport equation. SIAM J. Sci. Comput. **42**(6), A3677–A3703 (2020)
8. Engwer, C., Nüßing, A.: Geometric reconstruction of implicitly defined surfaces and domains with topological guarantees. ACM Trans. Math. Soft. **44**(2), Art. No. 14 (2018)
9. Giuliani, A.: A two-dimensional stabilized discontinuous Galerkin method on curvilinear embedded boundary grids. SIAM J. Sci. Comput. **44**(1), A389–A415 (2022)
10. Lukácová-Medvidová, M., Morton, K., Warnecke, G.: Finite volume evolution Galerkin methods for hyperbolic systems. SIAM J. Sci. Comput **26**(1), 1–30 (2004)
11. May, S., Streitbürger, F.: DoD stabilization for non-linear hyperbolic conservation laws on cut cell meshes in one dimension. Appl. Math. Comput. **419**, Art. 126854 (2022)
12. Schoeder, S., Sticko, S., Kreiss, G., Kronbichler, M.: High-order cut discontinuous Galerkin methods with local time stepping for acoustics. Int. J. Numer. Methods Eng. **121**(13), 2979–3003 (2020)
13. Streitbürger, F., Birke, G., Engwer, C., May, S.: Spectral and high order methods for partial differential equations ICOSAHOM 2020+1, vol. 137. Springer (2023)

# Numerical Simulation of a Barotropic Two-Phase Flow Model with Miscible Phases

**Jean Bussac and Khaled Saleh**

**Abstract** This paper addresses the numerical approximation of a compressible barotropic two-phase flow model with miscible conditions. The first phase is a liquid while the second phase corresponds to a gaseous mixture, which contains two components that share the same volume (for example vapor and air). The fluid dynamic is depicted by a Baer-Nunziato-type model, involving non conservative coupling terms. The core of the paper is the simulation of this model by a Suliciu relaxation scheme initially designed for immiscible mixtures. The numerical results illustrate the convergence of the scheme, its robustness for low volume fraction regimes and its computational cost efficiency.

**Keywords** Multiphase flow · Relaxation scheme · Miscible conditions

## 1 Introduction

This work concerns the simulation of the barotropic three-field two-phase Baer-Nunziato model, studied in [7]. This model corresponds to a mixture of a liquid phase ($l$) plus a gaseous phase containing two miscible gases: vapor ($v$) and an inert gas ($g$). In the nuclear industry, such flows may appear in accidental configurations such as the Loss of Coolant Accident (LOCA). In the one-dimensional framework, it is a seven equation system of partial differential equations involving non-conservative coupling terms, which reads

J. Bussac (✉)
Laboratoire de Mathématiques Jean Leray, Nantes Université, 44000 Nantes, France
e-mail: jean.bussac@univ-nantes.fr

K. Saleh
Université de Lyon, Institut Camille Jordan, 69622 Villeurbanne, France



E. Franck et al. (eds.), *Finite Volumes for Complex Applications X—Volume 2, Hyperbolic and Related Problems*, Springer Proceedings in Mathematics & Statistics 433,
https://doi.org/10.1007/978-3-031-40860-1_7

$$
\begin{aligned}
&\partial_t \alpha_l + u_l \partial_x \alpha_l = 0,\\
&\partial_t(\alpha_l\rho_l) + \partial_x(\alpha_l\rho_l u_l) = 0,\\
&\partial_t(\alpha_l\rho_l u_l) + \partial_x(\alpha_l\rho_l u_l^2 + \alpha_l p_l(\rho_l)) + p_v(\rho_v)\partial_x\alpha_v + p_g(\rho_g)\partial_x\alpha_g = 0,\\
&\text{and for } k = v, g:\\
&\partial_t(\alpha_k\rho_k) + \partial_x(\alpha_k\rho_k u_k) = 0,\\
&\partial_t(\alpha_k\rho_k u_k) + \partial_x(\alpha_k\rho_k u_k^2 + \alpha_k p_k(\rho_k)) - p_k(\rho_k)\partial_x\alpha_k = 0.
\end{aligned}
\tag{1}
$$

For $k \in \{l, v, g\}$, $\rho_k > 0$ and $u_k \in \mathbb{R}$ are respectively the density and velocity of field $k$. The pressure of field $k$ only depends on the corresponding field density and is given by a barotropic equation of state (see [2, 7]) $p_k : \rho_k \mapsto p_k(\rho_k)$ and we assume that $p_k'(\rho_k) > 0$. Finally $\alpha_k \in (0, 1)$ is the statistical fraction of field $k$. Since the vapor field and the inert gas share the same gaseous phase we have the saturation constraint $\alpha_v = \alpha_g = 1 - \alpha_l$.

**Remark 1** A mixture of three immiscible phases can be modeled by a similar set of PDEs. Assuming that the fields $v$ and $g$ do not share the same phase, the phase fraction of vapor is computed by an additional transport equation of velocity $u_l$ and $\alpha_g$ is determined by the saturation constraint $\alpha_l + \alpha_v + \alpha_g = 1$.

**Remark 2** Source terms can be added to the model (1) to account for mass and thermodynamical transfers.

**Proposition 1** *Denote* $\mathcal{U} := (\alpha_l, \alpha_l\rho_l, \alpha_l\rho_l u_l, \alpha_v\rho_v, \alpha_v\rho_v u_v, \alpha_g\rho_g, \alpha_g\rho_g u_g)$. *System* (1) *is weakly hyperbolic and admits the following real eigenvalues*

$$\sigma_1(\mathcal{U}) = u_l,\ \sigma_{2,3}(\mathcal{U}) = u_l \pm c_l,\ \sigma_{4,5}(\mathcal{U}) = u_v \pm c_v,\ \sigma_{6,7}(\mathcal{U}) = u_g \pm c_g,$$

*where* $c_k(\rho_k) = \sqrt{p_k'(\rho_k)}$ *is the speed of sound of field* $k$. *All the characteristic fields are genuinely non-linear excepting* $\sigma_1$ *which is linearly degenerate. The set of corresponding right eigenvectors spans* $\mathbb{R}^7$ *if and only if*

$$\alpha_l \neq 0,\ \alpha_v = \alpha_g \neq 0,\ |u_k - u_l| \neq c_k(\rho_k) \text{ for } k = g, v. \tag{2}$$

***Proof*** See [7].

The aim of this work is to extend the Suliciu relaxation scheme to the model (1). It is known in immiscible situations that the Suliciu relaxation scheme provides much better results than standard schemes in vanishing phase regimes ($\alpha_k \to 0$), which are quite important in practical applications [4, 8]. It is therefore desirable to have a Suliciu relaxation type scheme for the miscible case.

**Remark 3** The loss of hyperbolicity (2) can be due to the presence of a nearly vanishing phase $\alpha_k \to 0$, or to an interaction between the coupling wave $u_l$ and an acoustic wave $u_k \pm c_k$, $k = g, v$ (see [7]). This second resonant configuration is unlikely to happen in the concerned industrial applications (multiphase flows in nuclear reactors), where the flows have strongly subsonic relative velocities, i.e. a relative Mach number much smaller than one:

$$M_k = \frac{|u_l - u_k|}{c_k(\rho_k)} << 1 \ \text{ for } k = g, v.$$

Let the specific total energy of field $k$ be defined by $E_k = e_k + u_k^2/2$, where the specific internal energy of the field $k$ is such that $e_k'(\rho_k) = p_k(\rho_k)/\rho_k^2$. One can prove that the function $\mathcal{U} \mapsto (\alpha_k \rho_k E_k)(\mathcal{U})$ is (non strictly) convex (see [9]). The entropy weak solutions of system (1) are the weak solutions that satisfy the following inequality in the weak sense:

$$\partial_t\Big(\sum_{k=l,g,v} \alpha_k \rho_k E_k\Big) + \partial_x\Big(\sum_{k=l,g,v} (\alpha_k \rho_k E_k u_k + \alpha_k p_k(\rho_k) u_k)\Big) \leq 0. \tag{3}$$

For smooth solutions of (1), (3) is an equality.

## 2 Suliciu Relaxation Scheme

The solutions of (1) are formally recovered as the limit of the solutions of the following enlarged system, in the regime $\epsilon \to 0$:

$$\begin{aligned}
&\partial_t \alpha_l + u_l \partial_x \alpha_l = 0, \\
&\partial_t(\alpha_l \rho_l) + \partial_x(\alpha_l \rho_l u_l) = 0, \\
&\partial_t(\alpha_l \rho_l u_l) + \partial_x(\alpha_l \rho_l u_l^2 + \alpha_l \pi_l(\tau_l, \mathcal{T}_l)) + \pi_v(\tau_v, \mathcal{T}_v)\partial_x \alpha_v + \pi_g(\tau_g, \mathcal{T}_g)\partial_x \alpha_g = 0, \\
&\partial_t(\alpha_l \rho_l \mathcal{T}_l) + \partial_x(\alpha_l \rho_l \mathcal{T}_l u_l) = \frac{1}{\epsilon}\alpha \rho_l(\tau_l - \mathcal{T}_l), \\
&\text{and for } k = v, g: \\
&\partial_t(\alpha_k \rho_k) + \partial_x(\alpha_k \rho_k u_k) = 0, \\
&\partial_t(\alpha_k \rho_k u_k) + \partial_x(\alpha_k \rho_k u_k^2 + \alpha_k \pi_k(\tau_k, \mathcal{T}_k)) - \pi_k(\tau_k, \mathcal{T}_k)\partial_x \alpha_k = 0, \\
&\partial_t(\alpha_k \rho_k \mathcal{T}_k) + \partial_x(\alpha_k \rho_k \mathcal{T}_k u_k) = \frac{1}{\epsilon}\alpha \rho_k(\tau_k - \mathcal{T}_k).
\end{aligned} \tag{4}$$

The quantities $\mathcal{T}_k$ are additional unknowns approximating the specific volumes $\tau_k = 1/\rho_k$, and associated with linearizations of the pressure laws:

$$\pi_k(\tau_k, \mathcal{T}_k) = P_k(\mathcal{T}_k) + a_k^2(\mathcal{T}_k - \tau_k), \quad k = l, g, v,$$

where $\tau \mapsto P_k(\tau) = p_k(\tau^{-1})$. The quantities $a_k \in \mathbb{R}^+$ are the relaxation parameters and are chosen so as to satisfy the so-called subcharacteristic (or Whitham) condition $a_k^2 \geq -P_k'(\mathcal{T}_k)$. A Chapman-Enskog expansion of the solution of (4) shows that under Whitham's condition, the relaxation system (4) is a viscous approximation of (1).

**Proposition 2** *Denote $\mathcal{W} := (\mathcal{U}, \alpha_l\rho_l\mathcal{T}_l, \alpha_v\rho_v\mathcal{T}_v, \alpha_g\rho_g\mathcal{T}_g)$. The convective part of system* (4) *is weakly hyperbolic and admits the following real eigenvalues*

$$\sigma_{1,8,9,10}(\mathcal{W}) = u_l,\ \sigma_{2,3}(\mathcal{W}) = u_l \pm a_l\tau_l,\ \sigma_{4,5}(\mathcal{W}) = u_v \pm a_v\tau_v,\ \sigma_{6,7}(\mathcal{W}) = u_g \pm a_g\tau_g.$$

*All the characteristic fields linearly degenerate. The set of corresponding right eigenvectors spans $\mathbb{R}^7$ if and only if*

$$\alpha_l \neq 0,\ \alpha_v = \alpha_g \neq 0,\ |u_k - u_l| \neq a_k\tau_k \text{ for } k = g, v. \tag{5}$$

The Suliciu relaxation scheme consists in building an exact Riemann solver for the convective part of the enlarged system (4), which is (relatively) easy to compute because the system has only linearly degenerate fields. Such an exact Riemann solution was first computed in [3, 4] for two-phase flows and has been extended in [8] for immiscible multiphase barotropic flows. In [8], the transport equations for $\alpha_l$ and $\alpha_v$ are solved and by the volumic constraint and $\alpha_g$ is computed by enforcing the initialization $\alpha_g(t = 0) := 1 - \alpha_l(t = 0) - \alpha_v(t = 0)$. The extension to the barotropic three-field two-phase Baer-Nunziato model (1) is rather straightforward. It consists in solving the transport equation for $\alpha_l$ while $\alpha_g = \alpha_v$ are computed by enforcing the initialization:

$$\alpha_g(t = 0) = \alpha_v(t = 0) := 1 - \alpha_l(t = 0).$$

Since in both the miscible and immiscible cases, the phase fractions $\alpha_k$ satisfy the same transport equation of velocity $u_l$ and the overall Riemann solution is unchanged.

At the numerical level, a fractional step method is used. The first step is a time-advancing step using the exact solution of the Riemann problem for the convective part of (4). The second step consists in an instantaneous relaxation towards the equilibrium system (1) by imposing $\mathcal{T}_k = \tau_k$ in the solution obtained by the first step. The whole scheme can be written in the form of a non conservation finite volume scheme (see [8] for details). Denoting $\Delta t$ and $\Delta x$ the time and space steps and $\mathcal{U}_j^n$ a constant approximation of the solution at time $t^n = n\Delta t$ on $[x_{j-\frac{1}{2}}, x_{j+\frac{1}{2}}[$ where $x_{j+\frac{1}{2}} = (j + \frac{1}{2})\Delta x$ for all $j$ in $\mathbb{Z}$, the overall scheme takes the form of an explicit non conservative finite volume scheme:

$$\mathcal{U}_j^{n+1} = \mathcal{U}_j^n - \frac{\Delta t}{\Delta x}\left(\mathbf{F}^-(\mathcal{U}_j^n, \mathcal{U}_{j+1}^n) - \mathbf{F}^+(\mathcal{U}_{j-1}^n, \mathcal{U}_j^n)\right). \tag{6}$$

The time step $\Delta t$ is updated at each time iteration so as to satisfy the following CFL condition prescribed by the acoustic waves of the relaxation system (4):

$$\frac{\Delta t}{\Delta x} \max_{k=l,v,g} \max_{j\in\mathbb{Z}} \max\left\{|(u_k - a_k\tau_k)_j^n|, |(u_k + a_k\tau_k)_{j+1}^n|\right\} < \frac{1}{2}. \tag{7}$$

**Proposition 3** (Properties of the Suliciu relaxation scheme)

- *Under the CFL condition* (7), *the scheme preserves positive values of the phase fractions and densities.*
- *The discretizations of the partial masses and total mixture momentum are conservative.*
- *Under Whitham's condition, the numerical solution satisfies a discrete counterpart to the energy inequality* (3) *(see [8, (16)], [1]).*

## 3 Numerical Results

The relaxation scheme is compared to the non conservative Rusanov scheme [5], which is (to our knowledge) the only available scheme for this multiphase flow model. In the following we denote $\mathbb{U} = (\alpha_l, \rho_l, u_l, \rho_v, u_v, \rho_g, u_g)$ the vector of non conservative unknowns. We present two test-cases and plot the approximated solutions for a 100 cell mesh, which corresponds to a realistic $10^6$ cell mesh in a three-dimensional industrial context. The time step is chosen so as to impose a CFL number of 0.45, see (7)).

### 3.1 *Test-Case 1: Pure Contact Discontinuity*

The initial data is the following

$$\begin{aligned} \mathbb{U}_L =& (0.1, 3.0, 0.2, 0.3, 1.0, 0.4, 1.0), \\ \mathbb{U}_R =& (0.4, 2.0724212862957074, 0.2, 0.28339834471128433, \\ & 1.4702967632600488, 0.34584269153214842, 1.5879142504747157). \end{aligned} \tag{8}$$

We present the approximated solution in Fig. 1. The classical sonic wave oscillations are smaller for the relaxation scheme. The contact discontinuity is much less smeared with the relaxation scheme which is known to be less diffusive and therefore more accurate than Rusanov's scheme. We compute the discrete $L^1$-error between the approximate solution and the exact one at the final time $t = N\Delta t = 0.08$, normalized by the discrete $L^1$-norm of the exact solution. We observe the expected convergence rate of $\Delta x^{1/2}$ which is due to the presence of a contact discontinuity. Figure 2 also shows that it requires a hundred times more CPU time to obtain the same error with Rusanov's scheme (using more refined meshes) than with the relaxation scheme.

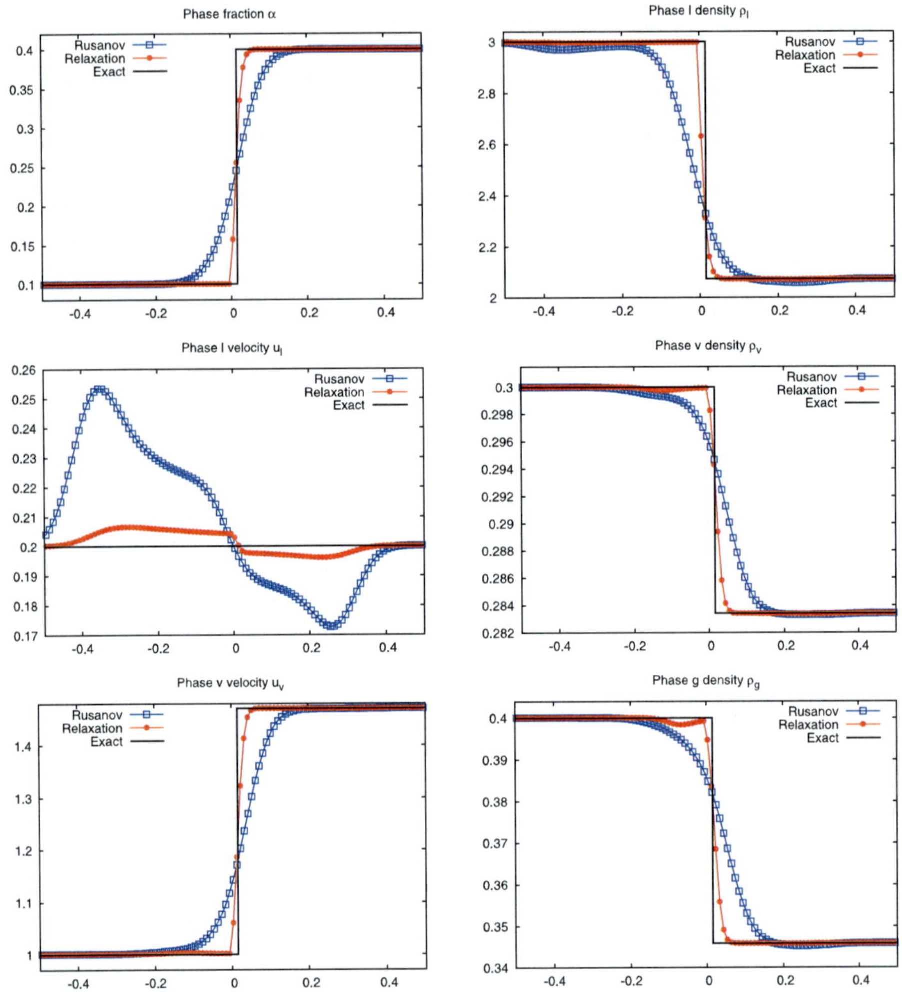


**Fig. 1** Test-case 1: solution for 100 cells at final time $t = 0.08$. Contact wave of velocity $u_l = 0.2$. Initial data given in (8). The $u_g$ plot has a similar profile to the $u_v$ one

## 3.2 Test-Case 2: Vanishing-Phase Test-Case with $u_v \pm c_v$ Shocks

In this test-case, the left initial data has no liquid phase, which corresponds to $\alpha_l = 0$ (in practice $\alpha_l = 10^{-10}$). We consider a shock-contact-shock wave configuration for the vapor field, while the other fields are only subject to the $u_l$ contact. The initial data is:

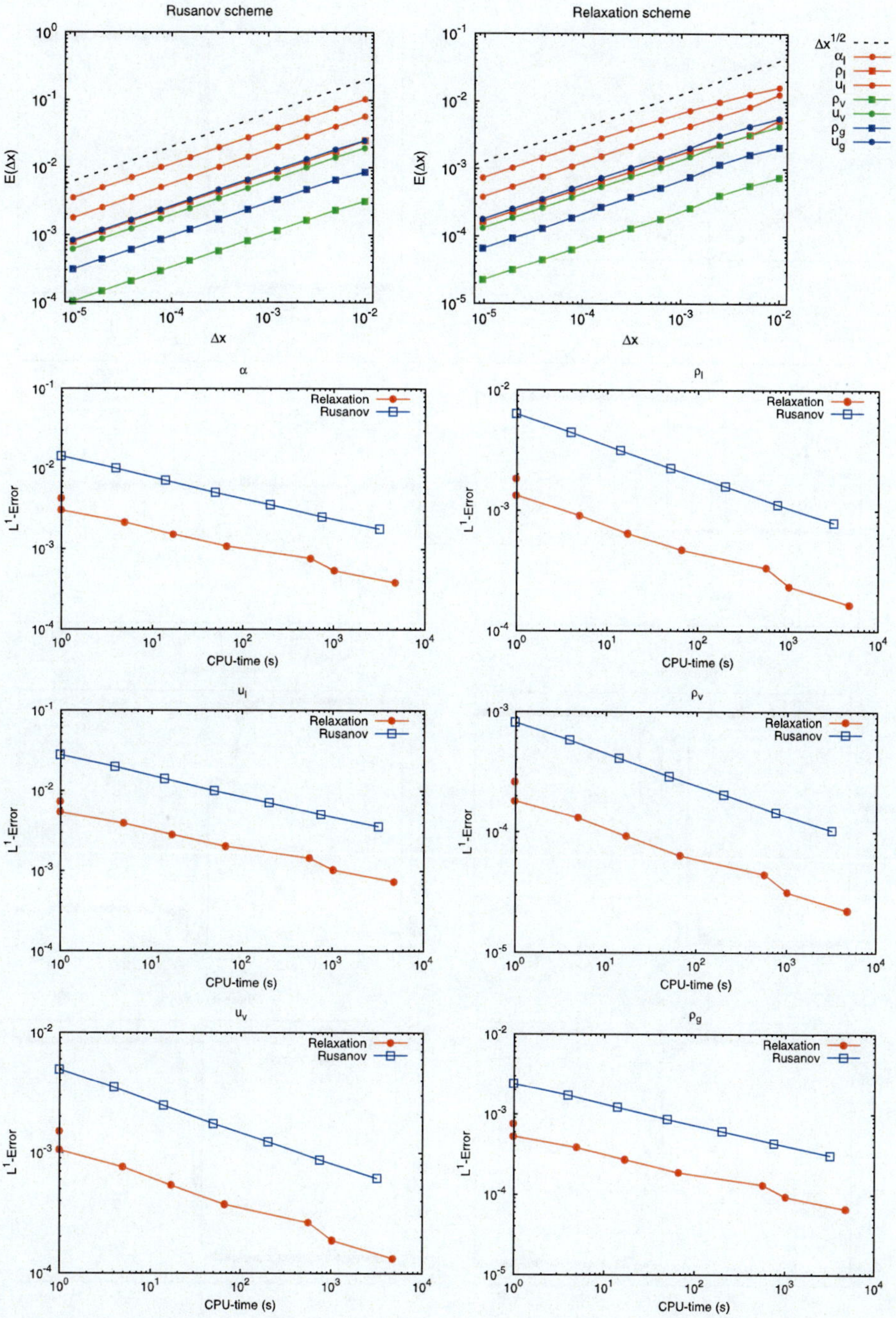


**Fig. 2** Test-case 1: $L^1$-error with respect to $\Delta x$ and $L^1$-error with respect to computational cost (in seconds). The $u_g$ plot is not represented since it is very similar to the $u_v$ plot

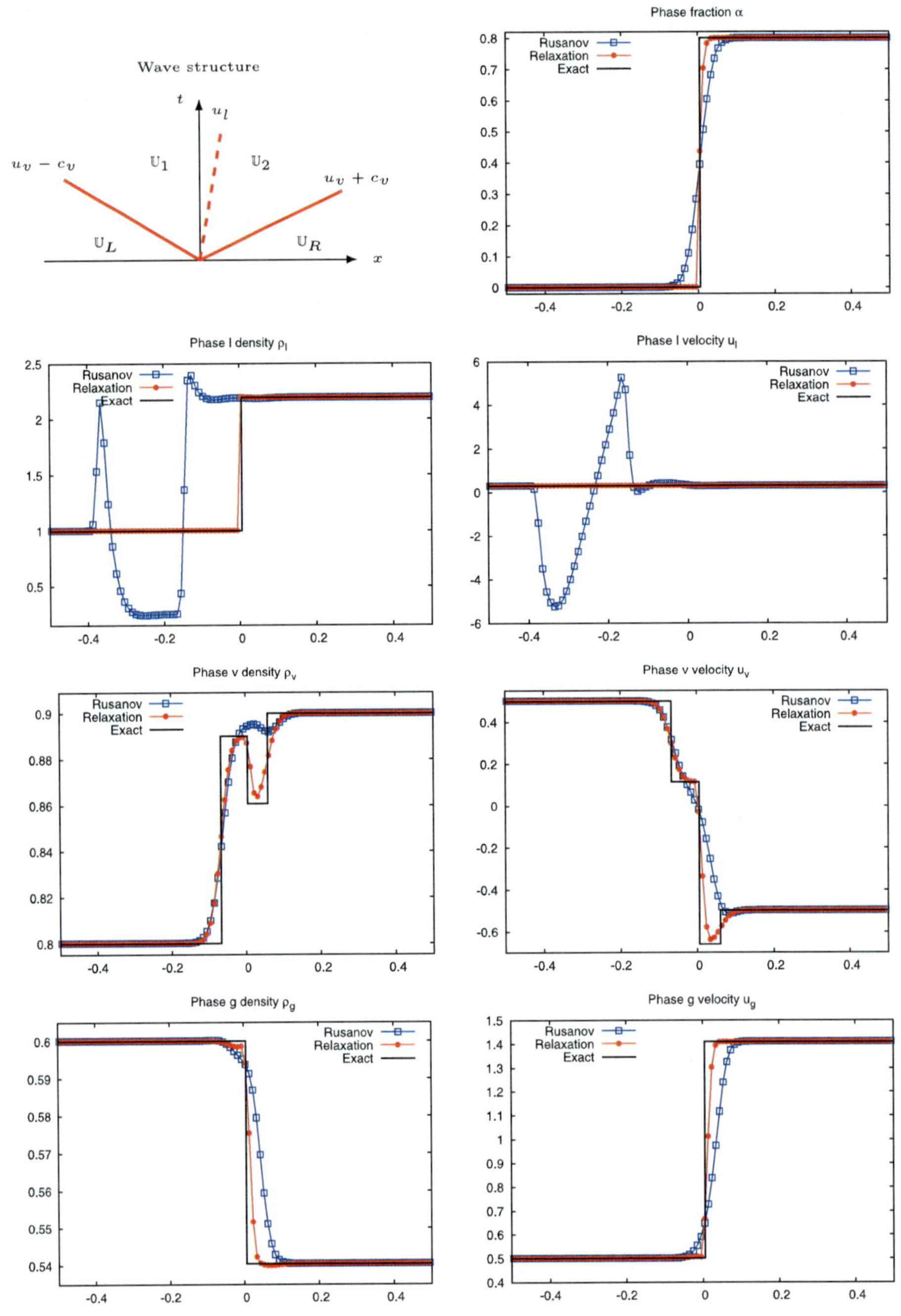


**Fig. 3** Test-case 2: solutions for 100 cells at final time $t = 0.02$. Contact wave of velocity $u_l = 0.3$, $u_v - c_v$-shock of velocity –3.3155813734376762 and $u_v + c_v$-shock of velocity 3.0689467170729379. Initial data given in (9)

$$
\begin{aligned}
\mathbb{U}_L &= (1.0.10^{-10}, 1.0, 0.3, 0.8, 0.5, 0.6, 0.5, ), \\
&\quad - 0.66086132955857257, 0.54048985831522922, 1.4101040855979630), \\
\mathbb{U}_R &= (0.8, 2.1958716914805883, 0.3, 0.9, \\
&\quad - 0.49798559025814237, 0.54048985831522922, 1.4101040855979630).
\end{aligned} \tag{9}
$$

The wave structure and approximated solutions are given in Fig. 3. On the one hand, the relaxation scheme follows the contact discontinuity much better than the Rusanov scheme and does not oscillate on the left constant state, which is a low fraction regime. On the other hand, close intermediate states of the vapor field are rather well captured by the relaxation scheme, even for a 100 cell mesh. On the contrary, due to its large diffusivity, the Rusanov scheme misses the intermediate state between the $u_l$-contact and the $(u_v + c_v)$-shock.

## 4 Conclusion

This work confirms the convincing behavior of this relaxation scheme already known for immiscible mixtures. For a given error on the solution, it performs much better than the Rusanov scheme in terms of CPU time cost (by a factor of order $10^2$), which is a discriminating feature in an industrial approach. Moreover, the low fraction regimes are approximated with robustness. In future works, it would be interesting to consider the thermodynamical relaxation by adding source terms to system (1). The three-phase four-field case [6] can be investigated.

## References

1. Bouchut, F.: Nonlinear Stability of Finite Volume Methods for Hyperbolic Conservation Laws: And Well-balanced Schemes for Sources. Springer Science & Business Media (2004)
2. Coquel, F., Gallouët, T., Hérard, J.-M., Seguin, N.C.R.: Acad. Sci. Paris **334**(10), 927–932 (2002)
3. Coquel, F., Hérard, J.-M., Saleh, K.: J. Comput. Phys. **330**, 401–435 (2017). Elsevier
4. Coquel, F., Hérard, J.-M., Saleh, K., Seguin, N.: ESAIM: M2AN **48**, 165–206 (2014). EDP Sciences
5. Gallouët, T., Hérard, J.-M, Seguin, N.: Math. Models Methods Appl. Sci. **14**(05), 663–700 (2004). World Scientific
6. Hérard, J.-M., Hurisse, O., Quibel, L.: ESAIM: M2AN **55** S251–S278 (2021)
7. Hérard, J.-M., Mathis, H.: ESAIM: M2AN **53**(4), 1373–1389 (2019). EDP Sciences
8. Saleh, K.: ESAIM: M2AN, Apr (2019). EDP Sciences
9. Saleh, K., Seguin, N.: ESAIM Proc. Surveys **69**, 70–78 (2020). EDP Sciences

# Local Characteristic Decomposition Based Central-Upwind Scheme for Compressible Multifluids

**Shaoshuai Chu and Alexander Kurganov**

**Abstract** In this paper, we introduce the local characteristic decomposition based central-upwind (LCD-CU) scheme for compressible multifluids. The scheme is implemented within the hybrid multifluid algorithm from [A. CHERTOCK, S. CHU, AND A. KURGANOV, J. Sci. Comput, 89(2021), Paper No. 48], according to which we use the level set method to track the position of material interfaces and replace the conservative compressible Euler equations with the pressure-based nonconservative ones in the vicinities of the interfaces. The LCD-CU scheme is used away from the interfaces and this helps to reduce the numerical dissipation in these areas. At the interfaces, we still use the path-conservative central-upwind scheme designed to accurately solve nonconservative hyperbolic systems. This leads to a substantially higher resolution, especially in the two-dimensional case.

**Keywords** Path-conservative scheme · Compressible multifluids · Local characteristic decomposition based central-upwind scheme · Hybrid algorithm

## 1 Introduction

This paper focuses on the development of a new numerical method for compressible multifluids, which are assumed to be immiscible. The governing equations in the two-dimensional (2-D) case read as

S. Chu
Department of Mathematics, Southern University of Science and Technology, Shenzhen 518055, China
e-mail: chuss2019@mail.sustech.edu.cn

A. Kurganov (✉)
Department of Mathematics, Shenzhen International Center for Mathematics, and Guangdong Provincial Key Laboratory of Computational Science and Material Design, Southern University of Science and Technology, Shenzhen 518055, China
e-mail: alexander@sustech.edu.cn

 
E. Franck et al. (eds.), *Finite Volumes for Complex Applications X—Volume 2, Hyperbolic and Related Problems*, Springer Proceedings in Mathematics & Statistics 433,
https://doi.org/10.1007/978-3-031-40860-1_8

$$\rho_t + (\rho u)_x + (\rho v)_y = 0, \tag{1}$$

$$(\rho u)_t + (\rho u^2 + p)_x + (\rho u v)_y = 0, \tag{2}$$

$$(\rho v)_t + (\rho u v)_x + (\rho v^2 + p)_y = 0, \tag{3}$$

$$E_t + [u(E + p)]_x + [v(E + p)]_y = 0, \tag{4}$$

where $x$ and $y$ are spatial variables, $t$ is time, and $\rho(x, y, t)$, $u(x, y, t)$, $v(x, y, t)$, $p(x, y, t)$, and $E(x, y, t)$ are the density, x- and y-velocities, pressure, and total energy, respectively.The system is completed through the following equations of state (EOS) for each of the fluid components:

$$p = (\gamma - 1)\Big[E - \frac{\rho}{2}(u^2 + v^2)\Big] - \gamma p_\infty, \tag{5}$$

where $\gamma$ represents the specific heat ratio and $p_\infty$ is the stiffness parameter, with $p_\infty = 0$ corresponding to the ideal gas case. In this paper, we consider a multifluid problem with two components and assume that $\gamma = \gamma_I$, $p_\infty = p_{\infty,I}$ and $\gamma = \gamma_{II}$, $p_\infty = p_{\infty,II}$ for the first and second fluid components, respectively.

The fluid components can be identified, for instance, by the level-set function $\phi$ whose zero level-set defines the interface between the fluid components; see [1, 3, 10, 16]. The function $\phi$ propagates with the fluid velocity and satisfies the following conservative equation:

$$(\rho\phi)_t + (\rho u\phi)_x + (\rho v\phi)_y = 0. \tag{6}$$

It is well-known that even when the initial data are smooth, solutions of (1)–(6) can produce complex nonsmooth wave patterns including shocks, rarefactions, and contact discontinuities. This makes it quite challenging to develop accurate and reliable numerical schemes for (1)–(6). A library of finite-volume (FV) methods for the 2-D hyperbolic systems of conservation laws have been proposed; see, e.g., the monographs [12, 15] and references therein. However, applying the single fluid FV methods to the multifluid system (1)–(6) may lead to significant pressure and velocity oscillations, which originate near the material interface and then typically spread all over the computational domain; see, e.g., [3, 13]. This occurs since in the cells where the interface is located, the fluids are artificially mixed, and the mixed cell average values often are nonphysical. Therefore, one needs to design special multifluid algorithms as it was done in, e.g., [1, 2, 4, 6, 8–10, 18–20].

In this paper, we construct a new numerical method for (1)–(6), which is, in fact, an improved version of the hybrid multifluid algorithm recently proposed in [8]. The key idea in [8] is to detect the vicinities of material interfaces and then to replace the conservative energy Eq. (4) there with a nonconservative evolution equation for the pressure $p$ (which, unlike the total energy $E$, remains continuous across the contact waves at the material interfaces):

$$p_t + (up)_x + (vp)_y = -\,[(\gamma - 1)p + \gamma p_\infty]\,u_x - [(\gamma - 1)p + \gamma p_\infty]\,v_y. \tag{7}$$

The resulting nonconservative system (1)–(3), (5)–(7) is then solved using the path-conservative central-upwind (PCCU) scheme introduced in [5]. Away from material interfaces, the original conservative system (1)–(6) was solved in [8] using the CU scheme from [14]. We now replace that CU scheme with the recently proposed local characteristic decomposition based central-upwind (LCD-CU) scheme from [7], which contains a substantially smaller amount of numerical dissipation and typically achieves an enhanced resolution; see the numerical results for the single fluid compressible Euler equations reported in [7, §5]. In this paper, we demonstrate how the implementation of the LCD-CU scheme can enhance the resolution achieved by the hybrid multifluid algorithm.

## 2 Hybrid Multifluid Algorithm

We follow the hybrid multifluid algorithm from [8]. Here, we briefly describe the main steps of the algorithm and focus on the novel features of the method designed in this paper.

*Step 1*. Given the set of discrete data for $\boldsymbol{V} = (\rho, u, v, p, \phi)^\top$, we determine the location of the material interfaces, which are assumed to be in the immediate vicinities of the FV cells in which $\phi$ changes sign (for the sake of simplicity, we use uniform Cartesian meshes).

*Step 2*. In the vicinities of the interfaces detected in *Step 1*, we discretize the non-conservative system (1)–(3), (5)–(7) using the second-order PCCU scheme.

*Step 3*. Away from the detected interface areas, we discretize the conservative system (1)–(6). Unlike the method from [8], where (1)–(6) was discretized using the CU scheme from [14], we now use the LCD-CU scheme, which was recently proposed in [7] but implemented here in a slightly different way since we use an alternative piecewise linear reconstruction procedure. As in [7], we perform the reconstruction in the local characteristic variable, but in order to switch to them, we now use the primitive variables $\boldsymbol{V}$ rather than the conservative ones $(\rho, \rho u, \rho v, E, \rho\phi)^\top$. This is done as follows. First, we rewrite (1)–(6) in terms of $\boldsymbol{V}$:

$$\boldsymbol{V}_t + \mathcal{A}\boldsymbol{V}_x + \mathcal{B}\boldsymbol{V}_y = \boldsymbol{0}, \tag{8}$$

where

$$\mathcal{A} = \begin{pmatrix} u & \rho & 0 & 0 & 0 \\ 0 & u & 0 & \frac{1}{\rho} & 0 \\ 0 & 0 & u & 0 & 0 \\ 0 & \gamma(p+p_\infty) & 0 & u & 0 \\ 0 & 0 & 0 & 0 & u \end{pmatrix} \quad \text{and} \quad \mathcal{B} = \begin{pmatrix} v & 0 & \rho & 0 & 0 \\ 0 & v & 0 & 0 & 0 \\ 0 & 0 & v & \frac{1}{\rho} & 0 \\ 0 & 0 & \gamma(p+p_\infty) & v & 0 \\ 0 & 0 & 0 & 0 & v \end{pmatrix}. \tag{9}$$

We then locally linearize $\mathcal{A}$ and $\mathcal{B}$ at each $x$- and $y$-cell interfaces and obtain the linearized matrices

$$\widehat{\mathcal{A}}=\begin{pmatrix}\widehat{u} & \widehat{\rho} & 0\ 0\ 0\\ 0 & \widehat{u} & 0\ \frac{1}{\widehat{\rho}}\ 0\\ 0 & 0 & u\ 0\ 0\\ 0 & \widehat{\gamma}(\widehat{p}+\widehat{p}_\infty) & 0\ \widehat{u}\ 0\\ 0 & 0 & 0\ 0\ \widehat{u}\end{pmatrix}\quad\text{and}\quad \widehat{\mathcal{B}}=\begin{pmatrix}v\ 0 & \rho & 0\ 0\\ 0\ \widehat{v} & 0 & 0\ 0\\ 0\ 0 & \widehat{v} & \frac{1}{\widehat{\rho}}\ 0\\ 0\ 0 & \widehat{\gamma}(\widehat{p}+\widehat{p}_\infty) & \widehat{v}\ 0\\ 0\ 0 & 0 & 0\ \widehat{v}\end{pmatrix}, \tag{10}$$

where $\widehat{\rho}=\sqrt{\rho_L\rho_R}$, $\widehat{u}=\frac{\sqrt{\rho_L}u_L+\sqrt{\rho_R}u_R}{\sqrt{\rho_L}+\sqrt{\rho_R}}$, $\widehat{v}=\frac{\sqrt{\rho_L}v_L+\sqrt{\rho_R}v_R}{\sqrt{\rho_L}+\sqrt{\rho_R}}$, $\widehat{p}=\frac{\sqrt{\rho_L}p_L+\sqrt{\rho_R}p_R}{\sqrt{\rho_L}+\sqrt{\rho_R}}$, $\widehat{\gamma}=\frac{\sqrt{\rho_L}\gamma_L+\sqrt{\rho_R}\gamma_R}{\sqrt{\rho_L}+\sqrt{\rho_R}}$, and $\widehat{p}_\infty=\frac{\sqrt{\rho_L}(p_\infty)_L+\sqrt{\rho_R}(p_\infty)_R}{\sqrt{\rho_L}+\sqrt{\rho_R}}$, where $(\cdot)_L$ and $(\cdot)_R$ stand for the left- and right-sided point values at the given cell interface. We then use the eigenvectors of the obtained constant matrices $\widehat{\mathcal{A}}$ and $\widehat{\mathcal{B}}$ to form the (locally defined) matrices $R$. We then introduce the local characteristic variables $\boldsymbol{\Gamma}:=R^{-1}\boldsymbol{V}$ at every cell interface (more precisely, in two cells on the left and two cells on the right of the interface), perform the piecewise linear reconstruction in $\boldsymbol{\Gamma}$ (in our numerical experiments, we have used the Minmod2 limiter, see, e.g., [12, 15], which determines the slopes of $\boldsymbol{\Gamma}$ at each cell adjusted to the material interface), use the obtained reconstructions to evaluate the point values of $\boldsymbol{\Gamma}$ at the cell interfaces, and then compute the corresponding point values of $\boldsymbol{V}$ using $\boldsymbol{V}=R\boldsymbol{\Gamma}$.

**Remark 1** Note that the nonconservative system (8)–(9) is different from (1)–(3), (5)–(7) and that it is used for the local characteristic decomposition only—not for the time evolution.

**Remark 2** We perform a piecewise linear reconstruction in the local characteristic variables since when we implement the LCD-CU scheme, these variables are available and it is well-known that reconstructing local characteristic variables typically leads to smaller or no oscillations.

## 3 Numerical Examples

In this section, we test the proposed second-order scheme on several numerical examples and compare the obtained results with those computed by the CU and mixed-order A-WENO schemes from [8] (this scheme is based on the fifth-order WENO-Z interpolation monotonized with the help of the second-order accurate piecewise linear reconstruction). In the rest of this section, we will refer to the proposed scheme as to LCD-CU scheme. In all of the examples below, we have solved the ODE systems using the three-stage third-order strong stability preserving Runge-Kutta method (see, e.g., [11]) with CFL number 0.3.

**Example 1** (*Water-Air Model Using the Stiff Equation of State*) In the first example taken from [3], we consider a gas-liquid multifluid system, in which the liquid

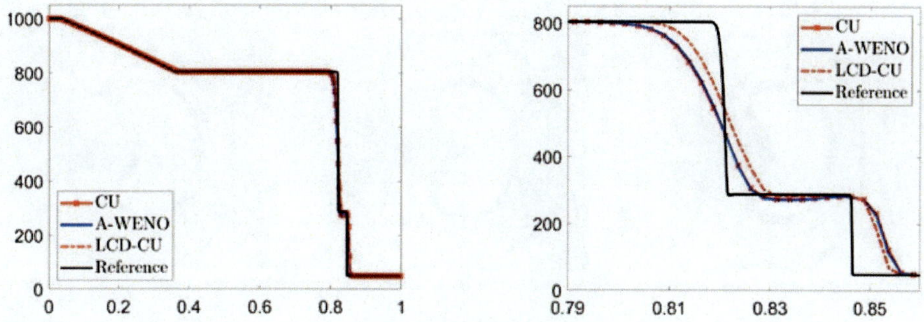


**Fig. 1** Example 1: Density $\rho$(left) and zoom at [0.79, 0.86] (right)

component is modeled by the stiff EOS (5) with $p_\infty \gg 1$. The initial conditions correspond to a severe water-air shock tube problem and given by

$$(\rho, u, p; \gamma, p_\infty)(x, 0) = \begin{cases} (1000, 0, 10^9; 4.4, 6 \cdot 10^8), & x < 0.7, \\ (50, 0, 10^5; 1.4, 0), & x > 0.7. \end{cases}$$

We compute the numerical solution until the final time $t = 0.00025$ on a uniform mesh with 400 cells. In Fig. 1, we plot the solution computed by the CU, mixed-order A-WENO, and LCD-CU schemes together with the corresponding reference solutions computed by the LCD-CU scheme on a much finer mesh with 8000 cells. As one can see, there is no oscillations in the numerical results and the density profile obtained by the LCD-CU scheme is slightly sharper than the one computed by the second-order CU and mixed-order A-WENO schemes; see zoom at the contact and shock waves area in Fig. 1 (right).

**Example 2** (*Shock Hitting Helium Bubble*) In the first 2-D example taken from [17], we consider the system (1)–(6) subject to the following initial conditions:

$$(\rho, u, v, p; \gamma, p_\infty)(x, y, 0) = \begin{cases} (4/29, 0, 0, 1; 5/3, 0), & x^2 + y^2 \le 0.0625, \\ (4/3, -0.3535, 0, 1.5; 1.4, 0), & x \ge 0.75, \\ (1, 0, 0, 1; 1.4, 0), & \text{otherwise}, \end{cases}$$

see [8, Fig. 3.6] for the sketch of the initial shock-bubble setting.

We compute the numerical results on the computational domain $[-3, 1] \times [-0.5, 0.5]$ by the second-order CU, second-order LCD-CU, and mixed-order A-WENO schemes on the uniform $2000 \times 500$ spatial mesh. In Fig. 2, we illustrate the results obtained at different time moments during the shock-bubble interaction process. Notice that the bubble changes its shape and propagates to the left, but in order to focus on the details of the bubble structure, we only zoom at $[\sigma, \sigma + 1] \times [-0.5, 0.5]$ square area containing the bubble ($\sigma$ is decreasing in time from –0.75 to –1.6). We plot Schlieren images of the magnitude of the density gradient field $|\nabla \rho|$. To this end, we have used the following shading function:

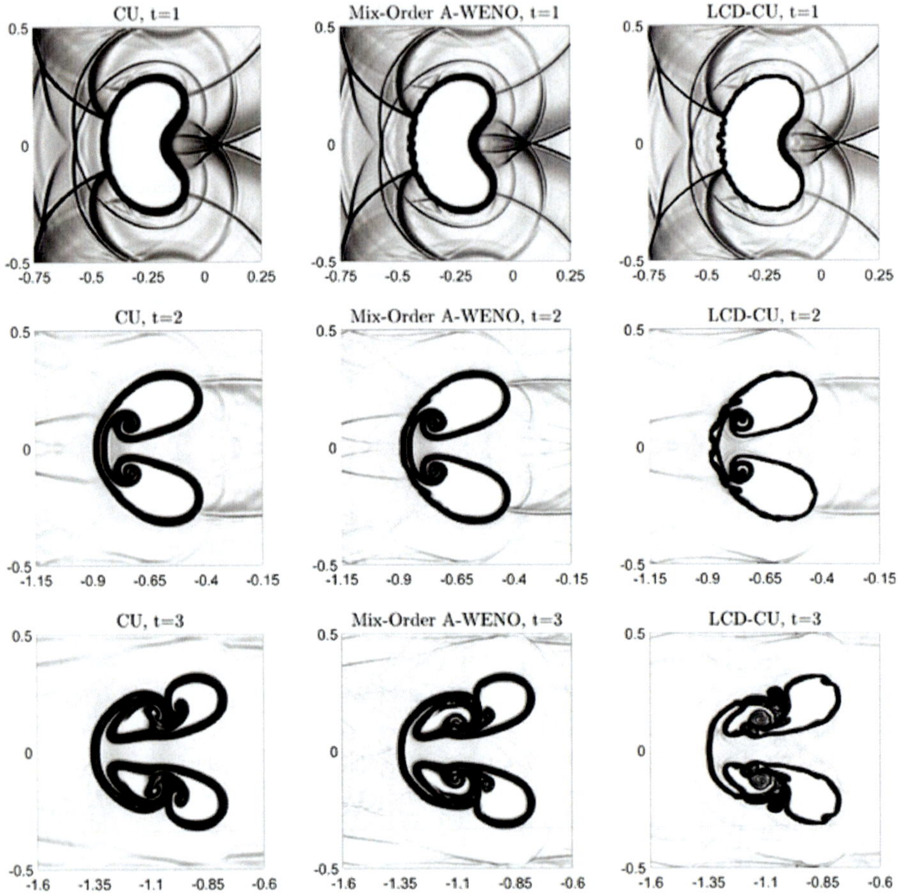


**Fig. 2** Example 2: Shock-Helium bubble interaction by the CU (left column), mixed-order A-WENO (middle column), and LCD-CU (right column) schemes at $t = 1, 2$, and 3

$$\exp\left(-\frac{K|\nabla\rho|}{\max(|\nabla\rho|)}\right), \quad K = 80,$$

where the numerical density derivatives are computed using standard central differencing.

As one can see from Fig. 2, the LCD-CU scheme can capture both the material interface and small features of the solution in a sharper manner than the second-order CU and mixed-order A-WENO schemes. The differences in the results computed by the studied schemes become more pronounced at larger times.

**Example 3** (*Shock Hitting R22 Bubble*) In the last example taken from [17], we consider another shock-bubble interaction problem with the following initial conditions:

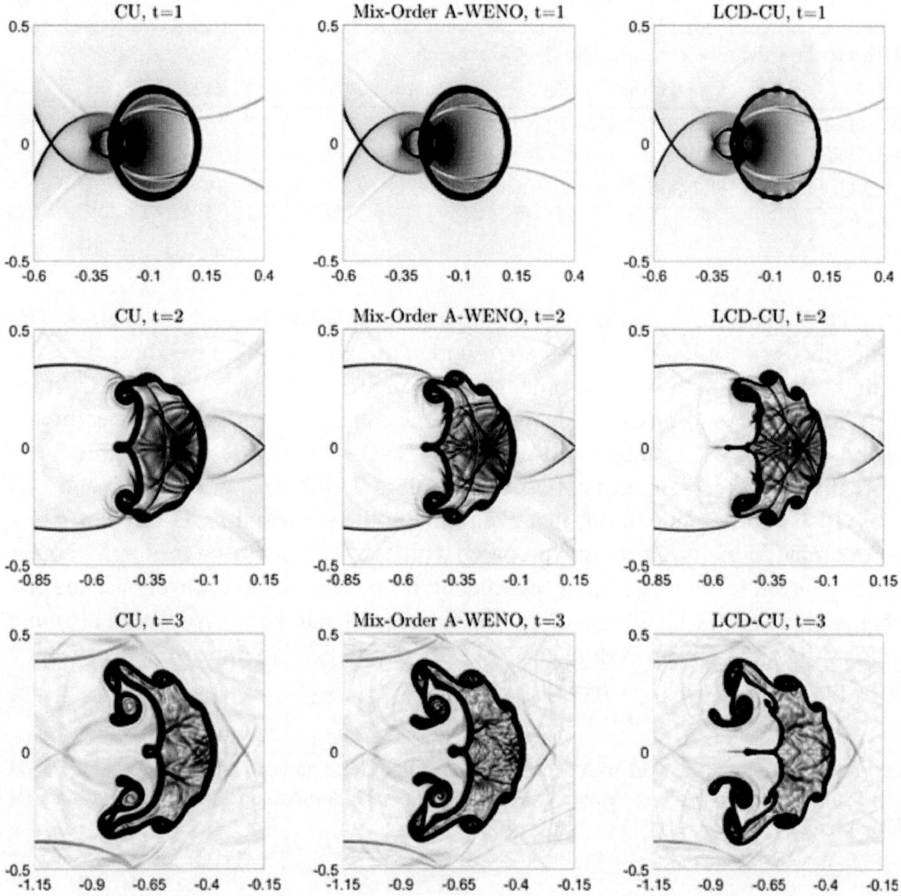


**Fig. 3** Example 3: Shock-R22 bubble interaction by the CU (left column), mixed-order A-WENO (middle column), and LCD-CU (right column) schemes at $t = 1$, 2, and 3

$$(\rho, u, v, p; \gamma, p_\infty)(x, y, 0) = \begin{cases} (3.1538, 0, 0, 1; 1.249, 0), & x^2 + y^2 \le 0.0625, \\ (4/3, -0.3535, 0, 1.5; 1.4, 0), & x \ge 0.75, \\ (1, 0, 0, 1; 1.4, 0), & \text{otherwise.} \end{cases}$$

We compute the numerical results on the same computational domain and meshes as in Example 2 and plot the Schlieren images of the magnitude of the density gradient field in Fig. 3. The numerical results show that the bubble changes its shape and propagates to the left, and in order to focus on the details of the bubble structure, we only zoom at $[\sigma, \sigma + 1] \times [-0.5, 0.5]$ square area containing the bubble ($\sigma$ is decreasing in time from –0.6 to –1.15). As one can see, the second-order LCD-CU scheme produces sharper numerical results than both the second-order CU and mixed-order A-WENO schemes.

**Table 1** Examples 1–3: CPU times consumed by the LCD-CU and mixed-order A-WENO schemes relative to the CPU time consumed by the CU scheme

| | Example 1 (%) | Example 2 (%) | Example 3 (%) |
|---|---|---|---|
| Mixed-order A-WENO | 282 | 297 | 280 |
| LCD-CU | 306 | 328 | 317 |

**Remark 3** We have measured the CPU times consumed by the studied schemes. The results obtained for Examples 1–3 are reported in Table 1, where we show the CPU time consumption of the second-order LCD-CU and mixed-order A-WENO schemes relative to the second-order CU scheme. As one can see, the times consumed by the second-order LCD-CU and mixed-order A-WENO schemes are around three times larger than times consumed by the second-order CU schemes in both 1-D and 2-D cases. It should be pointed out that even though the proposed LCD-CU scheme is more computationally expensive compared with its CU predecessor, the improvement in the achieved resolution is quite significant. If one tries to alternatively use the CU scheme on a finer mesh, the memory use may become an issue. We would also like to stress that the LCD procedure can be easily parallelized to improve the efficiency of the proposed LCD-CU scheme.

**Acknowledgements** The work of A. Kurganov was supported in part by NSFC grant 12171226 and by the fund of the Guangdong Provincial Key Laboratory of Computational Science and Material Design (No. 2019B030301001).

## References

1. Abgrall, R., Karni, S.: Ghost-fluids for the poor: a single fluid algorithm for multifluids. In: Hyperbolic Problems: Theory, Numerics, Applications. Birkhäuser, International Series of Numerical Mathematics, vols. 140, 141, pp. 1–10 (2001). https://doi.org/10.1007/978-3-0348-8370-2_1
2. Abgrall, R., Saurel, R.A.: Discrete equations for physical and numerical compressible multiphase mixtures. J. Comput. Phys. **186**, 361–396 (2003). https://doi.org/10.1016/S0021-9991(03)00011-1
3. Abgrall, R., Karni, S.: Computations of compressible multifluids. J. Comput. Phys. **169**, 594–623 (2001). https://doi.org/10.1006/jcph.2000.6685
4. Allaire, G., Clerc, S., Kokh, S.: A five-equation model for the simulation of interfaces between compressible fluids. J. Comput. Phys. **181**, 577–616 (2002). https://doi.org/10.1006/jcph.2002.7143
5. Castro Díaz, M.J., Kurganov, A., Morales de Luna, T.: Path-conservative central-upwind schemes for nonconservative hyperbolic systems. ESAIM Math. Model. Numer. Anal. **53**, 959–985 (2019). https://doi.org/10.1051/m2an/2018077
6. Cheng, J., Zhang, F., Liu, T.G.: A quasi-conservative discontinuous Galerkin method for solving five equation model of compressible two-medium flows. J. Sci. Comput. **85**, 12–35 (2020). https://doi.org/10.1007/s10915-020-01319-5

7. Chertock, A., Chu, S., Herty, M., Kurganov, A., Lukáčová-Medvid'ová, M.: Local characteristic decomposition based central-upwind scheme. J. Comput. Phys. **473**, Paper No. 111718 (2023). https://doi.org/10.1016/j.jcp.2022.111718
8. Chertock, A., Chu, S., Kurganov, A.: Hybrid multifluid algorithms based on the path-conservative central-upwind scheme. J. Sci. Comput. **89**, Paper No. 48 (2021). https://doi.org/10.1007/s10915-021-01656-z
9. Chertock, A., Karni, S., Kurganov, A.: Interface tracking method for compressible multifluids. M2AN Math. Model. Numer. Anal. **42**, 991–1019 (2008). https://doi.org/10.1051/m2an:2008036
10. Fedkiw, R.P., Aslam, T., Merriman, B., Osher, S.: A non-oscillatory Eulerian approach to interfaces in multimaterial flows (the ghost fluid method). J. Comput. Phys. **152**, 457–492 (1999). https://doi.org/10.1006/jcph.1999.6236
11. Gottlieb, S., Shu, C.- W., Tadmor, E.: Strong stability-preserving high-order time discretization methods. SIAM Rev. **43**, 89–112 (2001). https://doi.org/10.1137/S003614450036757X
12. Hesthaven, J.S.: Numerical methods for conservation laws: from analysis to algorithms. In: Computational Science & Engineering. SIAM, PA, (2018). https://doi.org/10.1137/1.9781611975109
13. Karni, S.: Hybrid multifluid algorithms. SIAM J. Sci. Comput. **17**, 1019–1039 (1996). https://doi.org/10.1137/S106482759528003X
14. Kurganov, A., Lin, C.-T.: On the reduction of numerical dissipation in central-upwind schemes. J. Comput. Phys. **2**, 141–163 (2007). https://doi.org/10.4236/ns.2010.28114
15. LeVeque, R.J.: Finite volume methods for hyperbolic problems. In: Cambridge Texts in Applied Mathematics. Cambridge University Press, Cambridge (2002). https://doi.org/10.1017/CBO9780511791253
16. Mulder, W., Osher, S., Sethian, J.A.: Computing interface motion in compressible gas dynamics. J. Comput. Phys. **100**, 209–228 (1992). https://doi.org/10.1016/0021-9991(92)90229-R
17. Quirk, J.J., Karni, S.: On the dynamics of a shock-bubble interaction. J. Fluid Mech. **318**, 129–163 (1996). https://doi.org/10.1017/S0022112096007069
18. Saurel, R., Abgrall, R.: A simple method for compressible multifluid flows. SIAM J. Sci. Comput. **21**, 1115–1145 (1999). https://doi.org/10.1137/S1064827597323749
19. Shyue, K.M.: An efficient shock-capturing algorithm for compressible multicomponent problems. J. Comput. Phys. **142**, 208–242 (1998). https://doi.org/10.1006/jcph.1998.5930
20. Wang, C.W., Shu, C.- W.: An interface treating technique for compressible multi-medium flow with Runge-Kutta discontinuous Galerkin method. J. Comput. Phys. **229**, 8823–8843 (2010). https://doi.org/10.1016/j.jcp.2010.08.012

# Simpson's Quadrature for a Nonlinear Variational Symplectic Scheme

**François Dubois and Juan Antonio Rojas-Quintero**

**Abstract** [01 June 2023]. We propose a variational symplectic numerical method for the time integration of dynamical systems issued from the least action principle. We assume a quadratic internal interpolation of the state between two time steps and we approximate the action in one time step by the Simpson's quadrature formula. The resulting scheme is nonlinear and symplectic. First numerical experiments concern a nonlinear pendulum and we have observed experimentally very good convergence properties.

**Keywords** Ordinary differential equations · Nonlinear pendulum · Numerical analysis

## 1 Introduction

The principle of least action is a key point for establishing evolution equations and partial differential equations, from classical to quantum mechanics and electromagnetisms [2, 12]. An important application of this principle is proposed with the finite element method and it is used for engineering applications since the 1950's. The principle of least action is also the starting point for the conception of symplectic numerical schemes for dynamical systems (see e.g. [5, 6, 11]). In particular, the Newmark scheme [7], which is very popular in structural dynamics, is symplectic. In [4], we have proposed a linear Simpson symplectic scheme that extends the previous works [6, 11] for quadratic interpolation.

F. Dubois (✉)
Laboratoire de Mathématiques d'Orsay, Faculté des Sciences d'Orsay, Université Paris-Saclay, Orsay, France
e-mail: francois.dubois@universite-paris-saclay.fr

LMSSC laboratory, Conservatoire National des Arts et Métiers, Paris, France

J. Antonio Rojas-Quintero
CONAHCYT - Tecnológico Nacional de México/I.T. Ensenada, Ensenada, BC 22780, Mexico

 
E. Franck et al. (eds.), *Finite Volumes for Complex Applications X—Volume 2, Hyperbolic and Related Problems*, Springer Proceedings in Mathematics & Statistics 433,
https://doi.org/10.1007/978-3-031-40860-1_9

In this contribution, we study a nonlinear Simpson symplectic scheme, as an alternative to Newmark's scheme. One motivation is to solve stiff problems in robotics [9]. We first recall in Sect. 2 the fundamental statements relative to Lagrangian and Hamiltonian mechanics and focus our attention on the nonlinear pendulum. Then the classical discrete dynamical system obtained directly with a second order discretization of the Lagrangian is presented in Sect. 3. Then we define and study in Sect. 4 a symplectic Simpson numerical scheme based on a quadrature with internal quadratic interpolation. First numerical results are presented in Sect. 5 and the exact solution of the nonlinear pendulum is recalled in the Annex.

## 2 Continuous Approach

We consider a dynamical system described by a state $q(t)$ of constant mass $m$ composed by a simple real variable to fix the ideas, and for $0 \leq t \leq T$. The continuous action $S_c$ introduces a Lagrangian $L$

$$L\Big(q, \frac{\mathrm{d}q}{\mathrm{d}t}\Big) = \frac{m}{2}\Big(\frac{\mathrm{d}q}{\mathrm{d}t}\Big)^2 - V(q) \tag{1}$$

and we have

$$S_c = \int_0^T L\Big(\frac{\mathrm{d}q}{\mathrm{d}t}, q(t)\Big)\, \mathrm{d}t. \tag{2}$$

The trajectories associated with the extremals of the action satisfy the Euler-Lagrange equations $\frac{\mathrm{d}}{\mathrm{d}t}\big(\frac{\partial L}{\partial(\frac{\mathrm{d}q}{\mathrm{d}t})}\big) = \frac{\partial L}{\partial q}$. With the Lagrangian proposed in (1), the differential equation

$$\frac{\mathrm{d}}{\mathrm{d}t}\Big(m\,\frac{\mathrm{d}q}{\mathrm{d}t}\Big) + \frac{\partial V}{\partial q} = 0 \tag{3}$$

of Newtonian mechanics is recovered. With the momentum $p \equiv \partial L / \partial(\frac{\mathrm{d}q}{\mathrm{d}t}) = m\,\frac{\mathrm{d}q}{\mathrm{d}t}$ and the Hamilton function $H(p, q) \equiv p\,\frac{\mathrm{d}q}{\mathrm{d}t} - L\big(q, \frac{\mathrm{d}q}{\mathrm{d}t}\big)$, we obtain the first order system of Hamilton's equations $\frac{\mathrm{d}p}{\mathrm{d}t} + \frac{\partial H}{\partial q} = 0,\ \frac{\mathrm{d}q}{\mathrm{d}t} - \frac{\partial H}{\partial p} = 0$. In the case of the Lagrangian function introduced in (1), we have $H(p, q) = \frac{1}{2m}\, p^2 + V(q)$ and

$$\frac{\mathrm{d}p}{\mathrm{d}t} + \frac{\partial V}{\partial q} = 0\,, \quad \frac{\mathrm{d}q}{\mathrm{d}t} - \frac{1}{m}\, p = 0. \tag{4}$$

In this contribution, we consider the case of the nonlinear pendulum that corresponds to the potential $V(q) = m\,\omega^2\,(1 - \cos q)$. Then $\frac{\mathrm{d}p}{\mathrm{d}t} + m\,\omega^2\, \sin q = 0$ and we have finally the second order dynamics $\frac{\mathrm{d}^2 q}{\mathrm{d}t^2} + \omega^2\, \sin q = 0$. An analytical solution of this problem is established in Annex. This exact solution is also shown in Fig. 1.

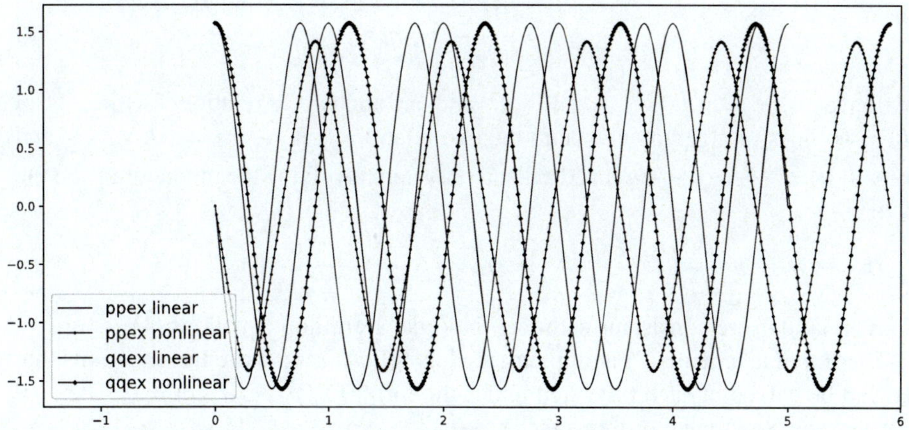


**Fig. 1** Typical evolution for five periods of an nonlinear pendulum satisfying the dynamics (4) with $V(q) = m\,\omega^2\,(1 - \cos q)$, $\omega = 2\,\pi$, $p(0) = 0$ and $q(0) = \frac{\pi}{2}$. We observe that the nonlinear period is not given by the linear evaluation $T = \frac{2\pi}{\omega}$ anymore but follows the relation (25)

## 3 Newmark Scheme

In this section we recall the classical Newmark scheme [7] which we use as the reference scheme for benchmarking purposes. A discretization of the relation (2) is obtained by splitting the interval $[0,\,T]$ into $N$ elements and we set $h = \frac{T}{N}$. At the discrete time $t_j = j\,h$, an approximation $q_j$ of $q(t_j)$ is introduced and a discrete form $S_d$ of the continuous action $S_c$ can be defined: $S_d = \sum_{j=1}^{N-1} L_N(q_j,\,q_{j+1})$. The discrete Lagrangian $L_N(q_\ell,\,q_r)$ is derived from the relation (1) with a centered finite difference approximation $\frac{\mathrm{d}q}{\mathrm{d}t} \simeq \frac{q_r - q_\ell}{h}$ and a midpoint quadrature formula $\int_0^h V\big(q(t)\big)\,\mathrm{d}t \simeq h\,V\big(\frac{q_\ell + q_r}{2}\big)$:

$$L_N(q_\ell,\,q_r) = \frac{m\,h}{2}\Big(\frac{q_r - q_\ell}{h}\Big)^2 - h\,V\Big(\frac{q_\ell + q_r}{2}\Big). \tag{5}$$

We observe that $S_d = \cdots + L_N(q_{j-1},\,q_j) + L_N(q_j,\,q_{j+1}) + \cdots$. Then the discrete Euler Lagrange equation is obtained when the action is stationary, $\delta S_d = 0$ for an arbitrary variation $\delta q_j$ of the discrete variable $q_j$. It takes the form

$$\frac{\partial L_N}{\partial q_r}(q_{j-1},\,q_j) + \frac{\partial L_N}{\partial q_\ell}(q_j,\,q_{j+1}) = 0. \tag{6}$$

Taking into account the relation (5), we obtain

$$\frac{q_{j+1} - 2\,q_j + q_{j-1}}{h^2} + \frac{1}{2\,m}\Big[V'_{j+1/2} + V'_{j-1/2}\Big] = 0 \tag{7}$$

with $V'_{j+1/2} \equiv \frac{\partial V}{\partial q}\big(\frac{q_j + q_{j+1}}{2}\big)$. Observe first that the scheme (7) is implicit. Secondly, this numerical scheme is clearly consistent with the second order differential equation (3) associated with the Lagrangian proposed in (1). The momentum $p_r$ is defined by

$$p_r = \frac{\partial L_N}{\partial q_r}(q_\ell,\, q_r). \tag{8}$$

We have $p_{j+1} = m\,\frac{q_{j+1}-q_j}{h} - \frac{h}{2}\,V'_{j+1/2}$ and an analogous relation for $p_j$. Then after two lines of algebra, we obtain the two relations $p_{j+1} - p_j = -h\,V'_{j+1/2}$ and $p_{j+1} + p_j = \frac{2m}{h}\,(q_{j+1} - q_j)$ and a discrete system involving the momentum and the state:

$$p_{j+1} - p_j + h\,\frac{\partial V}{\partial q}\Big(\frac{q_j + q_{j+1}}{2}\Big) = 0\,,\ q_{j+1} - q_j - \frac{h}{2\,m}\,(p_{j+1} + p_j) = 0. \tag{9}$$

These relations are consistent with the first order Hamilton equations (4).

The implicit scheme (9) is implemented as follows. We write the equations that have to be solved at each time step under the form $F_N(p_{j+1},\, q_{j+1}) = 0$. Then we consider the Newton algorithm $(p,\, q) \longrightarrow (p^*,\, q^*)$:

$$F_N(p,\, q) + \mathrm{d}F_N(p,\, q).(p^* - p,\, q^* - q) = 0$$

with the initialization $p = p_j,\ q = q_j$. Observe that the inverse of the Jacobian matrix can be evaluated easily:

$$\big(\mathrm{d}F_N(p,\, q)\big)^{-1} = \frac{1}{1 + \frac{h^2}{4m}\,V''(\frac{q_j+q}{2})}\begin{pmatrix} 1 & -\frac{h}{2}\,V''(\frac{q_j+q}{2}) \\ \frac{h}{2m} & 1 \end{pmatrix}.$$

In our experiments, this algorithm is converging to machine accuracy in five iterations, which is typical of Newton's algorithm.

By definition (see [11]), symplectic schemes are area-preserving. Therefore, it is sufficient to establish that the jacobian determinant $J \equiv \frac{\partial p_{j+1}}{\partial p_j}\,\frac{\partial q_{j+1}}{\partial q_j} - \frac{\partial p_{j+1}}{\partial q_j}\,\frac{\partial q_{j+1}}{\partial p_j}$ is equal to 1. We can show that Newmark's scheme (9) is symplectic. From (9), with $V''_{j+1/2} \equiv \frac{\partial V}{\partial q}(q_{j+1/2})$ and $q_{j+1/2} = \frac{1}{2}\,(q_j + q_{j+1})$, we have

$$\begin{aligned} \delta p_{j+1} + \tfrac{h}{2}\,V''_{j+1/2}\,\delta q_{j+1} &= \delta p_j - \tfrac{h}{2}\,V''_{j+1/2}\,\delta q_j \\ -\tfrac{h}{2m}\,\delta p_{j+1} + \delta q_{j+1} &= \tfrac{h}{2m}\,\delta p_j + \delta q_j. \end{aligned}$$

This system can be rewritten

$$A\begin{pmatrix}\delta p \\ \delta q\end{pmatrix}_{j+1} = B\begin{pmatrix}\delta p \\ \delta q\end{pmatrix}_{j}. \tag{10}$$

We have $\det A = 1 + \frac{h^2}{4m}\,V''_{j+1/2} = \det B$ and in consequence $J = 1$. Observe that although the Newmark scheme (9) is symplectic, it does not preserve the energy. While the quantity $H_j \equiv \frac{1}{2m}\,p_j^2 + V(q_j)$ remained constant for the linear version of the scheme (see [4]), it is not strictly constant in this nonlinear case.

## 4 Simpson's Quadrature with Quadratic Interpolation

Internal interpolation between 0 and $h$ is written in terms of quadratic finite elements (see e.g. [8]). For $0 \leq \theta \leq 1$, we first set

$$\varphi_0(\theta) = (1-\theta)\,(1-2\,\theta)\,,\ \ \varphi_{1/2}(\theta) = 4\,\theta\,(1-\theta)\,,\ \ \varphi_1(\theta) = \theta\,(2\,\theta - 1). \tag{11}$$

With $t = h\,\theta$, we consider the polynomial function

$$q(t) = q_\ell\,\varphi_0(\theta) + q_m\,\varphi_{1/2}(\theta) + q_r\,\varphi_1(\theta). \tag{12}$$

Then $q(0) = q_\ell$, $q(\frac{h}{2}) = q_m$ and $q(h) = q_r$ and the basis functions (11) are well adapted to these degrees of freedom. We have also $\frac{\mathrm{d}q}{\mathrm{d}t} = g_\ell\,(1-\theta) + g_r\,\theta$ with the derivatives $g_\ell$ and $g_r$ given by the relations

$$g_\ell = \frac{\mathrm{d}q}{\mathrm{d}t}(0) = \frac{1}{h}\big(-3\,q_\ell + 4\,q_m - q_r\big)\,,\ \ g_r = \frac{\mathrm{d}q}{\mathrm{d}t}(h) = \frac{1}{h}\big(q_\ell - 4\,q_m + 3\,q_r\big). \tag{13}$$

We remark also that

$$g_m = \frac{\mathrm{d}q}{\mathrm{d}t}\Big(\frac{h}{2}\Big) = \frac{1}{2}\,(g_\ell + g_r) = \frac{q_r - q_\ell}{h}. \tag{14}$$

Once the interpolation is defined in an interval of length $h$, we use it by splitting the range $[0,\,T]$ into $N$ pieces, and $h = \frac{T}{N}$. With $t_j = j\,h$, we set $q_j \simeq q(t_j)$ for $0 \leq j \leq N$ and $q_{j+1/2} \simeq q(t_j + \frac{h}{2})$ with $0 \leq j \leq N-1$. In the interval $[t_j,\,t_{j+1}]$, the function $q(t)$ is a polynomial of degree 2, represented by the relation (12) with $t = t_j + \theta\,h$, $q_\ell = q_j$, $q_m = q_{j+1/2}$ and $q_r = q_{j+1}$.

For the numerical integration of a regular function $\psi$ on the interval $[0,\,1]$, the Simpson method is very popular: $\int_0^1 \psi(\theta)\,\mathrm{d}\theta \simeq \frac{1}{6}\big[\psi(0) + 4\,\psi(\frac{1}{2}) + \psi(1)\big]$. This quadrature formula is exact up to polynomials of degree three. Then a discrete Lagrangian $L_S(q_\ell,\,q_m,\,q_r) \simeq \int_0^h \big[\frac{m}{2}\big(\frac{\mathrm{d}q}{\mathrm{d}t}\big)^2 - V(q)\big]\,\mathrm{d}t$ can be defined with the Simpson quadrature formula associated with an internal polynomial approximation $q(t)$ of degree 2 presented in (12):

$$L_S(q_\ell,\,q_m,\,q_r) = \frac{m\,h}{12}\,\big(g_\ell^2 + 4\,g_m^2 + g_r^2\big) - \frac{h}{6}\,\big(V_\ell + 4\,V_m + V_r\big) \tag{15}$$

with $V_\ell = V(q_\ell)$, $V_m = V(q_m)$ and $V_r = V(q_r)$. The discrete action $\Sigma_d$ for a motion $t \longmapsto q(t)$ between the initial time and a given time $T > 0$ is discretized with $N$ regular intervals and take the form $\Sigma_d = \sum_{j=1}^{N-1} L_S(q_j,\,q_{j+1/2},\,q_{j+1})$.

Euler-Lagrange equations, coming from the stationary action $\delta\Sigma_d = 0$, are first established for an arbitrary variation $\delta q_{j+1/2}$ of the internal degree of freedom in the interval $[t_j,\,t_{j+1}]$:

$$\frac{\partial L_S}{\partial q_m} = 0. \tag{16}$$

Due to the relations (13), (14), we first observe that $\frac{\partial g_\ell}{\partial q_m} = \frac{4}{h}$, $\frac{\partial g_m}{\partial q_m} = 0$ and $\frac{\partial g_r}{\partial q_m} = -\frac{4}{h}$. Then, due to the expression (15) of the discrete Lagrangian, we have, with $V'_m \equiv \frac{\partial V}{\partial q}(q_m)$,

$\frac{\partial L_S}{\partial q_m} = \frac{m\,h}{6}\,\big(g_\ell\,\frac{4}{h} + g_\ell\,\frac{4}{h}\big) - \frac{2}{3}\,h\,V'_m = \frac{2}{3}\,\frac{m}{h}\,\big(-4\,q_\ell + 8\,q_m - 4\,q_r\big) - \frac{2}{3}\,h\,V'_m$

$= \frac{16}{3} \frac{m}{h} \left[ q_m - \frac{1}{2} (q_\ell + q_r) - \frac{h^2}{8m} V'_m \right]$ and the Eq. (16) can be written

$$q_m - \frac{h^2}{8m} \frac{\partial V}{\partial q}(q_m) = \frac{1}{2} (q_\ell + q_r). \tag{17}$$

It defines implicitly the value $q_m$ at the middle of the interval as a function of the extremities $q_\ell$ and $q_r$. Under the form $\frac{4m}{h^2} (q_\ell - 2q_m + q_r) + \frac{\partial V}{\partial q}(q_m) = 0$, the relation (17) is clearly consistent with the differential equation (3).

The discrete action takes the form

$$\Sigma_d = \cdots + L_S(q_{j-1}, q_{j-1/2}, q_j) + L_S(q_j, q_{j+1/2}, q_{j+1}) + \ldots .$$

The variation of this discrete action is equal to zero. We have in consequence the following discrete Euler-Lagrange equations:

$$\frac{\partial L_S}{\partial q_r}(q_{j-1}, q_{j-1/2}, q_j) + \frac{\partial L_S}{\partial q_\ell}(q_j, q_{j+1/2}, q_{j+1}) = 0.$$

We obtain after some lines of elementary calculus

$$\frac{1}{h^2} \left( q_{j-1} - 2q_j + q_{j+1} \right) + \frac{1}{3m} \left( V'_{j-1/2} + V'_j + V'_{j+1/2} \right) = 0. \tag{18}$$

Due to the condition (16), the right momentum $p_r = \frac{\partial L_S}{\partial q_r}$ can be evaluated as follows:

$p_r = \frac{m\,h}{6} \left[ g_\ell \left( -\frac{1}{h} \right) + 4\, g_m \left( \frac{1}{h} \right) + g_r \left( \frac{3}{h} \right) \right] - \frac{h}{6} V'_r$

$= \frac{m}{6} \left[ -\frac{1}{h} \left( -3q_\ell + 4q_m - q_r \right) + \frac{1}{h} (q_r - q_\ell) + \frac{3}{h} \left( q_\ell - 4q_m + 3q_r \right) \right] - \frac{h}{6} V'_r$

$= \frac{m}{6h} \left[ 2q_\ell - 16q_m + 14q_r \right] - \frac{h}{6} V'_r = \frac{1}{6} \left[ \frac{2m}{h} \left( q_\ell - 8q_m + 7q_r \right) - h\, V'_r \right].$

We replace the intermediate value $q_m$ with the relation (17) and we have

$p_r = \frac{1}{6} \left[ \frac{2m}{h} \left( q_\ell - 8 \left( \frac{1}{2} (q_\ell + q_r) + \frac{h^2}{8m} V'_m \right) + 7q_r \right) - h\, V'_r \right]$

$= \frac{m}{h} \left( q_r - q_\ell \right) - \frac{h}{6} \left( 2V'_m + V'_r \right)$. In other words,

$$p_{j+1} = \frac{m}{h} \left( q_{j+1} - q_j \right) - \frac{h}{6} \left( 2V'_{j+1/2} + V'_{j+1} \right). \tag{19}$$

Similarly, we have $p_j = \frac{m}{h} \left( q_j - q_{j-1} \right) - \frac{h}{6} \left( 2V'_{j-1/2} + V'_j \right)$ and taking into account the discrete Euler-Lagrange equations (18), we find

$$p_j = \frac{m}{h} \left( q_{j+1} - q_j \right) + \frac{h}{6} \left( V'_j + 2V'_{j+1/2} \right). \tag{20}$$

From the relations (19) and (20), we deduce the discrete Hamiltonian dynamics

$$\begin{cases} p_{j+1} - p_j + \frac{h}{6} \left( V'_j + 4V'_{j+1/2} + V'_{j+1} \right) = 0 \\ q_{j+1} - q_j - \frac{h^2}{12m} \left( V'_{j+1} - V'_j \right) - \frac{h}{2m} \left( p_{j+1} + p_j \right) = 0 \end{cases} \tag{21}$$

We write the system (17), (21) under the form $F_S(q_{j+1/2}, p_{j+1}, q_{j+1}) = 0$. We have

$$
\mathrm{d}F_S(q_m,\,p,\,q) = \begin{pmatrix} 1-\frac{h\,\theta}{8} & 0 & -\frac{1}{2} \\ \frac{2}{3}\,m\,\theta & 1 & \frac{1}{6}\,m\,\varphi \\ 0 & -\frac{h}{2\,m} & 1-\frac{h\,\varphi}{12} \end{pmatrix}
$$

with $\theta \equiv \frac{h}{m}\,V''(q_m)$ and $\varphi \equiv \frac{h}{m}\,V''(q)$. After a formal calculation with the help of the free software "SageMath" [10], it is possible to explicit the inverse of this Jacobian matrix. At each time step, the numerical resolution of the nonlinear system of three equations $F_S(q_{j+1/2},\,p_{j+1},\,q_{j+1}) = 0$ is conducted with a Newton algorithm. As with Newmark's algorithm, we have observed machine precision convergence at the fifth iteration with the proposed scheme.

The Simpson scheme (17), (21) is symplectic. With the notations $V''_j \equiv \frac{\partial^2 V}{\partial q^2}(q_j)$ and $\widetilde{V}''_{j+1/2} \equiv \frac{2\,V''_{j+1/2}}{1-\frac{h^2}{8\,m}\,V''_{j+1/2}}$, we have by differentiation of the relations (21):

$$
\begin{aligned}
\delta p_{j+1} + \tfrac{h}{6}\,(V''_{j+1} + \widetilde{V}''_{j+1/2})\,\delta q_{j+1} &= \delta p_j - \tfrac{h}{6}\,(V''_j + \widetilde{V}''_{j+1/2})\,\delta q_j \\
-\tfrac{h}{2\,m}\,\delta p_{j+1} + \left(1 - \tfrac{h^2}{12\,m}\,V''_{j+1}\right)\delta q_{j+1} &= \tfrac{h}{2\,m}\,\delta p_j + \left(1 - \tfrac{h^2}{12\,m}\,V''_j\right)\delta q_j .
\end{aligned}
$$

As for the Newmark scheme, this system can be written with a relation (10). We have in this case $\det A = 1 + \frac{h^2}{12\,m}\,\widetilde{V}''_{j+1/2} = \det B$ and the relation

$$
\frac{\partial p_{j+1}}{\partial p_j}\,\frac{\partial q_{j+1}}{\partial q_j} - \frac{\partial p_{j+1}}{\partial q_j}\,\frac{\partial q_{j+1}}{\partial p_j} = 1
$$

is established.

## 5 First Numerical Experiments and Conclusions

We have implemented the Simpson symplectic scheme (17), (21) and have compared it with the Newmark scheme (9). Typical results for $N = 10$ meshes and one period are displayed in Fig. 2. They are compared with the exact solution presented in Fig. 1. Quantitative errors with the maximum norm are also presented in Table 1. An asymptotic order of convergence can be estimated for the momentum, the state and various energies.

In this work we have recalled Newmark's classical method which is very popular in some fields of the engineering sciences and is symplectic. We have proposed an alternative symplectic variational integrator based on Simpson's rule, to deal with nonlinear differential equations. The method was tested on the non-trivial nonlinear pendulum for which the analytical solution is known. We will be working on more complex problems in the future, starting with the symmetric spinning top. The authors thank the reviewers for their valuable comments.

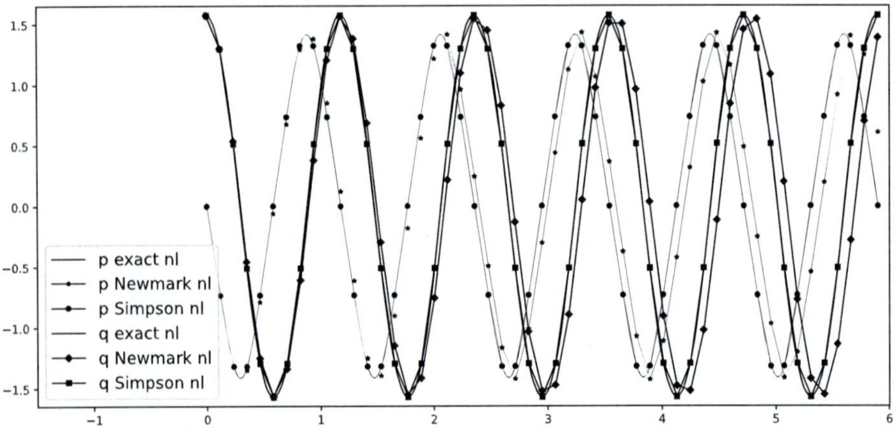


**Fig. 2** Comparing Newmark and Simpson schemes with 10 points per period and a total time of 5 periods. The exact solution is presented in Fig. 1. The errors for the Newmark scheme are clearly visible whereas the Simpson scheme is still very precise despite the reduced number of time steps

**Table 1** Errors in the maximum norm ($L_\infty$) for a simultation of a single period over all time steps. For a given discretization level, the Simpson scheme is more precise than the Newmark scheme by approximately three orders of magnitude. The orders of convergence are preliminary estimates

| | Number of meshes | 50 | 100 | 200 | Order |
|---|---|---|---|---|---|
| Newmark | Momentum | $2.93\ 10^{-2}$ | $7.32\ 10^{-3}$ | $1.83\ 10^{-3}$ | 2.0 |
| Symplectic Simpson | Momentum | $6.08\ 10^{-6}$ | $3.78\ 10^{-7}$ | $2.36\ 10^{-8}$ | 4.0 |
| Newmark | State | $5.26\ 10^{-3}$ | $1.31\ 10^{-3}$ | $3.29\ 10^{-4}$ | 2.0 |
| Symplectic Simpson | State | $1.05\ 10^{-6}$ | $6.51\ 10^{-8}$ | $4.06\ 10^{-9}$ | 4.0 |
| Newmark | Relative energy | $9.06\ 10^{-4}$ | $2.29\ 10^{-4}$ | $5.73\ 10^{-5}$ | 2.0 |
| Symplectic Simpson | Relative energy | $1.30\ 10^{-6}$ | $8.42\ 10^{-8}$ | $5.25\ 10^{-9}$ | 4.0 |

## Annex. Exact Solution of the Nonlinear Pendulum

In this Annex, we follow essentially the synthesis [3]. We consider an angle $\theta_0 \in (0,\ \pi)$ and the non linear pendulum problem

$$\frac{\mathrm{d}^2 q}{\mathrm{d}t^2} + \omega^2 \sin q = 0, \ \ q(0) = \theta_0, \ \ \frac{\mathrm{d}q}{\mathrm{d}t}(0) = 0. \tag{22}$$

It is easy to verify the conservation of energy: $\frac{\mathrm{d}}{\mathrm{d}t}\big[\frac{1}{2}\big(\frac{\mathrm{d}q}{\mathrm{d}t}\big)^2 + \omega^2\,(1-\cos q)\big] = 0$. We introduce the parameter $k \equiv \sin\big(\frac{\theta_0}{2}\big)$ and we have $0 < k < 1$. Due to the initial conditions in (22), we observe that $\frac{\mathrm{d}q}{\mathrm{d}t} < 0$ for small values of $t > 0$.

Then we have $\frac{\mathrm{d}q}{\mathrm{d}t} = -2\,\omega\,k\,\sqrt{1 - \frac{1}{k^2}\,\sin^2\big(\frac{q}{2}\big)}$. We introduce also the change of unknown defined by $\sin\varphi = \frac{1}{k}\,\sin\big(\frac{q}{2}\big)$. We observe that $\varphi(0) = \frac{\pi}{2}$. We have also $\frac{\mathrm{d}\varphi}{\mathrm{d}t} = -\omega\,\sqrt{1 - k^2\,\sin^2\varphi} = -\omega\,\cos\varphi$ and we deduce the differential relation

$$\omega\,\mathrm{d}t = -\frac{\mathrm{d}\varphi}{\sqrt{1 - k^2\,\sin^2\varphi}} \tag{23}$$

from the previous calculus.

We consider $F(a,\,m) \equiv \int_0^a \frac{\mathrm{d}\varphi}{\sqrt{1-m\,\sin^2\varphi}}$ the incomplete elliptic integral of the first kind and $K(m) \equiv \int_0^{\pi/2} \frac{\mathrm{d}\varphi}{\sqrt{1-m\,\sin^2\varphi}} = F(\frac{\pi}{2},\,m)$ the complete elliptic integral of the first kind presented e.g. in the book of Abramowitz and Stegun [1]. Then by integration of (23),

$$\omega\,t = -\int_{\pi/2}^{\varphi} \frac{\mathrm{d}\xi}{\sqrt{1 - k^2\,\sin^2\xi}} = K(k^2) - F(\varphi,\,k^2). \tag{24}$$

After a fourth of a period $T$, the parameter $\varphi$ changes from $\frac{\pi}{2}$ to zero and we obtain

$$\omega\,\frac{T}{4} = \int_0^{\pi/2} \frac{\mathrm{d}\varphi}{\sqrt{1 - k^2\,\sin^2\varphi}} = K(k^2). \tag{25}$$

The Jacobi amplitude $A(.,\,m)$ is defined (see e.g. [1]) as the reciprocal function of the incomplete elliptic integral of the first kind: $F(a,\,m) = u$ is equivalent to $a = A(u,\,m)$. Due to (24), we have $\varphi = A(K(k^2) - \omega\,t,\,k^2)$. Finally,

$$\sin\Big(\frac{q}{2}\Big) = k\,\sin\varphi\,,\quad \frac{\mathrm{d}q}{\mathrm{d}t} = -2\,\omega\,k\,\cos\varphi.$$

## References


1. Abramowitz, M., Stegun, I.A.: Handbook of Mathematical Functions with Formulas, Graphs, and Mathematical Tables. National Bureau of Standards, Applied Mathematics Series, vol. 55 (1964)
2. Arnold, V.: Mathematical Methods of Classical Mechanics. Springer (1974)
3. Chenciner, A.: Connaissez-vous le pendule? Gazette des Mathématiciens **86**, 21–27 (2000)
4. Dubois, F., Rojas-Quintero, J.A.: A variational symplectic scheme based on Simpson's quadrature. In: 6th conference on Geometric Science of Information, Saint-Malo (France) (2023)
5. Hairer, E., Wanner, G., Lubich, C.: Geometric Numerical Integration. Springer, Berlin Heidelberg (2006)
6. Kane, C., Marsden, J.E., Ortiz, M., West, M.: Variational integrators and the Newmark algorithm for conservative and dissipative mechanical systems. Int. J. Numer. Methods Eng. **49**, 1295–1325 (2000)
7. Newmark, N.M.: A method of computation for structural dynamics. J. Eng. Mech. Div. Am. Soc. Civ. Eng. **85**(EM3), 67–94 (1959)
8. Raviart, P.A., Thomas, J.M.: Introduction à l'analyse numérique des équations aux dérivées partielles. Masson, Paris (1983)

9. Rojas-Quintero, J.A., Villalobos-Chin, J., Santibanez, V.: Optimal control of robotic systems using finite elements for time integration of covariant control equations. IEEE Access **9**, 104980–105001 (2021)
10. SageMath, the Sage mathematics software system (Version 7.5.1). The Sage developers (2017). www.sagemath.org
11. Sanz-Serna, J.M.: Symplectic integrators for Hamiltonian problems: an overview. Acta Numerica **1**, 243–286 (1992)
12. Souriau, J.M.: Structure des systèmes dynamiques. Dunod, Paris (1970)

# An Active Flux Method for the Vlasov-Poisson System

**Yanick-Florian Kiechle, Erik Chudzik, and Christiane Helzel**

**Abstract** We derive an Active Flux method for the 1 + 1 dimensional Vlasov-Poisson system. During each time step one solution step of the Poisson problem is followed by one evolution step of the unsplit 1 + 1 dimensional Vlasov equation. The resulting method is third order accurate.

**Keywords** Active flux · Vlasov-poisson system · Plasma physics

## 1 Introduction

The Active Flux method is a finite volume method for hyperbolic conservation laws that was proposed by Eymann and Roe [3, 4, 7]. The original version of the Active Flux method, which is also used here, is third order accurate. While Roe and Eymann used unstructured triangular grids for their two-dimensional Active Flux method, we consider a Cartesian grid as in [1, 2, 6]. Our approach for the Vlasov-Poisson system relies on this grid structure.

The Active Flux method has a compact stencil in space and time. In contrast to classical finite volume methods, the stencil is enriched by adding point values located along grid cell interfaces. The point values are updated using truly multi-dimensional evolution formulas. The dense and compact stencil of the method gives accurate approximations even on coarse grids, while following the more physical approach, that one uses as little information as possible outside the true domain of dependence as pointed out by Roe [9]. Compared to third order discontinuous Galerkin methods the Active Flux method also gives comparable accuracy at substantially reduced cost [8]. Therefore, this method is an interesting candidate for the high-dimensional Vlasov equation.

The Vlasov-Poisson system models the evolution of a collisionless plasma. It consists of a kinetic equation in space and velocity for the plasma particles, which self

Y.-F. Kiechle · E. Chudzik (✉) · C. Helzel
Institute of Mathematics, Heinrich-Heine University Düsseldorf, Dusseldorf, Germany
e-mail: erik.chudzik@hhu.de

 
E. Franck et al. (eds.), *Finite Volumes for Complex Applications X—Volume 2, Hyperbolic and Related Problems*, Springer Proceedings in Mathematics & Statistics 433,
https://doi.org/10.1007/978-3-031-40860-1_10

interact via an electrostatic force [5]. Here we present the first Active Flux method for the Vlasov-Poisson system. We restrict our considerations to the $1+1$-dimensional case. During each time step the Poisson problem is solved using a higher order finite difference method. The Vlasov equation can be evolved in time using characteristic theory for the update of the point values and a finite volume approach for the update of the cell average values. In order to obtain third order accuracy the electric field needs to be prescribed as a function in space and time. While the solution of the Poisson problem provides the electric field at discrete times only, we use several moments of the Vlasov equation to derive a time dependent representation of the electric field. Our numerical results confirm third order accuracy of the resulting method.

## 2 The Mathematical Model

In physical applications one may be interested in the dynamics of a non-relativistic collisionless plasma with $S$ different plasma species. The Vlasov equation describes the evolution of a probability density of each species. Since we consider collisionless plasma, the particles do not interact by collision, but are connected exclusively indirectly by the electromagnetic field, which is why the Vlasov equation is coupled to Maxwell's equations. One obtains the Vlasov-Maxwell system. If we now make simplifying assumptions on the magnetic field as well as on physical constants and consider only one dynamical evolving and one stationary species, we obtain the Vlasov-Poisson system. See [5] for a more detailed derivation.

We now consider the $1+1$ dimensional Vlasov-Poisson system on the domain $(x, v, t) \in \Omega = (-L, L) \times \mathbb{R} \times \mathbb{R}^+$

$$\partial_t f + v\partial_x f + E(x,t)\partial_v f = 0 \tag{1}$$

$$\Delta\phi(x,t) = \rho(x,t) - \rho_0, \tag{2}$$

$$E(x,t) = \nabla\phi(x,t). \tag{3}$$

Here $f = f(x, v, t)$ is a probability density function, which depends on time $t$, macroscopic space $x$ and particle velocity $v$ and interacts via an electrostatic force. $E$ represents the electric field and $\phi$ is the electric potential. Furthermore we require periodic boundary conditions in $x$

$$f(-L, v, t) = f(L, v, t), \quad \phi(-L, t) = \phi(L, t) \quad \text{and} \quad E(-L, t) = E(L, t).$$

The total and background densities are defined by

$$\rho(x,t) := \int_{-\infty}^{\infty} f(x, v, t)\, dv, \quad \rho_0 := \frac{1}{2L}\int_{-L}^{L} \rho(x,t)\, dx. \tag{4}$$

## 3 Active Flux for Vlasov Poisson

First we construct a $(x, v)$-grid consisting of cells $C_{i,j} = [x_{i-\frac{1}{2}}, x_{i+\frac{1}{2}}] \times [v_{j-\frac{1}{2}}, v_{j+\frac{1}{2}}]$ with constant mesh widths $\Delta x = x_{i+\frac{1}{2}} - x_{i-\frac{1}{2}}$ and $\Delta v = v_{j+\frac{1}{2}} - v_{j-\frac{1}{2}}$ $\forall i, j$. Furthermore we have $x_{\frac{1}{2}} = -L$, $x_{N_x+\frac{1}{2}} = L$ and, considering a bounded domain in the velocity component, $v_{\frac{1}{2}} = -v_{max}, v_{N_v+\frac{1}{2}} = v_{max}$. In Fig. 1 we illustrate the grid used for the Poisson problem as well as a mesh cell for the the approximation of the Vlasov equation. Note that the Cartesian grid Active Flux method uses point values along the grid cell boundary as well as cell average values as degrees of freedom. More details and in particular the basis functions used for the reconstruction can be found in [1, 6].

During each time step, we solve the Poisson problem on the domain $(-L, L)$ using a grid with mesh width $\Delta x/2$. We use a fourth order accurate finite difference method. To obtain a unique solution of the Poisson problem we computed a solution to the problem

$$\begin{aligned} \Delta\phi(x,t) &= \rho(x,t) - \rho_0 \\ \phi(-L,t) &= \phi(L,t) = 0, \end{aligned}$$

assuming that $\rho$ and $\rho_0$ are given. We can then compute the electric field $E(x, t_n)$ at all grid points using higher order finite difference formulas. Now assume we have at least third order accurate approximations of the electric field at the current time $t_n$ at all points $x_{i-\frac{1}{2}}, x_i, x_{i+\frac{1}{2}}$.

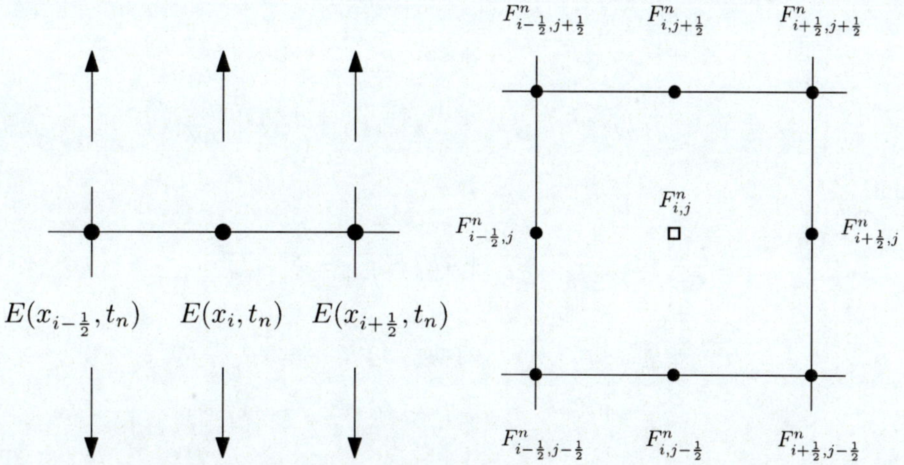


**Fig. 1** Left: One-dimensional spatial grid cell of the electric field. The arrows are to illustrate the integration over the velocity space V to obtain $\rho$. Right: Configuration of degrees of freedom in cell $C_{i,j}$

The interesting part is the update of the Vlasov equation, which will now be described in more detail. To simplify the presentation we write the equation in the form

$$\begin{aligned} f_t + h(f)_x + g(f)_v &= 0 \qquad (x, v, t) \in (-L, L) \times (-v_{max}, v_{max}) \times \mathbb{R}^+, \\ f(x, v, 0) &= f_0(x, v) \qquad (x, v) \in (-L, L) \times (-v_{max}, v_{max}). \end{aligned}$$

with initial values $f_0(x, v)$ which are periodic in $x$.

Here $f : (-L, L) \times (-v_{max}, v_{max}) \times \mathbb{R}^+ \to \mathbb{R}$ is the conserved quantity and $h, g : \mathbb{R} \to \mathbb{R}$ are flux functions, which have the form $h(f(x, v, t)) = vf(x, v, t)$ and $g(f(x, v, t)) = E(x, t)f(x, v, t)$.

The numerical flux over a vertical edge $x_{i+\frac{1}{2}} \times [v_{j-\frac{1}{2}}, v_{j+\frac{1}{2}}]$ of the cell $C_{i,j}$, i.e.

$$H_{i+\frac{1}{2},j} \approx \frac{1}{\Delta t \Delta v} \int_{t_n}^{t_{n+1}} \int_{v_{j-\frac{1}{2}}}^{v_{j+\frac{1}{2}}} h(f(x_{i+\frac{1}{2}}, v, t))\, dv\, dt$$

as well as the flux over the horizontal edge $[x_{i-\frac{1}{2}}, x_{i+\frac{1}{2}}] \times v_{j+\frac{1}{2}}$ of $C_{i,j}$, i.e.

$$G_{i,j+\frac{1}{2}} \approx \frac{1}{\Delta t \Delta x} \int_{t_n}^{t_{n+1}} \int_{x_{i-\frac{1}{2}}}^{x_{i+\frac{1}{2}}} g(f(x, v_{j+\frac{1}{2}}, t))\, dx\, dt$$

will be approximated using Simpson's rule. We obtain

$$\begin{aligned} H_{i+\frac{1}{2},j} = \frac{1}{36} \Big( & v_{j-\frac{1}{2}} F^n_{i+\frac{1}{2},j-\frac{1}{2}} + 4v_j F^n_{i+\frac{1}{2},j} + v_{j+\frac{1}{2}} F^n_{i+\frac{1}{2},j+\frac{1}{2}} \\ & + 4\; v_{j-\frac{1}{2}} F^{n+\frac{1}{2}}_{i+\frac{1}{2},j-\frac{1}{2}} + 16 v_j F^{n+\frac{1}{2}}_{i+\frac{1}{2},j} + 4 v_{j+\frac{1}{2}} F^{n+\frac{1}{2}}_{i+\frac{1}{2},j+\frac{1}{2}} \\ & + \; v_{j-\frac{1}{2}} F^{n+1}_{i+\frac{1}{2},j-\frac{1}{2}} + 4v_j F^{n+1}_{i+\frac{1}{2},j} + v_{j+\frac{1}{2}} F^{n+1}_{i+\frac{1}{2},j+\frac{1}{2}} \Big) \end{aligned}$$

and

$$\begin{aligned} G_{i,j+\frac{1}{2}} = \frac{1}{36} \Big( & E^n_{i-\frac{1}{2}} F^n_{i-\frac{1}{2},j+\frac{1}{2}} + 4E^n_i F^n_{i,j+\frac{1}{2}} + E^n_{i+\frac{1}{2}} F^n_{i+\frac{1}{2},j+\frac{1}{2}} \\ + \\ & , 4\; E^{n+\frac{1}{2}}_{i-\frac{1}{2}} F^{n+\frac{1}{2}}_{i-\frac{1}{2},j+\frac{1}{2}} + 16 E^{n+\frac{1}{2}}_i F^{n+\frac{1}{2}}_{i,j+\frac{1}{2}} + 4E^{n+\frac{1}{2}}_{i+\frac{1}{2}} F^{n+\frac{1}{2}}_{i+\frac{1}{2},j+\frac{1}{2}} \\ & + \; E^{n+1}_{i-\frac{1}{2}} F^{n+1}_{i-\frac{1}{2},j+\frac{1}{2}} + 4E^{n+1}_i F^{n+1}_{i,j+\frac{1}{2}} + E^{n+1}_{i+\frac{1}{2}} F^{n+1}_{i+\frac{1}{2},j+\frac{1}{2}} \Big). \end{aligned}$$

Here $E_i^{n+\frac{\tau}{\Delta t}}$ and $F_{i,j}^{n+\frac{\tau}{\Delta t}}$ denote the approximations of the electric field $E(x_i, t_n + \tau)$ and the distribution function $f(x_i, v_j, t_n + \tau)$ for $i = \frac{1}{2}, 1, \ldots, N_x + \frac{1}{2}$, $j = \frac{1}{2}, 1, \ldots, N_y + \frac{1}{2}$ and $\tau \in \{0, \frac{\Delta t}{2}, \Delta t\}$.

The values of $v$ arising in the numerical flux $H_{i+\frac{1}{2},j}$ are known values on our $(x, v)$-grid. However the values of the electric field arising in the numerical flux $G_{i,j+\frac{1}{2}}$ at later times need to be computed. The previously mentioned solution of the Poisson problem on a grid with mesh width $\Delta x/2$ provides accurate approximations of $E^n_{i-\frac{1}{2}}, E^n_i, E^n_{i+\frac{1}{2}}$ for all grid cells. In order to get a third order accurate approximation of the electric field at the intermediate and new time level we use a Taylor series expansion of the form

$$
\begin{aligned}
E(x, t_n + \tau) &= E(x, t_n) + (t_n + \tau - t_n)E_t(x, t_n) \\
&\qquad + \frac{1}{2!}(t_n + \tau - t_n)^2 E_{tt}(x, t_n) + \mathcal{O}(\tau^3) \\
&= E(x, t_n) + \tau E_t(x, t_n) + \frac{1}{2!}\tau^2 E_{tt}(x, t_n) + \mathcal{O}(\tau^3).
\end{aligned}
$$

Note that a second order accurate approximation of $E_t$ and a first order accurate approximation of $E_{tt}$ are sufficient since

$$
\begin{aligned}
E(x, t_n + \tau) &= \underbrace{E(x, t_n)}_{=\mathcal{O}(\tau^3)} + \tau \underbrace{E_t(x, t_n)}_{=\mathcal{O}(\tau^2)} + \frac{1}{2!}\tau^2 \underbrace{E_{tt}(x, t_n)}_{=\mathcal{O}(\tau)} + \mathcal{O}(\tau^3) \\
&= \mathcal{O}(\tau^3).
\end{aligned}
$$

We illustrate the computation of $E_t(x, t_n)$. From (2) and (3) one obtains

$$\partial_x E(x, t) = \partial_{xx}\phi(x, t) = \rho(x, t) - \rho_0 \tag{5}$$

as well as

$$\partial_t \partial_x E(x, t) = \partial_t(\rho(x, t) - \rho_0) = \partial_t \rho(x, t). \tag{6}$$

Integrating (1) over $V := [-v_{max}, v_{max}]$, we obtain

$$
\begin{aligned}
0 &= \int_V \partial_t f(x, v, t)\, dv + \int_V \partial_x v f(x, v, t)\, dv + \int_V \partial_v(E(x, t) f(x, v, t))\, dv \\
&= \underbrace{\partial_t \int_V f(x, v, t)\, dv}_{=\rho(x,t)} + \underbrace{\partial_x \int_V v f(x, v, t)\, dv}_{=\rho u(x,t)} + \int_V \partial_v(E(x, t) f(x, v, t))\, dv \\
&= \partial_t \rho(x, t) + \partial_x \rho u(x, t) + E(x, t) \underbrace{\int_V \partial_v f(x, v, t)\, dv}_{=0}.
\end{aligned}
$$

The last integral vanishes due to the fact that we assume that $f$ vanishes for large $v$. Thus, we obtain

$$\partial_t \rho(x,t) = -\partial_x \rho u(x,t) \tag{7}$$

for $(x,t) \in [-L, L] \times \mathbb{R}^+$. Using (7) and interchanging the order of derivatives in (6) we obtain

$$\partial_x E_t(x,t) = -\partial_x \rho u(x,t_n) \Rightarrow E_t(x,t) = -\rho u(x,t),$$

i.e. a representation of $E_t(x,t)$. Similarly one can calculate the representation for $E_{tt}(x,t)$, see [5] for details,

$$E_{tt}(x,t_n) = \partial_x \mathcal{E}(x,t_n) - \rho(x,t_n)E(x,t_n),$$

with $\rho$ as defined in (4) and

$$\rho u(x,t_n) = \int_V v f(x,v,t_n)\, dv \tag{8}$$

$$\mathcal{E}(x,t_n) = \int_V v^2 f(x,v,t_n)\, dv, \tag{9}$$

where $V = [-v_{max}, v_{max}]$ is. The integrals on the right hand side of (4), (8) and (9) can be computed from the degrees of freedom of the Active Flux method using Simpson's rule. For $\mathcal{E}$, for example, we compute

$$\mathcal{E}^n_{i-\frac{1}{2}} = \frac{\Delta v}{6} \sum_{j=1}^{N_v} (v^2_{j-\frac{1}{2}} F^n_{i-\frac{1}{2},j-\frac{1}{2}} + 4v_j^2 F^n_{i-\frac{1}{2},j} + v^2_{j+\frac{1}{2}} F^n_{i-\frac{1}{2},j+\frac{1}{2}})$$

$$\begin{aligned}\mathcal{E}^n_i = \frac{\Delta v}{6} \sum_{j=1}^{N_v} (v^2_{j-\frac{1}{2}} F^n_{i,j-\frac{1}{2}} &+4v_j^2 \frac{1}{16}(36F^n_{i,j} - 4(F^n_{i,j-\frac{1}{2}} + F^n_{i+\frac{1}{2},j} + F^n_{i,j+\frac{1}{2}} + F^n_{i-\frac{1}{2},j})\\ &-(F^n_{i-\frac{1}{2},j-\frac{1}{2}} + F^n_{i+\frac{1}{2},j-\frac{1}{2}} + F^n_{i+\frac{1}{2},j+\frac{1}{2}} + F^n_{i-\frac{1}{2},j+\frac{1}{2}})) + v^2_{j+\frac{1}{2}} F^n_{i,j+\frac{1}{2}})\end{aligned}$$

for $i = 1, \ldots, N_x$. For the approximation of $\mathcal{E}_{i-\frac{1}{2}}$ we can directly use the point value degrees of freedom. The computation of $\mathcal{E}_i$ requires point values of $f$ at the midpoints of a grid cell. These point values need to be computed from the Active Flux reconstruction. The other moments of $f$ are approximated analogously.

With this procedure we obtain approximations of the electric field at the intermediate and new time level with sufficient accuracy.

Furthermore, we need to compute the point values of $f$ at the intermediate and new time level which arise in the numerical fluxes $G_{i,j+\frac{1}{2}}$ and $H_{i+\frac{1}{2},j}$. These can be obtained by using the characteristics

$$x'(t) = v(t)$$
$$v'(t) = E(x(t), t)$$

of the Vlasov equation. Let us, for example, consider the approximation of $F^{n+\frac{1}{2}}_{i-\frac{1}{2},j-\frac{1}{2}} \approx f(x_{i-\frac{1}{2}}, v_{j-\frac{1}{2}}, t_n + \Delta t/2)$ using the Active Flux reconstruction at time $t_n$. In this case we solve the ode for the characteristics with initial values $x(t_{n+\frac{1}{2}}) = x_{i-\frac{1}{2}}$, $v(t_{n+\frac{1}{2}}) = v_{j-\frac{1}{2}}$ backwards in time for a time step of length $\Delta t/2$, as introduced in [2]. It turns out that a second order accurate ode solver is sufficient. By using Heun's method the footpoint $(x^n, v^n)$ of the characteristics can be computed as

$$\begin{aligned} x^* &= x_{i-\frac{1}{2}} - \tfrac{\Delta t}{2} v_{j-\frac{1}{2}}, & v^* &= v_{j-\frac{1}{2}} - \tfrac{\Delta t}{2} E(x_{i-\frac{1}{2}}, t_{n+\frac{1}{2}}), \\ x^n &= x_{i-\frac{1}{2}} - \tfrac{\Delta t}{4}(v_{j-\frac{1}{2}} + v^*), & v^n &= v_{j-\frac{1}{2}} - \tfrac{\Delta t}{4}(E(x_{i-\frac{1}{2}}, t_{n+\frac{1}{2}}) + E(x^*, t_n)). \end{aligned}$$

We already explained how to express the electric field at grid points of our spatial grid as a function of time. Now we however need to evaluate the electric field at a point $x^*$ which does not lie on the spatial grid. If we restrict the time step such that

$$\text{CFL} := \max\left(\frac{\max_j |v_j| \Delta t}{\Delta x}, \frac{\max_i |E(x_i, t)| \Delta t}{\Delta v}\right) \le 1,$$

then the point $x^*$ is located in the left or right neighbouring grid cell. We use Lagrangian interpolation in order to evaluate the electric field at $x^*$. For $v^* < 0$, for example, we interpolate

$$\left(x_{i-\frac{1}{2}}, E^{n+\tau}_{i-\frac{1}{2}}\right), \left(x_i, E^{n+\tau}_i\right), \left(x_{i+\frac{1}{2}}, E^{n+\tau}_{i+\frac{1}{2}}\right)$$

and obtain

$$E(x^*, t_n + \tau) = \sum_{k \in \{i-\frac{1}{2}, i, i+\frac{1}{2}\}} E^{n+\tau}_k L_k(x^*)$$

where $L_k$ are the Lagrangian basis polynomials. It is worth mentioning that these computations to evaluate the electric field at $x^*$ can be carried out not only with the electric field at current times but at every time $t_n + \tau$ for $\tau \in [0, \Delta t]$ as it is required in different ode solvers like the classical Runge-Kutta 4 method. $E(x, t_n + \tau)$ for $\tau \in [0, \Delta t]$ can be computed analogously via the Taylor series expansion in (5).

Once the footpoints of the characteristic curve are computed we obtain $F^{n+\frac{1}{2}}_{i-\frac{1}{2},j-\frac{1}{2}}$ by evaluating the piecewise quadratic Active Flux reconstruction, that interpolates the point values and preserves the cell average value at time $t_n$, at the point $(x^n, v^n)$. In an analogous way we can compute all the other point values of $f$ at the intermediate and final time.

After carrying out these steps for every cell $C_{i,j}$ the new cell average values of $f$ are finally computed using the finite volume update

$$F_{i,j}^{n+1} = F_{i,j}^{n} - \frac{\Delta t}{\Delta x}\left(H_{i+\frac{1}{2},j} - H_{i-\frac{1}{2},j}\right) - \frac{\Delta t}{\Delta y}\left(G_{i,j+\frac{1}{2}} - G_{i,j-\frac{1}{2}}\right).$$

## 4 Numerical Results

We consider the two stream instability, a classical test problem for numerical methods for the Vlasov-Poisson system, which was also studied in [5].

The initial values have the form

$$f(x, v, 0) = \frac{v^2}{\sqrt{8\pi}}\left(2 - \cos\left(\frac{x}{2}\right)\right)e^{-\frac{v^2}{2}}.$$

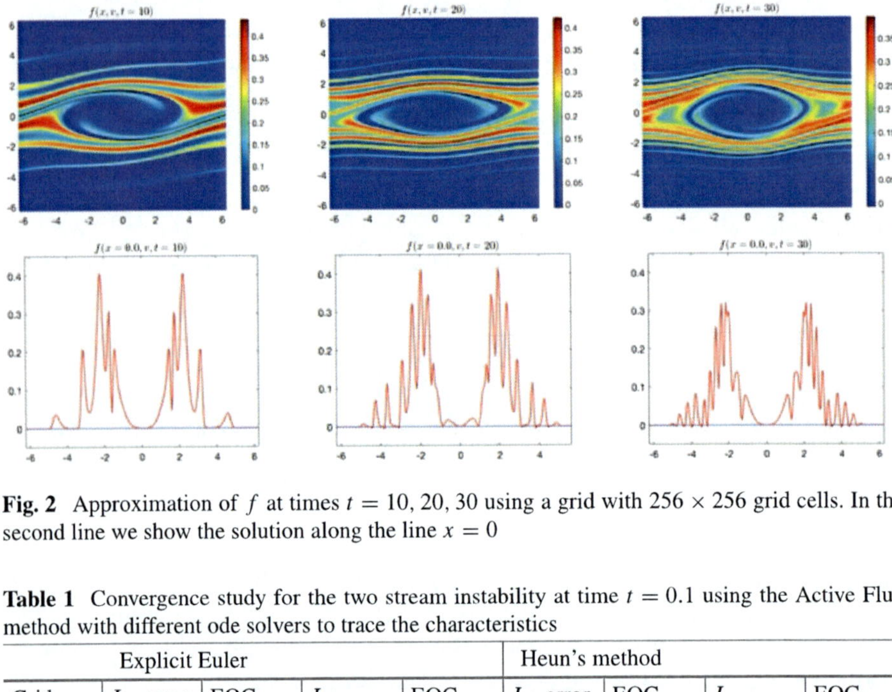

**Fig. 2** Approximation of $f$ at times $t = 10, 20, 30$ using a grid with $256 \times 256$ grid cells. In the second line we show the solution along the line $x = 0$

**Table 1** Convergence study for the two stream instability at time $t = 0.1$ using the Active Flux method with different ode solvers to trace the characteristics

| | Explicit Euler | | | | Heun's method | | | |
|---|---|---|---|---|---|---|---|---|
| Grid | $L_1$-error | EOC | $L_\infty$-error | EOC | $L_1$-error | EOC | $L_\infty$-error | EOC |
| $128^2$ | $8.0047 \cdot 10^{-5}$ | – | $5.2311 \cdot 10^{-6}$ | – | $6.5572 \cdot 10^{-5}$ | – | $4.7229 \cdot 10^{-6}$ | – |
| $256^2$ | $1.6305 \cdot 10^{-5}$ | 2.2955 | $1.0540 \cdot 10^{-6}$ | 2.3113 | $9.3983 \cdot 10^{-6}$ | 2.8026 | $6.7712 \cdot 10^{-7}$ | 2.8022 |
| $512^2$ | $3.8017 \cdot 10^{-6}$ | 2.1006 | $2.2940 \cdot 10^{-7}$ | 2.1999 | $1.2499 \cdot 10^{-6}$ | 2.9106 | $9.2055 \cdot 10^{-8}$ | 2.8789 |

The Vlasov-Poisson problem is solved on the domain $(x, v) \in [-2\pi, 2\pi] \times [-2\pi, 2\pi]$ with double periodic boundary conditions. Numerical results for $f$ at different times are shown in Fig. 2.

In Table 1 we show results of an experimental convergence study, where we compared a highly resolved computation for $f$ at time $t = 0.1$, computed on a grid with $2048 \times 2048$ grid cells, with approximations computed on several coarser grids. The test confirms the expected third order convergence rate. If we use the explicit Euler method to trace the characteristics, then the experimentally observed order of convergence is reduced to two.

Future work will concentrate on a positivity preserving approach and on extensions to higher dimensions.

## References

1. Barsukow, W., Hohm, J., Klingenberg, C., Roe, P.L.: The active flux scheme on Cartesian grids and its low Mach number limit. J. Sci. Comput. **81**, 594–622 (2019)
2. Calhoun, D., Chudzik, E., Helzel, C.: The Cartesian grid Active Flux method with adaptive mesh refinement. J. Sci. Comput. **94** (2023)
3. Eymann, T.A., Roe, P.L.: Active Flux Schemes for Systems. AIAA 2011–3840 (2011)
4. Eymann, T.A., Roe, P.L.: Multidimensional active flux schemes. AIAA Conference Paper, June (2013)
5. Rossmanith, J.A., Seal, D.C.: A positivity-preserving high-order semi-Lagrangian discontinuous Galerkin scheme for the Vlasov-Poisson equations. J. Comput. Phys. **230**, 6203–6232 (2011)
6. Helzel, C., Kerkmann, D., Scandurra, L.: A new ADER method inspired by the active flux method. J. Sci. Comput. **80**, 1463–1497 (2019)
7. Roe, P.: Is discontinuous reconstruction really a good idea? J. Sci. Comput. **73**, 1094–1114 (2017)
8. Roe, P., Maeng, J., Fan, D.: Comparing Active Flux and Discontinuous Galerkin Methods for Compressible Flow. AIAA Aerospace Sciences Meeting, January (2018)
9. Roe, P.: Designing CFD methods for bandwidth - A physical approach. Comput. Fluids **214**, 104774 (2021)

# On Thermodynamically Compatible Finite Volume Schemes for Overdetermined Hyperbolic Systems

**Michael Dumbser, Saray Busto, and Andrea Thomann**

**Abstract** We discuss a general class of thermodynamically compatible finite volume schemes for the discretization of overdetermined, thermodynamically compatible first order hyperbolic systems. The new schemes were introduced in [4, 11]. *By construction* they satisfy an *entropy inequality* and are *nonlinearly stable* in the energy norm. A peculiar feature of the new approach is that the *entropy inequality/entropy inequalities* are discretized *directly*, while *total energy conservation* is achieved as a mere *consequence* of the thermodynamically compatible discretization of all the other equations. The schemes can be applied to a very general class of nonlinear systems of hyperbolic partial differential equations, including both, conservative and non-conservative products, as well as potentially stiff algebraic relaxation source terms, provided that the underlying system is *overdetermined* and therefore satisfies an additional extra conservation law. The new methodology is applied to four particular thermodynamically compatible systems: the equations of ideal magnetohydrodynamics (MHD), the unified first order hyperbolic model of continuum mechanics of Godunov, Peshkov and Romenski (GPR), the first order hyperbolic model for turbulent shallow water flows of Gavrilyuk et al. and to the compressible two-fluid model of Romenski et al., which is endowed with *two entropy inequalities*. To the best knowledge of the authors, up to now there exists no standard finite volume scheme that is able to solve hyperbolic PDE systems with more than one entropy inequality. This feature can be considered as the main advantage of our scheme over existing standard FV methods. We present a set of numerical results in order to show their potential, efficiency and practical applicability.

M. Dumbser (✉)
Department of Civil, Environmental and Mechanical Engineering, University of Trento, Via Mesiano 77, I-38123 Trento, Italy
e-mail: michael.dumbser@unitn.it

S. Busto
Departamento de Matemática Aplicada I, Universidade de Vigo, Campus As Lagoas Marcosende s/n, E-36310 Vigo, Spain

A. Thomann
Université de Strasbourg, CNRS, Inria, IRMA, 67000 Strasbourg, France

 
E. Franck et al. (eds.), *Finite Volumes for Complex Applications X—Volume 2, Hyperbolic and Related Problems*, Springer Proceedings in Mathematics & Statistics 433,
https://doi.org/10.1007/978-3-031-40860-1_11

**Keywords** Overdetermined thermodynamically compatible hyperbolic systems · Thermodynamically compatible finite volume schemes (HTC) · Discrete entropy inequality · Nonlinear stability in the energy norm · Applications to magnetohydrodynamics · Continuum mechanics · Turbulent shallow water and compressible two-fluid flows

## 1 Governing Equations

The overdetermined systems of nonlinear hyperbolic, thermodynamically compatible partial differential equations considered in this work, i.e. the equations of ideal magnetohydrodynamics (MHD) [3], the unified first order hyperbolic model of continuum mechanics of Godunov, Peshkov and Romenski (GPR model) [6, 8], the hyperbolic model of turbulent shallow water flows (TSW) of Gavrilyuk et al. [5, 7] and the compressible two-fluid model of Romenski et al. [9, 10], can be cast into the general form of a nonlinear hyperbolic system of balance laws with parabolic vanishing viscosity regularization,

$$\partial_t \mathbf{q} + \partial_k \mathbf{f}_k(\mathbf{q}) + \mathbf{B}_k(\mathbf{q})\partial_k \mathbf{q} - \partial_m (\epsilon \partial_m \mathbf{q}) = \mathbf{P} + \mathbf{S}(\mathbf{q}), \tag{1}$$

and they are all endowed with an *extra conservation law* for the total energy density of the form

$$\frac{\partial \mathcal{E}}{\partial t} + \partial_k F_k - \partial_m (\epsilon \partial_m \mathcal{E}) = 0. \tag{2}$$

Here, $\mathbf{q}$ is the state vector, the flux tensor is denoted by $\mathbf{f}_k(\mathbf{q})$ and $\mathbf{B}_k(\mathbf{q})\partial_k \mathbf{q}$ includes the non-conservative products. The parabolic regularization terms read $\partial_m (\epsilon \partial_m \mathbf{q})$, with the vanishing viscosity parameter $\epsilon > 0$ and the associated entropy production term contained in $\mathbf{P}$. If present in the model, the (potentially stiff) algebraic relaxation source terms are contained in $\mathbf{S}(\mathbf{q})$. In the extra conservation law (2) the total energy flux is denoted by $F_k = F_k(\mathbf{q})$.

The *main field* or *thermodynamic dual variables* are defined as $\mathbf{p} = \partial_{\mathbf{q}} \mathcal{E}$ and therefore one trivially has $\mathbf{p} \cdot \partial_t \mathbf{q} = \partial_t \mathcal{E}$. Thus, for thermodynamic compatibility of the full system (1) with (2), the following identity must hold:

$$\mathbf{p} \cdot \partial_k \mathbf{f}_k(\mathbf{q}) + \mathbf{p} \cdot \mathbf{B}_k(\mathbf{q})\partial_k \mathbf{q} = \partial_k F_k. \tag{3}$$

Moreover, we also have

$$\mathbf{p} \cdot \mathbf{P} + \mathbf{p} \cdot \partial_m (\epsilon \partial_m \mathbf{q}) = \partial_m (\epsilon \partial_m \mathcal{E}), \tag{4}$$

i.e. compatibility of the non-negative entropy production term with the dissipative parabolic vanishing viscosity terms. Finally, the relation

$$\mathbf{p} \cdot \mathbf{S}(\mathbf{q}) = 0 \tag{5}$$

is required for the thermodynamic compatibility of the algebraic source terms, if present in the system.

## 2 Numerical Method and Results

In this section we briefly summarize the thermodynamically compatible finite volume schemes introduced in [4, 11]. For further details, the reader is referred to [4, 11]. The computational domain $\Omega \subset \mathbb{R}^d$ in $d$ space dimensions is paved by a general mesh of *orthogonal* polygonal/polyhedral control volumes $\Omega^\ell$. The common edge/face of two neighboring polygons/polyhedra $\Omega^\ell$ and $\Omega^\imath$ is denoted by $\partial\Omega^{\ell\imath} = \Omega^\ell \cap \Omega^\imath$ and $N_\ell$ is the set of neighbors of element $\Omega^\ell$. The semi-discrete finite volume scheme for the discretization of the general PDE system (1) with extra conservation law (2) reads for each control volume $\Omega^\ell$ as follows:

$$\begin{aligned}\frac{\partial \mathbf{q}^\ell}{\partial t} = &-\sum_{\imath \in N_\ell} \frac{|\partial\Omega^{\ell\imath}|}{|\Omega^\ell|}\left(\mathcal{F}^{\ell\imath} \cdot \mathbf{n}^{\ell\imath} + \mathcal{D}\left(\mathbf{q}^\ell, \mathbf{q}^\imath\right) \cdot \mathbf{n}^{\ell\imath}\right) \\ &+\sum_{\imath \in N_\ell} \frac{|\partial\Omega^{\ell\imath}|}{|\Omega^\ell|}\left(\mathcal{G}\left(\mathbf{q}^\ell, \mathbf{q}^\imath\right) \cdot \mathbf{n}^{\ell\imath} + \mathbf{P}\left(\mathbf{q}^\ell, \mathbf{q}^\imath\right)\right) + \mathbf{S}(\mathbf{q}^\ell),\end{aligned} \tag{6}$$

where $\mathbf{n}^{\ell\imath}$ is the unit normal vector pointing from element $\Omega^\ell$ to element $\Omega^\imath$ and thus $\mathbf{n}^{\imath\ell} = -\mathbf{n}^{\ell\imath}$. The two-point numerical flux in normal direction $\mathcal{F}^{\ell\imath} \cdot \mathbf{n}^{\ell\imath}$, the fluctuations $\mathcal{D}\left(\mathbf{q}^\ell, \mathbf{q}^\imath\right) \cdot \mathbf{n}^{\ell\imath}$, the dissipative terms $\mathcal{G}\left(\mathbf{q}^\ell, \mathbf{q}^\imath\right) \cdot \mathbf{n}^{\ell\imath}$, with the related entropy production terms $\mathbf{P}\left(\mathbf{q}^\ell, \mathbf{q}^\imath\right)$, and the algebraic source terms $\mathbf{S}(\mathbf{q}^\ell)$ must verify the compatibility conditions

$$\begin{aligned}\mathbf{p}^\ell \cdot \left(\mathcal{F}^{\ell\imath} \cdot \mathbf{n}^{\ell\imath} - \mathbf{f}_k^\ell n_k^{\ell\imath}\right) + \mathbf{p}^\imath \cdot \left(\mathbf{f}_k^\imath n_k^{\ell\imath} - \mathcal{F}^{\ell\imath} \cdot \mathbf{n}^{\ell\imath}\right)& \\ +\mathbf{p}^\ell \cdot \mathcal{D}\left(\mathbf{q}^\ell, \mathbf{q}^\imath\right) \cdot \mathbf{n}^{\ell\imath} + \mathbf{p}^\imath \cdot \mathcal{D}\left(\mathbf{q}^\imath, \mathbf{q}^\ell\right) \cdot \mathbf{n}^{\imath\ell} &= \left(F_k^\imath - F_k^\ell\right) n_k^{\ell\imath},\end{aligned} \tag{7}$$

$$\mathbf{p}^\ell \cdot \mathbf{S}\left(\mathbf{q}^\ell\right) = 0. \tag{8}$$

Since the source term satisfies $\mathbf{p} \cdot \mathbf{S} = 0$ the condition (8) holds trivially pointwise and therefore nothing special needs to be done in order to assure compatibility of the algebraic source terms. Concerning the construction of a thermodynamically compatible numerical flux we start with a simple *central discretization* of the numerical flux and of the non-conservative product, i.e. we choose

$$\tilde{\mathcal{F}}^{\ell\imath} \cdot \mathbf{n}^{\ell\imath} = \frac{1}{2}\left(\mathbf{f}_k^\ell + \mathbf{f}_k^\imath\right) n_k^{\ell\imath} \tag{9}$$

and

$$\mathcal{D}\left(\mathbf{q}^{\ell},\mathbf{q}^{r}\right)\cdot\mathbf{n}^{\ell r}=\frac{1}{2}\mathbf{B}_k(\bar{\mathbf{q}})n_k^{\ell r}\left(\mathbf{q}^{r}-\mathbf{q}^{\ell}\right),\quad \bar{\mathbf{q}}=\frac{1}{2}\left(\mathbf{q}^{r}+\mathbf{q}^{\ell}\right). \tag{10}$$

This discretization is in general *not yet compatible* with the extra conservation law (2) and therefore the numerical flux (9) needs to be subsequently *corrected* in order to achieve thermodynamic compatibility. Imposing condition (7) the resulting *thermodynamically compatible* **Abgrall flux** [1, 2] reads

$$\mathcal{F}^{\ell r}\cdot\mathbf{n}^{\ell r}=\tilde{\mathcal{F}}^{\ell r}\cdot\mathbf{n}^{\ell r}-\alpha^{\ell r}\left(\mathbf{p}^{r}-\mathbf{p}^{\ell}\right), \tag{11}$$

with the scalar correction factor

$$\alpha^{\ell r}=\frac{\left(F_k^{r}-F_k^{\ell}\right)n_k^{\ell r}+\left(\tilde{\mathcal{F}}^{\ell r}\cdot\mathbf{n}^{\ell r}\right)\cdot\left(\mathbf{p}^{r}-\mathbf{p}^{\ell}\right)-\left(\mathbf{p}^{r}\cdot\mathbf{f}_k^{r}-\mathbf{p}^{\ell}\cdot\mathbf{f}_k^{\ell}\right)n_k^{\ell r}}{\left(\mathbf{p}^{r}-\mathbf{p}^{\ell}\right)^2} -\frac{\left(\mathbf{p}^{r}+\mathbf{p}^{\ell}\right)\cdot\mathcal{D}\left(\mathbf{q}^{\ell},\mathbf{q}^{r}\right)\cdot\mathbf{n}^{\ell r}}{\left(\mathbf{p}^{r}-\mathbf{p}^{\ell}\right)^2}. \tag{12}$$

It is easy to check that the Abgrall flux (11) together with (9), (10) and (12) satisfies (7) *by construction*, since the correction factor $\alpha^{\ell r}$ is obtained by imposing condition (7) on the numerical flux (11).

Following [2, 3, 5, 6] the viscous part of the numerical flux reads

$$\mathcal{G}\left(\mathbf{q}^{\ell},\mathbf{q}^{r}\right)=\epsilon^{\ell r}\frac{\mathbf{q}^{r}-\mathbf{q}^{\ell}}{\delta^{\ell r}}=\epsilon^{\ell r}\frac{\Delta\mathbf{q}^{\ell r}}{\delta^{\ell r}},\quad \delta^{\ell r}=\left\|\mathbf{x}^{r}-\mathbf{x}^{\ell}\right\|. \tag{13}$$

For the MHD equations and the GPR model [4], the related entropy production term reads

$$\mathbf{P}\left(\mathbf{q}^{\ell},\mathbf{q}^{r}\right)=\left(0,\mathbf{0},\Pi^{\ell r},\mathbf{0},\mathbf{0}\right)^{T}, \tag{14}$$

where the only non-vanishing contribution is in the PDE for the entropy density and is given by

$$\Pi^{\ell r}=\frac{1}{2}\epsilon^{\ell r}\frac{\Delta\mathbf{q}^{\ell r}}{T^{\ell}}\,\partial^2_{\mathbf{qq}}\tilde{\mathcal{E}}^{\ell r}\frac{\Delta\mathbf{q}^{\ell r}}{\delta^{\ell r}}, \tag{15}$$

with the temperature $T^{\ell}=\frac{(\rho^{\ell})^{\gamma-1}}{(\gamma-1)c_v}e^{\frac{s^{\ell}}{c_v}}$ for the ideal gas equation of state. Instead, for the turbulent shallow water model [4, 5, 7] and the compressible two-fluid model [9–11] the corresponding entropy production terms are given in [4] and [11], respectively.

In all cases above, the Roe matrix of the Hessian of the energy potential, which appears in the production term (15), reads

$$\partial^2_{\mathbf{qq}} \tilde{\mathcal{E}}^{\ell r} = \int_0^1 \partial^2_{\mathbf{qq}} \mathcal{E}\left(\tilde{\boldsymbol{\psi}}(s)\right) ds =: \left(\partial^2_{\mathbf{pp}} \tilde{L}^{\ell r}\right)^{-1}. \tag{16}$$

Following [2, 5, 6] the former definition is based on the simple straight line segment path in $\mathbf{q}$ variables

$$\tilde{\boldsymbol{\psi}}(s) = \mathbf{q}^{\ell} + s\left(\mathbf{q}^{r} - \mathbf{q}^{\ell}\right), \qquad 0 \le s \le 1. \tag{17}$$

By construction, the Roe matrix $\partial^2_{\mathbf{qq}} \tilde{\mathcal{E}}^{\ell r}$ satisfies the Roe property

$$\partial^2_{\mathbf{qq}} \tilde{\mathcal{E}}^{\ell r} \cdot \left(\mathbf{q}^{r} - \mathbf{q}^{\ell}\right) = \left(\mathbf{p}^{r} - \mathbf{p}^{\ell}\right). \tag{18}$$

This allows to rewrite jumps in terms of the conservative variables $\mathbf{q}$ into jumps in terms of the main field variables (thermodynamic dual variables) $\mathbf{p}$. Throughout this paper the path integral in (16) is approximated numerically at the aid of a Gauss-Legendre quadrature formula using three quadrature points. For a detailed analysis on the influence of the quadrature error the reader is referred to [6]. Following [4] and [11] it is possible to prove a discrete cell-entropy inequality and nonlinear stability in the energy norm. It is possible to show via numerical experiments that with classical Runge-Kutta time discretization (method of lines) the proposed method is second order accurate in space and time on uniform Cartesian meshes. In the following we show some applications to the ideal MHD equations, see Fig. 1, to the GPR model of continuum mechanics, see Fig. 2 and to the compressible two-fluid model, see Fig. 3. We also show the results obtained for the non-conservative turbulent shallow water system [7] in the presence of shock waves, see Fig. 4. For more test cases and further details on the computational setup and for detailed discussion of the results and a comparison with available reference solutions, see [4] and [11], respectively.

**Fig. 1** HTC finite volume scheme applied to the MHD equations, see [4]. Left: Orszag-Tang vortex system. Right: MHD rotor problem

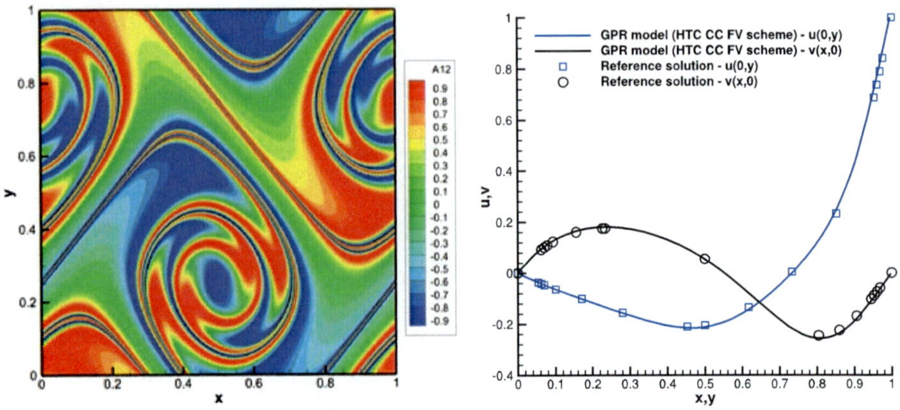


**Fig. 2** HTC finite volume scheme applied to the GPR model, see [4]. Left: double shear layer test. Right: lid-driven cavity

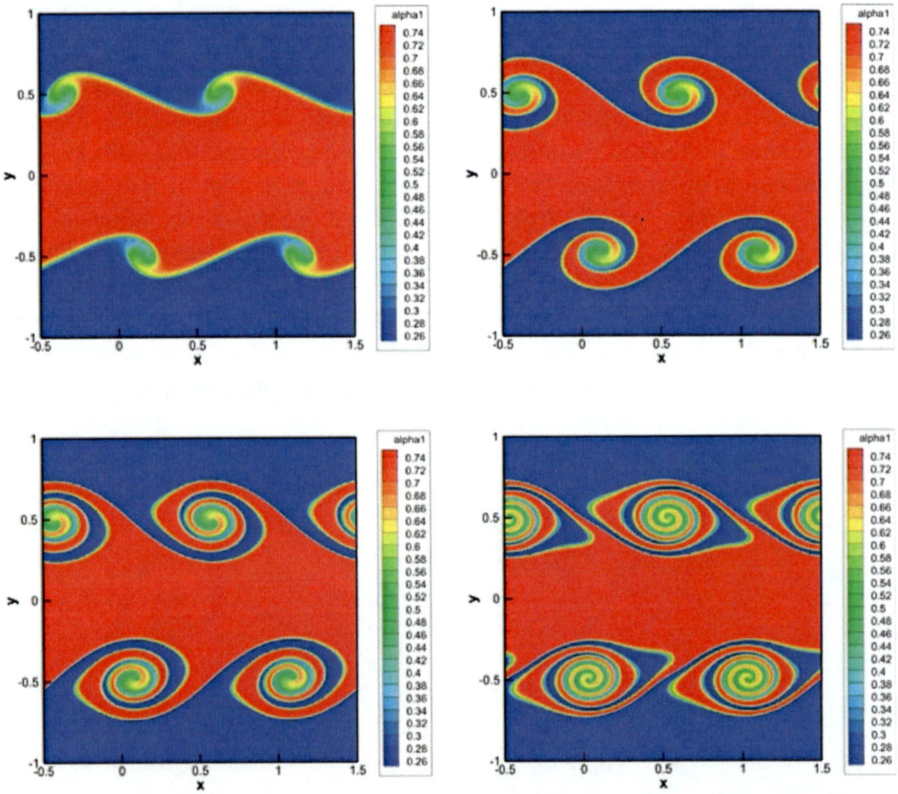


**Fig. 3** HTC finite volume scheme applied to the compressible two-fluid model, see [11]. Two-fluid Kelvin-Helmholtz instability at times $t \in \{3, 4, 5, 6\}$

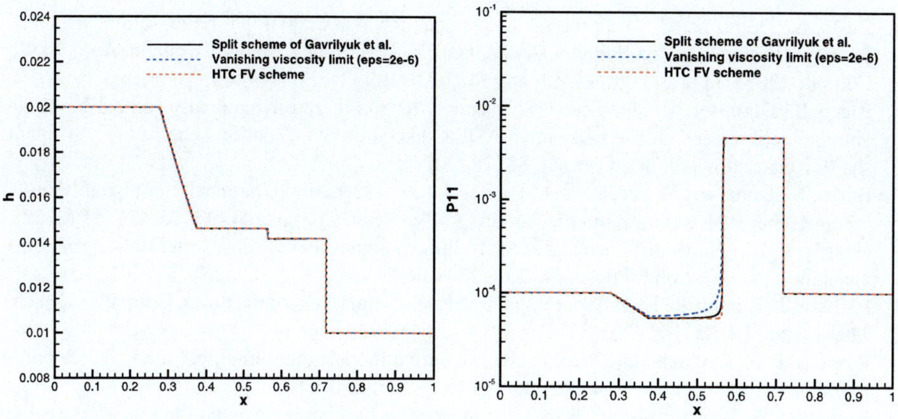


**Fig. 4** HTC finite volume scheme applied to a Riemann problem in the non-conservative turbulent shallow water model, see [7]. Left: water depth. Right: Reynolds stress tensor component $P_{11}$

## 3 Conclusion

In [4, 11] we have introduced a simple, efficient and very general family of thermodynamically compatible finite volume schemes based on the *Abgrall flux* recently presented in Abgrall et al. [2]. The thermodynamically compatible numerical flux (11) with (12) also directly takes into account the non-conservative products present in the governing PDE system. The new approach is totally universal and applies to general overdetermined first order hyperbolic system of balance laws of the very general form (1) that satisfies an extra conservation law of the type (2). Unlike in previous work, the numerical flux is not based on path integrals that require expensive and error prone numerical quadrature, but it instead relies on a very simple correction based on the jumps in the main field variables. A key feature of our scheme is that the total energy conservation is obtained as a consequence of the compatible discretization, while entropy-like variables are discretized directly. For more details, the interested reader is referred to [4, 11] and references therein.

## References

1. Abgrall, R.: A general framework to construct schemes satisfying additional conservation relations. Application to entropy conservative and entropy dissipative schemes. J. Comput. Phys. **372**, 640–666 (2018)
2. Abgrall, R., Busto, S., Dumbser, M.: A simple and general framework for the construction of thermodynamically compatible schemes for computational fluid and solid mechanics. Appl. Math. Comput. **440**, 127629 (2023)
3. Busto, S., Dumbser, M.: A new thermodynamically compatible finite volume scheme for magnetohydrodynamics. SIAM J. Numer. Anal. **61**, 343–364 (2023)

4. Busto, S., Dumbser, M.: A new class of general, efficient and simple finite volume schemes for overdetermined thermodynamically compatible hyperbolic systems. Commun. Appl. Math. Comput. (2023, in press). https://doi.org/10.1007/s42967-023-00307-4
5. Busto, S., Dumbser, M., Gavrilyuk, S., Ivanova, K.: On thermodynamically compatible finite volume methods and path-conservative ADER discontinuous Galerkin schemes for turbulent shallow water flows. J. Sci. Comput. **88**, 28 (2021)
6. Busto, S., Dumbser, M., Peshkov, I., Romenski, E.: On thermodynamically compatible finite volume schemes for continuum mechanics. SIAM J. Sci. Comput. **44**, A1723–A1751 (2022)
7. Gavrilyuk, S., Ivanova, K., Favrie, N.: Multi-dimensional shear shallow water flows: problems and solutions. J. Comput. Phys. **366**, 252–280 (2018)
8. Peshkov, I., Romenski, E.: A hyperbolic model for viscous Newtonian flows. Continuum Mech. Thermodyn. **28**, 85–104 (2016)
9. Romenski, E., Drikakis, D., Toro, E.: Conservative models and numerical methods for compressible two-phase flow. J. Sci. Comput. **42**, 68–95 (2010)
10. Romenski, E., Resnyansky, A., Toro, E.: Conservative hyperbolic formulation for compressible two-phase flow with different phase pressures and temperatures. Q. Appl. Math. **65**, 259–279 (2007)
11. Thomann, A., Dumbser, M.: Thermodynamically compatible discretization of a compressible two-fluid model with two entropy inequalities. J. Sci. Comput. **97**, 9 (2023)

# A Scheme Using the Wave Structure of Second-Moment Turbulent Models for Incompressible Flows

Martin Ferrand, Jean-Marc Hérard, Thomas Norddine, and Simon Ruget

**Abstract** We focus herein on the analysis of the one-dimensional Riemann problem arising from the convective subset of a second-moment turbulent non-conservative model for incompressible flows. The sketch of proof of existence and uniqueness of the solution is given, assuming a set of approximate jump conditions. Some first numerical simulations applying for the Finite Volume method are given and compared with another scheme classically used in CFD codes. This suggests to implement standard projection schemes to cope with the complete model.

**Keywords** Second-moment turbulent closure · Incompressible turbulent flows · Hyperbolic systems · Riemann problem

## 1 Introduction

Unsteady Reynolds-averaged Navier–Stokes models are widely used in the industrial framework in order to predict incompressible or compressible flows (see e.g. [8]). Second-moment tensors also arise when modelling shallow-water models (see [4, 5]). All these models contain first-order non-conservative terms which require special attention (see e.g. [2]). This work is highly motivated by the occurrence of non-physical oscillations in many practical situations (see e.g. [9]).

M. Ferrand (✉) · J.-M. Hérard · T. Norddine · S. Ruget
CEREA, École des Ponts, EDF R & D, Chatou, France
e-mail: martin.ferrand@edf.fr

EDF R & D, 6, quai Watier, 78400 Chatou, France



E. Franck et al. (eds.), *Finite Volumes for Complex Applications X—Volume 2, Hyperbolic and Related Problems*, Springer Proceedings in Mathematics & Statistics 433,
https://doi.org/10.1007/978-3-031-40860-1_12

## 2 Second-Moment Turbulence Models

The system of equations is composed of mass conservation (1a), momentum conservation (1b) and second-moment transport equation (1c):

$$\operatorname{div}\left(\overline{\boldsymbol{u}}\right) = 0, \tag{1a}$$

$$\partial_t \overline{\boldsymbol{u}} + \left(\overline{\boldsymbol{u}} \cdot \nabla\right) \overline{\boldsymbol{u}} + \mathbf{div}\left(\mathbf{R}\right) - \mathbf{div}\left(\overline{\boldsymbol{\Sigma}}\right) = \mathbf{0}, \tag{1b}$$

$$\partial_t \mathbf{R} + \left(\overline{\boldsymbol{u}} \cdot \nabla\right) \mathbf{R} + \mathbf{R} \cdot \nabla \overline{\boldsymbol{u}} + \nabla \overline{\boldsymbol{u}}^T \cdot \mathbf{R} + \mathbf{div}\left(\mathbf{G_R}\right) - \boldsymbol{\Phi} = \mathbf{0}, \tag{1c}$$

with $\overline{\boldsymbol{\Sigma}} := -\frac{\overline{p}}{\rho_0}\mathbf{1} + \nu_0\left(\nabla\overline{\mathbf{u}} + \nabla\overline{\mathbf{u}}^T\right)$ the mean stress tensor per mass unit, $\overline{p}$ the mechanical pressure. The Reynolds stress tensor $\mathbf{R} := \overline{\mathbf{u}' \otimes \mathbf{u}'}$ is symmetric and it is also expected to be positive and half-definite (called realizability constraint), owing to its intrinsic form (see [8]). Models for $\mathbf{G_R} := \overline{\mathbf{u}' \otimes \mathbf{u}' \otimes \mathbf{u}'}$ usually rely on a classical first gradient assumption $\mathbf{G_R} = -C_R \nu_t \left(\mathbf{R},\ T\right) \nabla \mathbf{R}$, where $C_R$ is a positive constant, $\nu_t$ is the turbulent viscosity, and $T$ is the time scale related to turbulent dissipation. Moreover $\boldsymbol{\Phi} := \overline{\mathbf{u}' \otimes \mathbf{div}\left(\boldsymbol{\Sigma}'\right) + \mathbf{div}\left(\boldsymbol{\Sigma}'\right) \otimes \mathbf{u}'}$ is classically split into two parts:

$$\boldsymbol{\Phi} = \boldsymbol{\Phi}^r\left(\nabla\overline{\boldsymbol{u}},\ \mathbf{R}\right) + \boldsymbol{\Phi}^s\left(\mathbf{R},\ T\right), \tag{2}$$

where the so-called rapid contribution verifies: $\boldsymbol{\Phi}^r(\mathbf{0}, \mathbf{R}) = \mathbf{0}$ and $\boldsymbol{\Phi}^r(\nabla\overline{\boldsymbol{u}},\ \mathbf{0}) = 0$. Moreover $\boldsymbol{\Phi}^r$ is linear with respect to $\nabla\overline{\boldsymbol{u}}$ (see e.g. [8] for a review of the turbulence models). The following result is useful to examine the realizability of the Reynolds stress tensor.

**Lemma 1** *Assume that the governing equation of a symmetric second-rank tensor* $\mathbf{R} \in \mathbb{R}^3 \times \mathbb{R}^3$ *is:*

$$\partial_t \mathbf{R} + \left(\overline{\boldsymbol{u}} \cdot \nabla\right) \mathbf{R} + \mathbf{R} \cdot \mathbf{H} + \mathbf{H}^T \cdot \mathbf{R} = \mathbf{0}.$$

*Let* $\delta_R$ *be the determinant of tensor* $\mathbf{R}$. *Then the governing equation of* $\delta_R$ *is:*

$$\partial_t \delta_R + \left(\overline{\boldsymbol{u}} \cdot \nabla\right) \delta_R + 2tr(\mathbf{H})\delta_R = 0.$$

A proof can be found in appendix 1 of [7]. $\mathbf{H}$ is deduced from the terms $\mathbf{R} \cdot \nabla\overline{\boldsymbol{u}} + \nabla\overline{\boldsymbol{u}}^T \cdot \mathbf{R}$, $\boldsymbol{\Phi}^r$ and $\boldsymbol{\Phi}^s$, and the boundedness of the trace of $\mathbf{H}$ must be examined to ensure positive values of $\delta_R$.

## 3 A First-Order Time Scheme

Starting with initial conditions $(\overline{\boldsymbol{u}}^n,\ \mathbf{R}^n)$ and using suitable boundary conditions, $(\overline{\boldsymbol{u}}^{n+1},\ \mathbf{R}^{n+1})$ arises from the following first-order time scheme:

$$\operatorname{div}\left(\overline{\boldsymbol{u}}^{n+1}\right) = 0, \tag{3a}$$

$$\frac{\overline{\boldsymbol{u}}^{n+1} - \overline{\boldsymbol{u}}^{n}}{\Delta t} + \nabla \frac{\overline{p}^{n+1}}{\rho_0} - \mathbf{div}\left(\nu_0 \nabla \overline{\mathbf{u}}^{n+1}\right) = -\left(\left(\overline{\boldsymbol{u}} \cdot \nabla\right) \overline{\boldsymbol{u}} + \mathbf{div}\,(\mathbf{R})\right)^n, \tag{3b}$$

$$\frac{\mathbf{R}^{n+1} - \mathbf{R}^{n}}{\Delta t} - \mathbf{div}\left(C_R \nu_t \nabla \mathbf{R}^{n+1}\right) - \boldsymbol{\Phi}^s\left(\mathbf{R}^{n+1}, T\right) = -\left(\left(\overline{\boldsymbol{u}} \cdot \nabla\right) \mathbf{R}\right)^n - \left(\mathbf{R} \cdot \nabla \overline{\boldsymbol{u}} + \nabla \overline{\boldsymbol{u}}^T \cdot \mathbf{R}\right)^n + \boldsymbol{\Phi}^r\left(\nabla \overline{\boldsymbol{u}}, \mathbf{R}\right)^n. \tag{3c}$$

A Stokes-like step has been applied in the scheme (3a) and (3b) (see, among others, [1, 12]). The space discretization associated with the right-hand-side of (3b), (3c) is derived from the analysis of the following evolution system (4):

$$\partial_t \overline{\boldsymbol{u}} + \left(\overline{\boldsymbol{u}} \cdot \nabla\right) \overline{\boldsymbol{u}} + \mathbf{div}\,(\mathbf{R}) = \mathbf{0}, \tag{4a}$$

$$\partial_t \mathbf{R} + \left(\overline{\boldsymbol{u}} \cdot \nabla\right) \mathbf{R} + \mathbf{R} \cdot \nabla \overline{\boldsymbol{u}} + \nabla \overline{\boldsymbol{u}}^T \cdot \mathbf{R} - \boldsymbol{\Phi}^r\left(\nabla \overline{\boldsymbol{u}}, \mathbf{R}\right) = \mathbf{0}. \tag{4b}$$

## 4 Analysis of the Evolution Step

In the following, we restrict to the simple Lumley's model which does not contain so-called rapid terms: $\Phi^{\mathbf{r}} = \mathbf{0}$. We also restrict from now on to the two-dimensional framework to ease the presentation. We first recall that the latter system (4) is hyperbolic if the Reynolds stress tensor is realizable (which is guaranteed, owing to Lemma 1, since in that case we have: $tr\mathbf{H} = \operatorname{div}\overline{\mathbf{u}} = 0$). Equation (4) is invariant under frame rotation, so we can rewrite it in the $(\mathbf{n}, \boldsymbol{\tau})$ reference frame (where $\mathbf{n}, \boldsymbol{\tau}$ are some unit orthogonal vectors), defining $x_n := \mathbf{x} \cdot \mathbf{n}$, $u_n := \overline{\boldsymbol{u}} \cdot \mathbf{n}$, $u_\tau := \overline{\boldsymbol{u}} \cdot \boldsymbol{\tau}$, $R_{nn} := \mathbf{n}^T \mathbf{R} \mathbf{n}$, $R_{n\tau} := \mathbf{n}^T \mathbf{R} \boldsymbol{\tau}$ and $R_{\tau\tau} := \boldsymbol{\tau}^T \mathbf{R} \boldsymbol{\tau}$. We now define the evolution step in the $\mathbf{n}-$direction, by neglecting the transverse derivatives, which yields:

$$\partial_t u_n + u_n \partial_n u_n + \partial_n R_{nn} = 0, \tag{5a}$$

$$\partial_t u_\tau + u_n \partial_n u_\tau + \partial_n R_{n\tau} = 0, \tag{5b}$$

$$\partial_t R_{nn} + u_n \partial_n R_{nn} + 2 R_{nn} \partial_n u_n = 0, \tag{5c}$$

$$\partial_t R_{n\tau} + u_n \partial_n R_{n\tau} + R_{nn} \partial_n u_\tau + R_{n\tau} \partial_n u_n = 0, \tag{5d}$$

$$\partial_t R_{\tau\tau} + u_n \partial_n R_{\tau\tau} + 2 R_{n\tau} \partial_n u_\tau = 0. \tag{5e}$$

We set $\mathbf{W} := (u_n,\ u_\tau,\ R_{nn},\ R_{\tau\tau},\ R_{n\tau})$. We assume from now on that initial and boundary conditions comply with the strictly realizability constraint, which will imply that $\mathbf{R}$ remains strictly realizable. Once more, system (5) preserves the realizability, which implies that $R_{nn} > 0$ whatever the unit vector $\mathbf{n}$ is.

**Proposition 1** (Hyperbolic evolution step) *System (5) is hyperbolic. It admits five real eigenvalues which are:*

$$\begin{aligned} \lambda_1 &= u_n - \sqrt{2R_{nn}}, & \lambda_2 &= u_n - \sqrt{R_{nn}}, & \lambda_3 &= u_n, \\ \lambda_4 &= u_n + \sqrt{R_{nn}}, & \lambda_5 &= u_n + \sqrt{2R_{nn}}. \end{aligned} \tag{6}$$

*and the associated right eigenvectors span* $\mathbb{R}^5$. *Fields associated with eigenvalues* $\lambda_{2,3,4}$ *are linearly degenerate (LD). The 1-field and 5-field are genuinely non linear (GNL).*

We refer to [6] for a proof. Note that this can be extended to the three-dimensional framework. Right eigenvectors are:

$$\mathbf{r}_1 = \left(1,\ \frac{R_{n\tau}}{R_{nn}},\ -\sqrt{2R_{nn}},\ -\frac{\sqrt{2}R_{n\tau}^2}{(R_{nn})^{\frac{3}{2}}},\ -\frac{\sqrt{2}R_{n\tau}}{\sqrt{R_{nn}}}\right)^T, \tag{7a}$$

$$\mathbf{r}_2 = \left(0,\ 1,\ 0,\ -\frac{2R_{n\tau}}{\sqrt{R_{nn}}},\ -\sqrt{R_{nn}}\right)^T, \tag{7b}$$

$$\mathbf{r}_3 = (0,\ 0,\ 0,\ 1,\ 0)^T, \tag{7c}$$

$$\mathbf{r}_4 = \left(0,\ 1,\ 0,\ \frac{2R_{n\tau}}{\sqrt{R_{nn}}},\ \sqrt{R_{nn}}\right)^T, \tag{7d}$$

$$\mathbf{r}_5 = \left(1,\ \frac{R_{n\tau}}{R_{nn}},\ \sqrt{2R_{nn}},\ \frac{\sqrt{2}R_{n\tau}^2}{(R_{nn})^{\frac{3}{2}}},\ \frac{\sqrt{2}R_{n\tau}}{\sqrt{R_{nn}}}\right)^T. \tag{7e}$$

Riemann invariants $I_R^i$ of the $i$-th field ($f \in I_R^i$, $\nabla_{\mathbf{W}} f \cdot \mathbf{r}_i(\mathbf{W}) = 0$) are:

$$I_R^1 = \left\{u_n + \sqrt{2R_{nn}},\ \frac{R_{n\tau}}{R_{nn}},\ R_{\tau\tau} - \frac{R_{n\tau}^2}{R_{nn}},\ u_\tau + R_{n\tau}\sqrt{\frac{2}{R_{nn}}}\right\}, \tag{8a}$$

$$I_R^2 = \left\{u_n,\ R_{nn},\ u_\tau + \frac{R_{n\tau}}{\sqrt{R_{nn}}},\ R_{\tau\tau}R_{nn} - R_{n\tau}^2\right\}, \tag{8b}$$

$$I_R^3 = \left\{u_n,\ u_\tau,\ R_{nn},\ R_{n\tau}\right\}, \tag{8c}$$

$$I_R^4 = \left\{u_n,\ R_{nn},\ u_\tau - \frac{R_{n\tau}}{\sqrt{R_{nn}}},\ R_{\tau\tau}R_{nn} - R_{n\tau}^2\right\}, \tag{8d}$$

$$I_R^5 = \left\{u_n - \sqrt{2R_{nn}},\ \frac{R_{n\tau}}{R_{nn}},\ R_{\tau\tau} - \frac{R_{n\tau}^2}{R_{nn}},\ u_\tau - R_{n\tau}\sqrt{\frac{2}{R_{nn}}}\right\}. \tag{8e}$$

Due to non-conservative terms in (5), shock relations cannot be derived in a classical way. Following [2] **approximate** shock relations can nevertheless be proposed.

These relations are valid in the limit of weak shocks. Assuming a linear path from left to right with respect to the variable $\mathbf{W}$, these relations write:

$$\sigma[u_n] = \widehat{u_n}[u_n] + [R_{nn}], \tag{9a}$$

$$\sigma[u_\tau] = \widehat{u_n}[u_\tau] + [R_{n\tau}], \tag{9b}$$

$$\sigma[R_{nn}] = 2\widehat{R_{nn}}[u_n] + \widehat{u_n}[R_{nn}], \tag{9c}$$

$$\sigma[R_{n\tau}] = \widehat{R_{n\tau}}[u_n] + \widehat{R_{nn}}[u_\tau] + \widehat{u_n}[R_{n\tau}], \tag{9d}$$

$$\sigma[R_{\tau\tau}] = 2\widehat{R_{n\tau}}[u_\tau] + \widehat{u_n}[R_{\tau\tau}], \tag{9e}$$

where $[x] := x_r - x_l$ denotes jump and $\widehat{x} := \frac{x_r + x_l}{2}$ is the arithmetic mean between left and right states of the discontinuity travelling at speed $\sigma$.

**Theorem 1** (Existence and uniqueness of the solution) *Considering approximate shock relations (9), there exists a unique self-similar realizable solution ($\mathbf{W}(x_n, t) = \mathbf{W}(\frac{x_n}{t})$, with $\mathbf{W} = (u_n, u_\tau, R_{nn}, R_{\tau\tau}, R_{n\tau})$) to the Riemann problem associated with system (5) supplemented with strictly realizable initial conditions $\mathbf{W}(x_n, 0) = \mathbf{W}^L$ for $x_n < 0$ and $\mathbf{W}(x_n, 0) = \mathbf{W}^R$ for $x_n > 0$, if and only if the following condition holds:*

$$u_n^R - u_n^L < \left(\sqrt{2R_{nn}^L} + \sqrt{2R_{nn}^R}\right). \tag{10}$$

As in the compressible framework, the Riemann problem structure of system (4) allows to specify wall, symmetry and inlet/outlet boundary conditions.

***Proof*** The proof follows the main guidelines recalled in [11] and detailed in [7]. We consider two initial realizable states $\mathbf{W}^R$, $\mathbf{W}^L$. We look for the four intermediate states labelled $I, II, III, IV$ from left to right separating the 5 waves.

**Step 0: Preliminary calculation.** To connect left and right states $\mathbf{W}^{l,r}$ across the 1st and 5th waves we set $z := \frac{R_{nn}^r}{R_{nn}^l} > 0$. To select the physical solution for shocks (9), we use $[u_n] = u_n^r - u_n^l \leq 0$ (see [7]).

For the 1-shock connection ($z > 1$):

$$\begin{aligned} &R_{nn}^r = zR_{nn}^l; \quad u_n^r = u_n^l + \tfrac{1-z}{\sqrt{z+1}}\sqrt{R_{nn}^l}; \quad u_\tau^r = u_\tau^l + \tfrac{1-z}{\sqrt{z+1}}\tfrac{R_{n\tau}^l}{\sqrt{R_{nn}^l}}; \\ &R_{n\tau}^r = zR_{n\tau}^l; \quad R_{\tau\tau}^r = R_{\tau\tau}^l + (z-1)\tfrac{(R_{n\tau}^l)^2}{R_{nn}^l}. \end{aligned} \tag{11}$$

For the 5-shock connection ($0 < z < 1$):

$$\begin{aligned} &R_{nn}^r = zR_{nn}^l; \quad u_n^r = u_n^l + \tfrac{z-1}{\sqrt{z+1}}\sqrt{R_{nn}^l}; \quad u_\tau^r = u_\tau^l + \tfrac{z-1}{\sqrt{z+1}}\tfrac{R_{n\tau}^l}{\sqrt{R_{nn}^l}}; \\ &R_{n\tau}^r = zR_{n\tau}^l; \quad R_{\tau\tau}^r = R_{\tau\tau}^l + (z-1)\tfrac{(R_{n\tau}^l)^2}{R_{nn}^l}. \end{aligned} \tag{12}$$

To connect the left and right states through the 1-rarefaction wave (respectively 5-rarefaction wave) we use the Riemann invariants (8a) (respectively (8e)).

**Step 1: Solution in terms of $u_n$ and $R_{nn}$ variables.** A glance at (8b)–(8d) shows that $u_n$ and $R_{nn}$ are constant across the three LD waves. First we can focus only on $u_n$ and $R_{nn}$ variables. We denote their intermediate values by $u_n^\sharp = u_n^I = u_n^{II} = u_n^{III} = u_n^{IV}$ and $R_{nn}^\sharp = R_{nn}^I = R_{nn}^{II} = R_{nn}^{III} = R_{nn}^{IV}$, and link these to initial states $\mathbf{W}^{R,L}$:

$$\begin{cases} u_n^L = u_n^\sharp - h_1(z_1)\sqrt{2R_{nn}^L}, \text{ with } z_1 := \dfrac{R_{nn}^\sharp}{R_{nn}^L} > 0, \\ u_n^R = u_n^\sharp + h_5(z_5)\sqrt{2R_{nn}^R}, \text{ with } z_5 := \dfrac{R_{nn}^R}{R_{nn}^\sharp} > 0, \end{cases} \tag{13}$$

$h_{1,5}(z)$ are defined using (8a), (8e), (11), (12), following [11]. Note that $z_1 z_5 = \frac{R_{nn}^R}{R_{nn}^L}$. We set:

$$\Psi(z_5) := u_n^R - u_n^L - \sqrt{2R_{nn}^R}h_5(z_5) - \sqrt{2R_{nn}^L}h_1(z_1(z_5)) = 0. \tag{14}$$

Solving (14) gives $z_5$, hence $R_{nn}^\sharp$ and $u_n^\sharp$. The solution is unique since $\Psi$ is a strictly monotonic function. Its lower and upper bounds enable to conclude that $u_n^\sharp$ and $R_{nn}^\sharp$ exist if (10) holds. $z_1$ and $z_5$ enable to compute the remaining components $(u_\tau, R_{\tau\tau}, R_{n\tau})$ of intermediate states $I$ and $IV$.

**Step** 2: **Solution in terms of $u_\tau$, $R_{n\tau}$, and $R_{\tau\tau}$ variables.** Let $u_\tau^\star = u_\tau^{II} = u_\tau^{III}$ and $R_{n\tau}^\star = R_{n\tau}^{II} = R_{n\tau}^{III}$, given by:

$$\begin{cases} u_\tau^\star + \dfrac{R_{n\tau}^\star}{\sqrt{R_{nn}^\sharp}} = u_\tau^I + \dfrac{R_{n\tau}^I}{\sqrt{R_{nn}^\sharp}}, \\ u_\tau^\star - \dfrac{R_{n\tau}^\star}{\sqrt{R_{nn}^\sharp}} = u_\tau^{IV} - \dfrac{R_{n\tau}^{IV}}{\sqrt{R_{nn}^\sharp}}. \end{cases} \tag{15}$$

Then, deduce $R_{\tau\tau}^{II}$ and $R_{\tau\tau}^{III}$ thanks to $(R_{nn}^\sharp R_{\tau\tau}^{II} - R_{n\tau}^{\star,2}) = (R_{nn}R_{\tau\tau} - R_{n\tau}^2)^I$ and $(R_{nn}^\sharp R_{\tau\tau}^{III} - R_{n\tau}^{\star,2}) = (R_{nn}R_{\tau\tau} - R_{n\tau}^2)^{IV}$.

## 5 Numerical Scheme for the Evolution System (4)

The objective of the present work is to obtain a stable numerical scheme (unlike those usually used, which may use partial upwinding which respect to the material mean velocity). Therefore we restrict the presentation to low order schemes such as Rusanov scheme. Other approximate Godunov schemes and MUSL techniques might also be considered.

For the sake of brevity the scheme is presented in a one dimensional framework. The evolution step (4) can be written (using the incompressibility constraint):

$$\frac{\partial \mathbf{W}}{\partial t} + \frac{\partial (u_n \mathbf{W})}{\partial x_n} + \mathbf{C}(\mathbf{W}) \frac{\partial \mathbf{W}}{\partial x_n} = \mathbf{0}, \tag{16}$$

$\mathbf{W}_i^{n+1}$ will be computed using the following Rusanov-like [10] scheme:

$$\Delta x_i (\mathbf{W}_i^{n+1} - \mathbf{W}_i^n) + \Delta t^n \left( \mathcal{F}^n_{i+\frac{1}{2}} - \mathcal{F}^n_{i-\frac{1}{2}} \right) + \Delta t^n \mathcal{N}_i^n = \mathbf{0}, \tag{17}$$

where the time step $\Delta t^n$ satisfies the CFL condition $\max_i \left( |\lambda^n_{i+\frac{1}{2}}| \right) \cdot \frac{\Delta t^n}{\Delta x_i} \leq \frac{1}{2}$, $\Delta x_i$ denotes the size of cell $i$. The numerical flux $\mathcal{F}^n_{i+\frac{1}{2}}$ and the non-conservative contribution $\mathcal{N}_i^n$ are:

$$\mathcal{F}^n_{i+\frac{1}{2}} := (u_n)^n_{i+\frac{1}{2}} \frac{\mathbf{W}_i^n + \mathbf{W}_{i+1}^n}{2} - \frac{\lambda^n_{i+\frac{1}{2}}}{2} (\mathbf{W}_{i+1}^n - \mathbf{W}_i^n), \tag{18a}$$

$$\mathcal{N}_i^n := \mathbf{C}(\mathbf{W}_i^n) \left( \frac{\mathbf{W}_{i+1}^n - \mathbf{W}_{i-1}^n}{2} \right), \tag{18b}$$

with $\lambda^n_{i+\frac{1}{2}} = \max_k \max_{l=i,i+1} \left( |\lambda_k (u_n \mathbf{I} + \mathbf{C}) (\mathbf{W}_l^n)| \right)$, and where the discrete divergence free condition on $(u_n)^n_{i+\frac{1}{2}}$ holds.

Unlike scheme (17) and (18), standard schemes only consider material convection upwinding for stabilization: $\lambda^n_{i+\frac{1}{2}} = |(u_n)^n_{i+\frac{1}{2}}|$.

The following test case describes a flow in the vicinity of a wall, while prescribing shear stress. The initial states $\mathbf{W}^{L,R}$ and the intermediate states $(I, II, III, IV)$ arising in the 1-dimensional Riemann problem associated with (5) are given in Table 1. Computations have been performed using code_saturne finite volume platform. The CFL parameter is set to 0.5 and the meshes contain from 250 to 32000 cells. The

**Table 1** Analytical solution for the tangential flow in the vicinity of a wall

| Zone | $L$ | $I$ | $II$ | $III$ | $IV$ | $R$ |
|---|---|---|---|---|---|---|
| $u_n$ | $u_n(=0)$ | $u_n$ | $u_n$ | $u_n$ | $u_n$ | $u_n(=0)$ |
| $u_\tau$ | $u_\tau^L(=1)$ | $u_\tau^L$ | $\frac{u_\tau^L+u_\tau^R}{2} + \frac{R_{n\tau}^L - R_{n\tau}^R}{2\sqrt{R_{nn}}}$ | $\frac{u_\tau^L+u_\tau^R}{2} + \frac{R_{n\tau}^L - R_{n\tau}^R}{2\sqrt{R_{nn}}}$ | $u_\tau^R$ | $u_\tau^R(=1)$ |
| $R_{nn}$ | $R_{nn}(=0.5)$ | $R_{nn}$ | $R_{nn}$ | $R_{nn}$ | $R_{nn}$ | $R_{nn}(=0.5)$ |
| $R_{n\tau}$ | $R_{n\tau}^L(=0.4)$ | $R_{n\tau}^L$ | $\frac{\left(u_\tau^L - u_\tau^R\right)\sqrt{R_{nn}}}{2} + \frac{R_{n\tau}^L + R_{n\tau}^R}{2}$ | $\frac{\left(u_\tau^L - u_\tau^R\right)\sqrt{R_{nn}}}{2} + \frac{R_{n\tau}^L + R_{n\tau}^R}{2}$ | $R_{n\tau}^R$ | $R_{n\tau}^R(=0.5)$ |
| $R_{\tau\tau}$ | $R_{\tau\tau}^L(=0.3)$ | $R_{\tau\tau}^L$ | $R_{\tau\tau}^L + \frac{(R_{n\tau}^{II})^2 - (R_{n\tau}^L)^2}{R_{nn}}$ | $R_{\tau\tau}^R + \frac{(R_{n\tau}^{III})^2 - (R_{n\tau}^R)^2}{R_{nn}}$ | $R_{\tau\tau}^R$ | $R_{\tau\tau}^R(=0.4)$ |

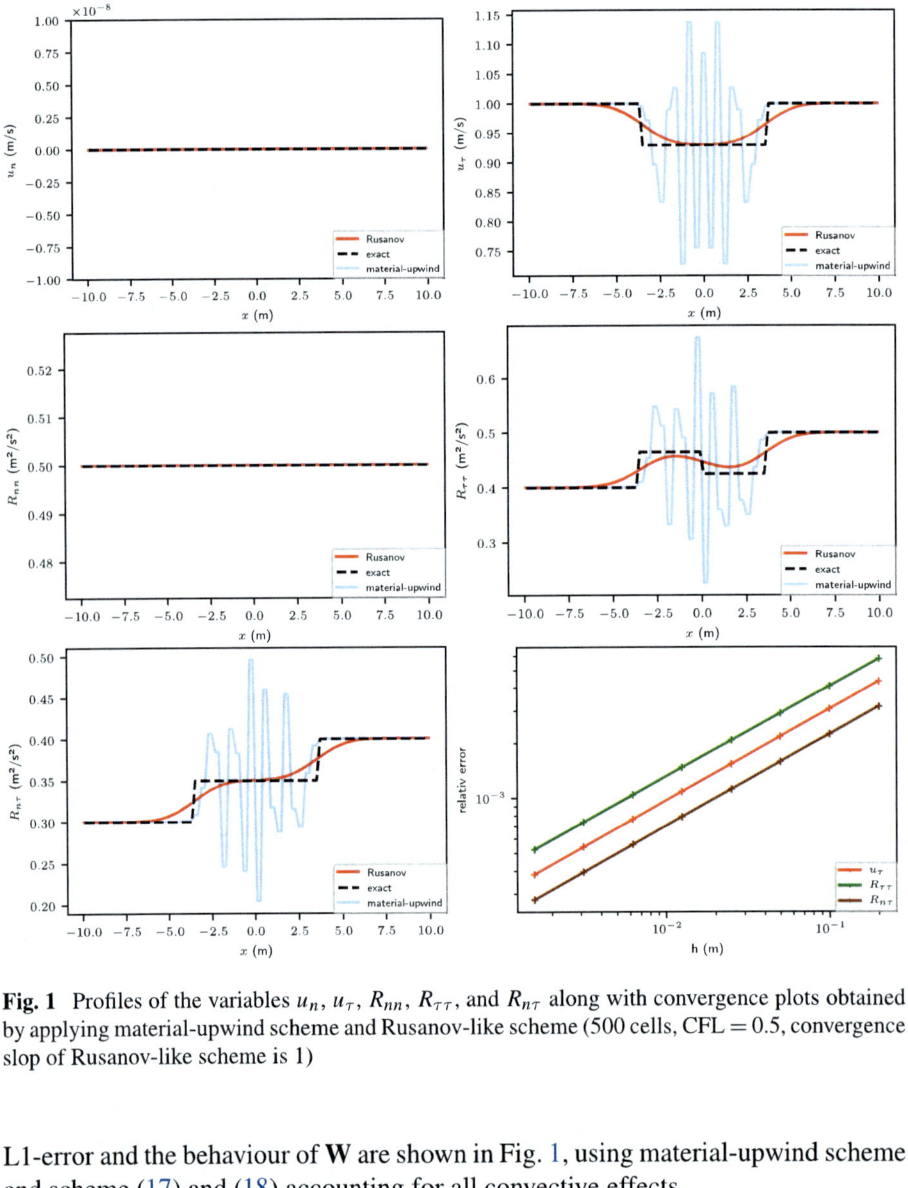


**Fig. 1** Profiles of the variables $u_n$, $u_\tau$, $R_{nn}$, $R_{\tau\tau}$, and $R_{n\tau}$ along with convergence plots obtained by applying material-upwind scheme and Rusanov-like scheme (500 cells, CFL = 0.5, convergence slop of Rusanov-like scheme is 1)

L1-error and the behaviour of $\mathbf{W}$ are shown in Fig. 1, using material-upwind scheme and scheme (17) and (18) accounting for all convective effects.

Of course, more accurate Riemann solvers can be investigated. This methodology can be applied to projection step methods or Uzawa algorithms. This strategy is also suitable for anisothermal flows or mean species transport equations with second-moment turbulence closures (see [3]). Eventually, more complex models involving non-zero contribution $\boldsymbol{\Phi}^r\ (\nabla \overline{\boldsymbol{u}},\ \mathbf{R})$ may be considered, using the same approach (see for instance appendix 2 in [7]).

## References


1. Chorin, A.J.: Numerical solution of the Navier-Stokes equations. Math. Comput. **22**(104), 745–762 (1968)
2. Dal Maso, G., Lefloch, P., Murat, F.: Definition and weak stability of nonconservative products. Journal de mathématiques pures et appliquées **74**(6), 483–548 (1995)
3. Ferrand, M., Hérard, J.-M., Norddine, T., Ruget, S.: Stable schemes for second-moment turbulent models for incompressible flows. In preparation
4. Gavrilyuk, S., Gouin, H.: Geometric evolution of the Reynolds stress tensor. Int. J. Eng. Sci. **59**, 65–73 (2012)
5. Gavrilyuk, S., Gouin, H.: Variational formulation for models of shear shallow water flows and ideal turbulence. Int. J. Non-Linear Mech. **119**, 103312 (2020)
6. Hérard, J.-M.: Basic analysis of some second moment closures part I: incompressible isothermal turbulent flows. Theor. Comput. Fluid Dyn. **6**(4), 213–233 (1994)
7. Hérard, J.-M.: Basic analysis of some second moment closures part II : incompressible turbulent flows including buoyant effects (1994). https://hal.archives-ouvertes.fr/hal-02007060
8. Pope, S. B.: Turbulent Flows. Cambridge university press (2000)
9. Richards, P.J., Norris, S.E.: Appropriate boundary conditions for computational wind engineering models revisited. J. Wind Eng. Ind. Aerodyn. **99**(4), 257–266 (2011)
10. Rusanov, V.V.: The calculation of the interaction of non-stationary shock waves with barriers. Zhurnal Vychislitel'noi Matematiki i Matematicheskoi Fiziki, Russ. Acad. Sci., Branch Math. Sci. **1**(2), 267–279 (1961)
11. Smoller, J.: Shock Waves and Reaction-diffusion Equations, vol. 258. Springer Science & Business Media (2012)
12. Temam, R.: Une méthode d'approximation de la solution des équations de Navier-Stokes. Bulletin de la Société Mathématique de France **96**, 115–152 (1968)

# Study of a Numerical Scheme with Transport-Acoustic Operator Splitting on a Staggered Mesh

**Ahmad El Halabi, Pierre Fernier, Thomas Galie, Samuel Kokh, and Khaled Saleh**

**Abstract** We present an approximation of the Euler equations by the means of an explicit finite volume method on a staggered grid scheme. The discretization strategy relies on a two-step algorithm that accounts separately for the acoustic and material transport effects in a similar way to the Lagrange-Projection methods.

**Keywords** Euler equations · Staggered mesh · Finite volume methods · Operator splitting

## 1 Introduction

Staggered grid discretizations have been originally designed for the approximation of incompressible flows. Since the seminal work [6], this approach has been extended to compressible models, the main objective being to simulate flows which may be both in highly compressible (large Mach number) and nearly incompressible (low Mach

A. E. Halabi · T. Galie (✉)
Service de Thermo-hydraulique et de Mécanique des Fluides, CEA, Université Paris-Saclay, 91191 Gif-sur-Yvette, France
e-mail: thomas.galie@cea.fr

A. E. Halabi
e-mail: ahmad.elhalabi@cea.fr

P. Fernier
Sorbonne Université, Paris, France

S. Kokh
Service de Génie Logiciel pour la Simulation, CEA, Université Paris-Saclay, 91191 Gif-sur-Yvette, France
e-mail: samuel.kokh@cea.fr

K. Saleh
CNRS UMR 5208, Institut Camille Jordan, Université Lyon 1, Université de Lyon, 43 bd 11 novembre 1918, 69622 Villeurbanne cedex, France
e-mail: saleh@math.univ-lyon1.fr

 
E. Franck et al. (eds.), *Finite Volumes for Complex Applications X—Volume 2, Hyperbolic and Related Problems*, Springer Proceedings in Mathematics & Statistics 433,
https://doi.org/10.1007/978-3-031-40860-1_13

number) regimes. The staggered location of the unknown over the mesh provides for Cartesian meshes a natural mean of evaluation of fluxes and gradients. However, as the original method is not conservative, the computation of shocks can be difficult.

Over the past years many contributions have suggested means to restore conservativity, enhance the stability of the method as well as extensions to unstructured meshes (see [3–5, 7, 8] and the references therein).

In this study, we present a fully conservative explicit staggered discretization for approximating one-dimensional compressible flows. The method relies on a separate treatment of acoustic and transport phenomena by means of an operator splitting that shares similarities with the Lagrange-Projection method as in [4], for instance. However one of the main difference of our scheme compared to the latter is the use of a relaxation approach for the linearization of the pressure gradient (see [1] for details). Thus this new scheme follows the ideas developed for the colocated method of [2], our final goal being the simulation of multiphase flows near the incompressible regime on general meshes.

## 2 Flow Equations and Acoustic/Transport Operator Splitting Approximation

We consider in this work a one-dimensional flow that is governed by the classic compressible Euler equations. We note $\rho$, $u$ and $e$ respectively the density, velocity and specific internal energy of the fluid. We suppose that the pressure $P$ is given by an Equation Of State (EOS) of the form $P = \mathcal{P}^{eos}(\rho, e)$. The sound velocity $c$ is classically defined by $c^2 = \partial_\rho \mathcal{P}^{eos} + (P/\rho^2)\partial_e \mathcal{P}^{eos}$ and is supposed to be real-valued. If $E = e + u^2/2$ denotes the specific total energy, the equation of motion reads in a one-dimensional setting

$$\rho_t + (\rho u)_x = 0, \qquad (\rho u)_t + (\rho u^2 + P)_x = 0, \qquad (\rho E)_t + (\rho E u + P u)_x = 0. \tag{1}$$

Following [2], we remark that smooth solutions of (1) verify

$$\begin{aligned}
&\rho_t && + \rho u_x && + u \rho_x && = 0, && \text{(2a)}\\
&(\rho u)_t && + \rho u u_x + P_x && + u(\rho u)_x && = 0, && \text{(2b)}\\
&(\rho E)_t && + \rho E u_x + (P u)_x && + u(\rho E)_x && = 0. && \text{(2c)}
\end{aligned}$$

This suggests to consider the subsystems

$$\rho_t + \rho u_x = 0, \quad (\rho u)_t + \rho u u_x + P_x = 0, \quad (\rho E)_t + \rho E u_x + (P u)_x = 0 \tag{3}$$

and

$$\rho_t + u \rho_x = 0, \quad (\rho u)_t + u(\rho u)_x = 0 \quad (\rho E)_t + u(\rho E)_x = 0. \tag{4}$$

As in [2], we propose to approximate the solutions of (1) by successively approximating the solution of (3) and (4). Before going any further, let us briefly recall properties of the subsystems (3) and (4). The acoustic system (3) reads equivalently

$$\rho\,(1/\rho)_t - u_x = 0, \qquad \rho u_t + P_x = 0, \qquad \rho E_t + (Pu)_x = 0, \tag{5}$$

and its associated Jacobian matrix possesses three eigenvalues $(-c, 0, c)$, so that it is strictly hyperbolic. The waves associated with the characteristic velocities $\pm c$ are genuinely nonlinear while the stationary wave is linearly degenerate. The transport system (4) is hyperbolic as it has diagonal form with $u$ as triple characteristic velocity.

## 3 Staggered Discretization

We consider a discretization of the real line into (primal) cells $K_i = (x_{i-1/2}, x_{i+1/2})$ with $i \in \mathbb{Z}$ and a series of instant $t^{n+1} = t^n + \delta t^n$ for $n \in \mathbb{N}$, where $\Delta x_i = x_{i+1/2} - x_{i-1/2} > 0$ is a space step and $\delta t^n > 0$ is the time step. The intervals $D_{i+1/2} = (x_i, x_{i+1})$ are referred to as dual cells, with $x_i = (x_{i+1/2} + x_{i-1/2})/2$ and $\Delta x_{i+1/2} = x_{i+1} - x_i$. For any fluid parameter $b$, we note $b_i^n$ and $b_{i+1/2}^n$ respective approximations of $\frac{1}{\Delta x_i}\int_{K_i} b(x, t^n)\mathrm{d}x$ and $\frac{1}{\Delta x_{i+1/2}}\int_{D_{i+1/2}} b(x, t^n)\mathrm{d}x$. If the vector $\mathcal{U}^n = \left(\rho_i^n, (\rho u)_{i+1/2}^n, (\rho E)_i^n\right)^T$ denotes a discrete approximate solution of (1) at instant $t^n$, the overall numerical scheme will first yield an updated value $\widetilde{\mathcal{U}}$ by integrating the acoustic system (3). Then by approximating the solution of the transport system (4) we will obtain the updated fluid parameters $\mathcal{U}^{n+1}$ at $t^{n+1}$. In the following, we will note $\lambda_k^n = \delta t^n/\Delta x_k$, for $n > 0$ and $k \in \mathbb{Z}/2$.

### *3.1 Approximation of the Acoustic Subsystem*

Using a similar strategy as in [2], we consider a Suliciu-type relaxation approximation of (5) that reads

$$r\,(1/\rho)_t - u_x = 0, \qquad r u_t + \pi_x = 0, \qquad r E_t + (\pi u)_x = 0, \tag{6a$_\nu$}$$

$$r_t = \nu(\rho - r), \qquad r\pi_t + a^2 u_x = \nu(P - \pi), \tag{6b$_\nu$}$$

where $a > 0$ is an approximation of $\rho c$ that needs to be chosen large enough in order to ensure stability. More precisely, $a$ must satisfy Whitham's condition which reads $a^2 > (\rho c)^2$ in order for system (6) to be a viscous approximation of system (3). Formally by considering system (6) in the limit $\nu \to \infty$, one retrieves the acoustic system (5). In practice, the regime $\nu \to \infty$ is obtained at each time step $t^n$ by classically enforcing $r_i^n = \rho_i^n$ and $\pi_i^n = \mathcal{P}^{eos}(\rho_i^n, e_i^n)$. Therefore, the equation on $r$ just expresses the fact that the density is constant over the acoustic step while the equation on $\pi$ is used as a linearization of the pressure gradient in the momentum equation. We discretize (6) with the following update formula

$$\rho_i^n\left(1/\widetilde{\rho}_i - 1/\rho_i^n\right) - \lambda_i^n\left(u_{i+1/2}^* - u_{i-1/2}^*\right) = 0, \tag{7a}$$

$$\rho_{i+1/2}^n\left(\widetilde{u}_{i+1/2} - u_{i+1/2}^n\right) + \lambda_{i+1/2}^n\left(\widetilde{\pi}_{i+1} - \widetilde{\pi}_i\right) = 0, \tag{7b}$$

$$\rho_i^n\left(\widetilde{E}_i - E_i^n\right) + \lambda_i^n\left(\widetilde{\pi}_{i+1/2}^* u_{i+1/2}^* - \widetilde{\pi}_{i-1/2}^* u_{i-1/2}^*\right) = 0, \tag{7c}$$

$$\rho_i^n\left(\widetilde{\pi}_i - \pi_i^n\right) + a_i^2\lambda_i^n\left(u_{i+1/2}^* - u_{i-1/2}^*\right) = 0, \tag{7d}$$

where $a_i = \rho_i^n c_i^n$ and

$$u_{i+1/2}^* = u_{i+1/2}^n - \frac{1}{2a_{i+1/2}}\left(\pi_{i+1}^n - \pi_i^n\right) \quad \text{with } a_{i+1/2} = \rho_{i+1/2}^n c_{i+1/2}^n, \tag{8a}$$

$$\widetilde{\pi}_{i+1/2}^* = \theta_{i+1/2}\widetilde{\pi}_i + \left(1 - \theta_{i+1/2}\right)\widetilde{\pi}_{i+1} \quad \text{with } \theta_{i+1/2} = \frac{1}{1 + \frac{\rho_i^n \Delta x_i}{\rho_{i+1}^n \Delta x_{i+1}}}. \tag{8b}$$

One can notice that $0 < \theta_{i+1/2} < 1$, $\forall i \in \mathbb{Z}$, so that the pressure at face $\widetilde{\pi}_{i+1/2}^*$ is a convex combination of the corresponding left and right relaxation pressures. The variables $\rho_{i+1/2}^n$, $c_{i+1/2}^n$ and $E_i^n$ are defined as follow:

$$\varphi_{i+1/2}^n = \frac{\Delta x_i \varphi_i^n + \Delta x_{i+1}\varphi_{i+1}^n}{2\Delta x_{i+1/2}}, \quad \text{for } \varphi \in \{\rho, c\}, \tag{9a}$$

$$E_i^n = e_i^n + \frac{1}{2}\left[\left(u_{i-1/2}^n\right)^2/2 + \left(u_{i+1/2}^n\right)^2/2\right]. \tag{9b}$$

## 3.2 Approximation of the Transport Subsystem

For more convenience, we will note in the following $\Delta\varphi_i^* = \varphi_{i+1/2}^* - \varphi_{i-1/2}^*$, for $\varphi \in \{u, \pi u\}$ and $\Delta\varphi_{i+1/2}^* = \varphi_{i+1}^* - \varphi_i^*$ for $\varphi \in \{\pi, u\}$ and $i \in \mathbb{Z}$. The discretization of the transport system (4) is performed following [2], more precisely we use the fact that $u\varphi_x = (u\varphi)_x - \varphi u_x$ for $\varphi \in \{\rho, \rho u, \rho E\}$. Thus, by writting $\delta\mathcal{U}^{n+1} = \mathcal{U}^{n+1} - \widetilde{\mathcal{U}}$, the update during the transport step writes

$$\delta\rho_i^{n+1} + \lambda_i^n\left(F_{i+1/2} - F_{i-1/2} - \widetilde{\rho}_i \Delta u_i^*\right) = 0, \tag{10a}$$

$$\delta(\rho u)_{i+1/2}^{n+1} + \lambda_{i+1/2}^n\left(F_{i+1}\widetilde{u}_{i+1}^{up} - F_i\widetilde{u}_i^{up} - \widetilde{(\rho u)}_{i+1/2}\Delta u_{i+1/2}^*\right) = 0, \tag{10b}$$

$$\delta(\rho E)_i^{n+1} + \lambda_i^n\left(F_{i+1/2}\widetilde{E}_{i+1/2}^{up} - F_{i-1/2}\widetilde{E}_{i-1/2}^{up} - \widetilde{(\rho E)}_i \Delta u_i^*\right) = 0. \tag{10c}$$

The numerical mass flux $F_{i+1/2}$ and total energy $\widetilde{E}^{up}_{i+1/2}$ are evaluated thanks to the classical upwind flux, more precisely

$$F_{i+1/2} = \widetilde{\rho}^{up}_{i+1/2} u^*_{i+1/2}, \quad \text{with} \quad (\widetilde{\rho}, \widetilde{E})^{up}_{i+1/2} = \begin{cases} (\widetilde{\rho}, \widetilde{E})_i, & \text{if } u^*_{i+1/2} > 0, \\ (\widetilde{\rho}, \widetilde{E})_{i+1}, & \text{otherwise.} \end{cases}$$

In (10b), the momentum fluxes $F_i$ are evaluated using an upwind procedure proposed in [8] that can be expressed as follow:

$$\widetilde{u}^{up}_i = \begin{cases} \widetilde{u}_{i-1/2}, & \text{if } F_i > 0 \\ \widetilde{u}_{i+1/2}, & \text{otherwise.} \end{cases} \quad \text{with } F_i = \frac{\Delta x_{i-1/2} F_{i-1/2} + \Delta x_{i+1/2} F_{i+1/2}}{2\Delta x_i}.$$

## 3.3 Overall Scheme

If we set $\widetilde{\rho}_{i+1/2} = \rho^n_{i+1/2} / \left(1 + \lambda^n_{i+1/2} \Delta u^*_{i+1/2}\right)$, the overall update of the conservative variable by the Staggered-Lagrange-Projection (SLP) takes the form of the following conservative scheme

$$\rho^{n+1}_i = \rho^n_i - \lambda^n_i \left(F_{i+1/2} - F_{i-1/2}\right), \tag{11a}$$

$$(\rho u)^{n+1}_{i+1/2} = (\rho u)^n_{i+1/2} - \lambda^n_{i+1/2} \left(F_{i+1} \widetilde{u}^{up}_{i+1} - F_i \widetilde{u}^{up}_i + \Delta \widetilde{\pi}_{i+1/2}\right), \tag{11b}$$

$$(\rho E)^{n+1}_i = (\rho E)^n_i - \lambda^n_i \left(F_{i+1/2} \widetilde{E}^{up}_{i+1/2} - F_{i-1/2} \widetilde{E}^{up}_{i-1/2} + \Delta \left(\widetilde{\pi}^* u^*\right)_i\right). \tag{11c}$$

Given the update value of $E^{n+1}_i$ and $u^{n+1}_{i+1/2}$ obtained by (11), the updated value $e^{n+1}_i$ of the internal energy is obtained thanks to (9b). The time step is set according to the relation

$$\delta t = K \min\left(\delta t_{ac}, \delta t_{tr}\right), \quad K \in (0, 1), \tag{12}$$

where $\delta t_{ac}$ and $\delta t_{tr}$ are given by

$$\delta t_{ac} = \min_{i \in \mathbb{Z}} \left( \frac{\Delta x_i}{\left(\Delta u^*_i\right)^-} \right) \quad \text{and} \quad \delta t_{tr} = \min_{i \in \mathbb{Z}} \left( \frac{\Delta x_i}{\left(u^*_{i+1/2}\right)^+ + \left(-u^*_{i-1/2}\right)^+} \right)$$

with $b^- = -\min(b, 0)$ and $b^+ = \max(b, 0)$. The time step $\delta t_{ac}$ ensures the positivity of $\widetilde{\rho}_i$ in (7a) while $\delta t_{tr}$ ensures the positivity of $\rho^{n+1}_i$ in (11a). One can show the following proposition.

**Proposition 1** *The SLP scheme (11) is conservative with respect to the density, momentum, total energy and under the condition* (12) *it preserves the positivity of the density.*

A proof of this proposition can be obtained following similar lines as in [2].

## 4 Numerical Results

We now test our numerical SLP scheme against classical tests of the literature and also display comparison with the MAC scheme presented in [8]. In the following, the fluid is supposed to be governed by the Perfect Gas EOS defined by $\mathcal{P}^{eos}(\rho, e) = (\gamma - 1)\rho e$, with $\gamma = 1.4$. All tests have been running with constant value $K = 0.1$ in the stability condition (12). Initial conditions of the first two tests are detailed in [10] and the third one in [9, problem case 3]. **Toro test 4** We propose to test the new SLP scheme against a first classical Riemann problem. This test represents the collision of two strong right and left shocks and consists of a left facing shock, a right travelling contact discontinuity and a right travelling shock wave. The computation was performed using a 500-cell mesh. We can see in Fig. 1 that the SLP scheme succeeds in capturing the right discontinuities while the MAC scheme suffers an overshoot near the shock. **Toro test 5** We now turn to a second classical Riemann problem. For this test, we also used a 500-cell grid. The test consists in a left rarefaction wave, a contact discontinuity and a strong right shock. Figure 2 displays a comparison between the SLP scheme and the MAC scheme. One can observe spurious oscillations in the MAC scheme in the intermediate state where the velocity vanishes and of maximal amplitude near the right shock. This behavior of the explicit MAC scheme is already known and is due to a lack of numerical dissipation in the

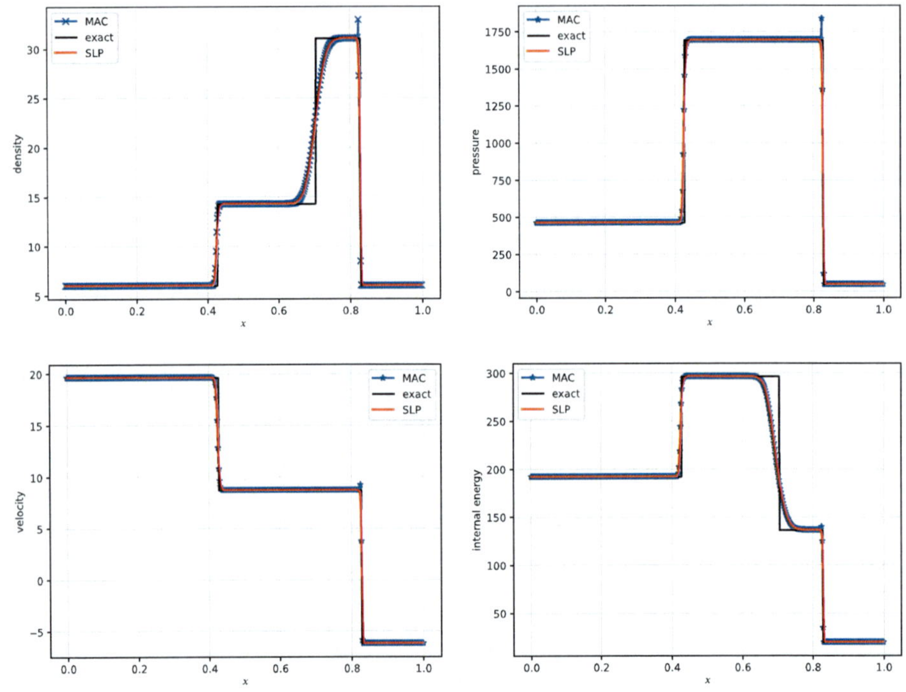


**Fig. 1** Results obtained for Toro test 4 at instant $t = 0.035$

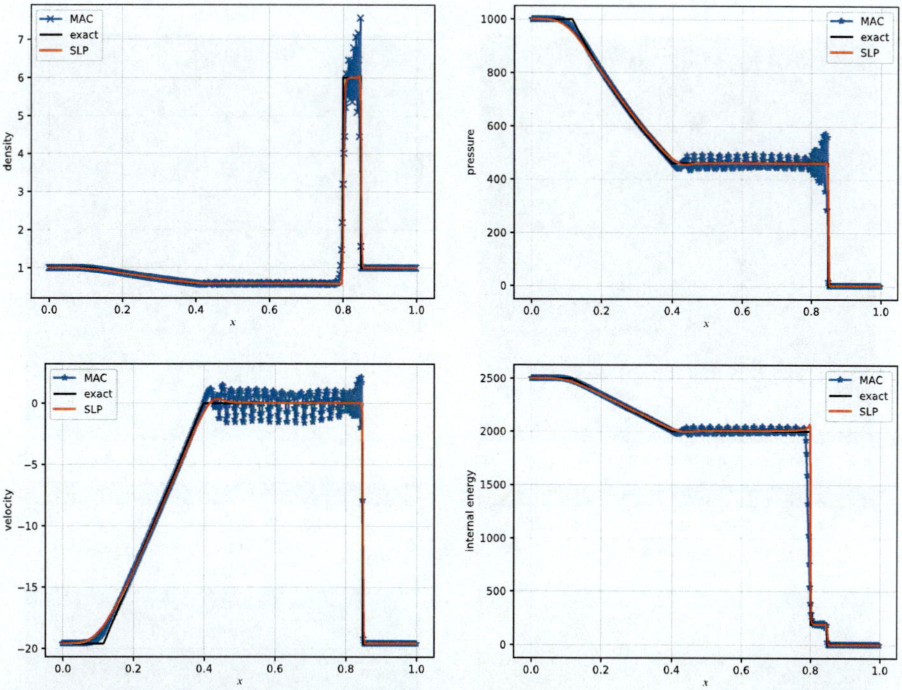


**Fig. 2** Results obtained for the Toro test 5 at instant $t = 0.012$

momentum equation when the velocity vanishes, as already pointed out in [8]. The SLP scheme does not suffer of such oscillations, mostly thanks to the extra diffusive term in the face velocity (8a).

### Two-Dimensional Riemann Problem

Using a directional splitting, the SLP scheme can be extended to multi-dimensional problems approximated over a Cartesian grid. We consider the two-dimensional Riemann problem of case 3 studied in [9] and performed using a $400 \times 400$-cell regular mesh. Results for both schemes are displayed in Fig. 3. Spurious oscillations appear for the MAC scheme mostly in the zero-velocity region of the test, which is caused by a lack of numerical dissipation as previously discussed. On the other hand, the new SLP scheme does not exhibit such behavior and presents good agreement with results shown in [9].

## 5 Conclusion

We proposed a fully conservative discretization of the compressible Euler equations over a staggered mesh using an explicit acoustic/transport operator splitting approach. The method shows promising results on both one-dimensional and two-dimensional

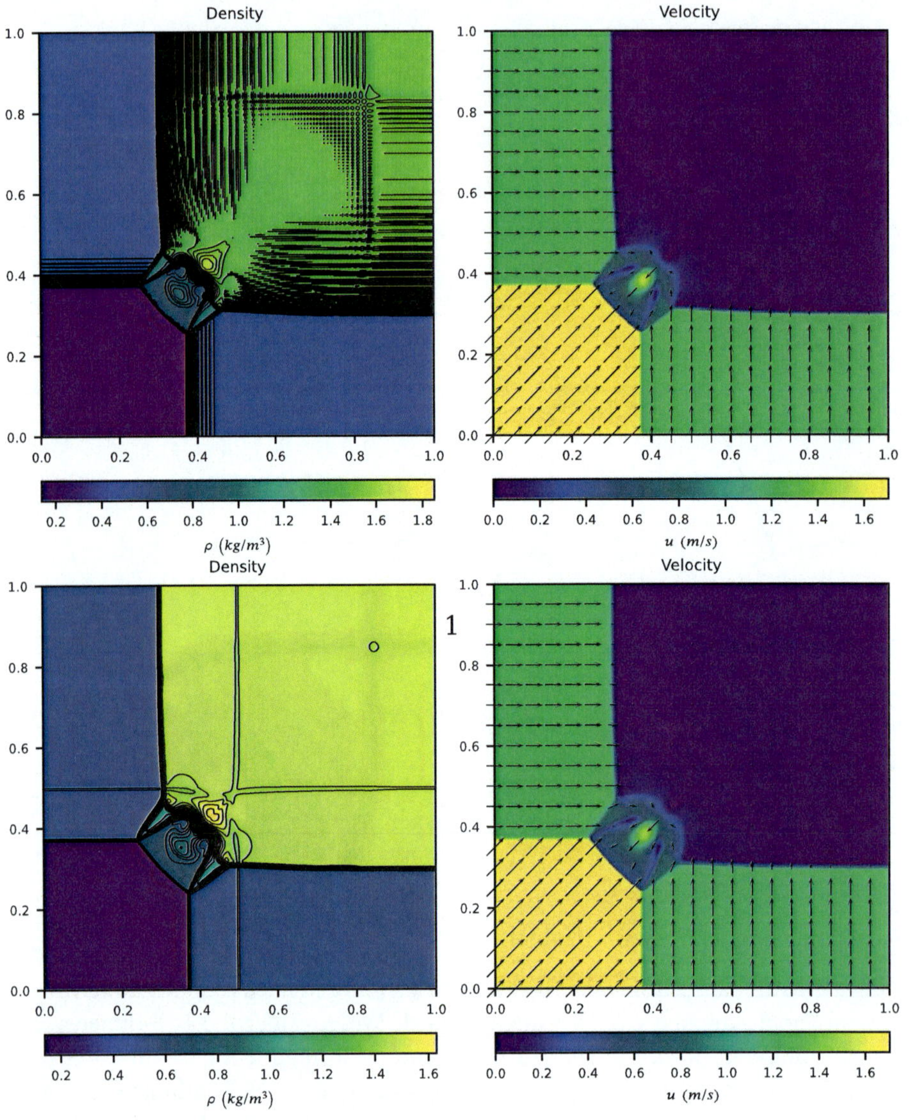


**Fig. 3** Two-dimensional Riemann problem: results for the MAC (above) and the SLP schemes (below) at instant $t = 0.3$

tests. Further works also include a more thorough study of the stability properties of the scheme and the extension to unstructured meshes.

## References

1. Bouchut, F.: Nonlinear Stability of Finite Volume Methods for Hyperbolic Conservation Laws, pp. 33–36. Birkhäuser, Basel (2004)
2. Chalons, C., Girardin, M., Kokh, S.: An all-regime Lagrange-Projection like scheme for the gas dynamics equations on unstructured meshes. Commun. Comput. Phys. **20**(1), 188–233 (2016)
3. Dakin, G., Després, B., Jaouen, S.: High-order staggered schemes for compressible hydrodynamics. Weak consistency and numerical validation. JCP **376**, 339–364 (2019)
4. Dakin, G., Jourdren, H.: High-order accurate Lagrange-remap hydrodynamic schemes on staggered cartesian grids. Comptes Rendus Mathématique **354**(2), 211–217 (2016)
5. Goudon, T., Llobell, J., Minjeaud, S.: A staggered scheme for the Euler equations. In: Springer Proceedings in Mathematics & Statistics, pp. 91–99. Springer International Publishing (2017)
6. Harlow, F., Amsden, A.: A numerical fluid dynamics calculation method for all flow speeds. JCP **8**(2), 197–213 (1971)
7. Herbin, R., Kheriji, W., Latché, J.-C.: Staggered schemes for all speed flows. ESAIM: Proc. **35**, 122–150 (2012)
8. Herbin, R., Latché, J.-C., Nguyen, T.T.: Consistent segregated staggered schemes with explicit steps for the isentropic and full Euler equations. ESAIM: Math. Modell. Numer. Anal. **52**(3), 893–944 (2018)
9. Liska, R., Wendroff, B.: Comparison of several difference schemes on 1d and 2d test problems for the Euler equations. SIAM J. Sci. Comput. **25**(3), 995–1017 (2003)
10. Toro, E.F.: Riemann Solvers and Numerical Methods for Fluid Dynamics, pp. 404–405. Springer, Berlin, Heidelberg (2009)

# Uncertainty Propagation of the Shock Position for Hyperbolic PDEs Using a Sensitivity Equation Method

**Camilla Fiorini**

**Abstract** In this work, we design a method to provide confidence intervals for the position of a discontinuity arising in hyperbolic conservation laws when some of the parameters of the problem are uncertain. This is based on a sensitivity equation method, therefore can be used only if the variance of the uncertain parameters is small. We illustrate how this method works on a simple one-dimensional conservation law: the inviscid Burgers equation. Multiple test cases are defined, some for which the analytical solution is known, in order to test the method.

**Keywords** Hyperbolic conservation laws · Shock position · Uncertainty propagation · Sensitivity equation method

## 1 Introduction

First order nonlinear hyperbolic partial differential equations (PDEs) are known to develop discontinuities, even when the initial condition is continuous. This adds a challenge if one wants to perform uncertainty propagation studies: a slight change in an input parameter can bring important changes in the output in the neighbourhood of the discontinuity. This translates to a high variance of the output close to the discontinuity [10]. If such variance is then used to compute confidence intervals for the output, this leads to artificially large intervals in the neighbourhood of the discontinuity, sometimes even providing negative values for variables such as the pressure or the density [4]. The use of modified sensitivities in [2, 4] solved the problem of negative values, but is equivalent to neglecting the variance of the position of the discontinuity. In this work, we use a sensitivity equation method to provide a first order estimate of this neglected quantity.

C. Fiorini (✉)
Laboratoire M2N, CNAM, Paris, France
e-mail: camilla.fiorini@lecnam.net
URL: http://camillafiorini.netlify.app/

 
E. Franck et al. (eds.), *Finite Volumes for Complex Applications X—Volume 2, Hyperbolic and Related Problems*, Springer Proceedings in Mathematics & Statistics 433,
https://doi.org/10.1007/978-3-031-40860-1_14

Here we consider a scalar one-dimensional conservation law, for the sake of simplicity. However, this approach can be easily extended to systems. The extension to higher dimension, on the other hand, is more difficult and beyond the scope of this work.

## 2 State Model

In this section, we describe the state model and its analytical solution: this is a well known problem and we use it mainly as an opportunity to introduce the notation. A scalar one dimensional conservation law can be written in the following form:

$$\begin{cases} \partial_t u(x,t) + \partial_x f(u(x,t)) = 0 & x \in \mathbb{R}, t > 0 \\ u(x,0) = g(x) & x \in \mathbb{R}, \end{cases} \tag{1}$$

where $u(x,t)$ is the conserved variable, $f$ the flux function and $g$ the initial condition. The method of characteristics can be used to solve analytically (1), and one can write the solution implicitly as follows:

$$u(x,t) = g(x - tf'(u(x,t))), \tag{2}$$

where $f' = \frac{df}{du}$. This method is valid as long as there is no intersection among the characteristics: in case of intersection a shock is generated and the solution $u(x,t)$ becomes discontinuous. We can define the *breaking time* $t_s$ as the smallest $t$ for which the characteristics intersect:

$$t_s = -\frac{1}{\min h'(x)}, \tag{3}$$

where $h(x) := f'(g(x))$.

The point $x_{s,0}$ from which the shock originates is known too:

$$x_{s,0} = \bar{x} + h(\bar{x})t_s, \text{ where } \bar{x} := \operatorname{argmin} h'(x).$$

If the initial data $g$ is discontinuous in a point $x_d$, two scenarios are possible: if $h(x_d^+) < h(x_d^-)$, the initial discontinuity is transported for all $t > 0$ ($t_s = 0$); otherwise the initial discontinuity is smoothed out and a rarefaction wave is generated.

Once the shock is generated, it moves along a curve $x_s(t)$ such that $\frac{dx_s}{dt} = v(t)$, which separates the plane $(x,t)$ into two parts, and in both of them the method of characteristics is valid. We denote with the superscript $+$ (respectively $-$) the quantities in the right (respectively left) part of the plane, i.e. $x > x_s(t)$ (respectively $x < x_s(t)$). The speed of the shock $v(t)$ can be computed using the Rankine–Hugoniot conditions:

$$v(t) = \frac{f(u^+(x_s(t),t)) - f(u^-(x_s(t),t))}{u^+(x_s(t),t) - u^-(x_s(t),t)}. \tag{4}$$

The position of the shock $x_s(t)$ can be then determined by solving the following ODE:

$$\begin{cases} \dfrac{dx_s}{dt} = v(t), \\ x_s(t_s) = x_{s,0}. \end{cases} \tag{5}$$

Finally, the solution of (1) can be written in the following compact form:

$$u(x,t) = u^+(x,t)H(x - x_s(t)) + u^-(x,t)H(x_s(t) - x), \tag{6}$$

where $u^\pm$ are obtained separately from (2) and $H$ is the Heaviside function.

# 3 Sensitivity Equation

## 3.1 Derivation of the Sensitivity Equation

In this section, we apply the continuous sensitivity equation (CSE) method to derive the sensitivity equations. First, we define the sensitivity with respect to a parameter of interest $a$ as the derivative of the state $u$ with respect to $a$ and we use the notation $u_a$:

$$u_a = \frac{\partial u}{\partial a}. \tag{7}$$

We now apply the CSE method and we differentiate the system (1) with respect to the parameter $a$, obtaining:

$$\begin{cases} \partial_a(\partial_t u(x,t)) + \partial_a(\partial_x f(u(x,t))) = 0 & x \in \mathbb{R},\ t > 0 \\ \partial_a u(x,0) = \partial_a g(x) & x \in \mathbb{R}, \end{cases}$$

and exchanging the derivatives in space and time with the ones with respect to $a$ one obtains the following equation and initial condition for the sensitivity:

$$\begin{cases} \partial_t u_a + \partial_x(f_a(u, u_a)) = 0 & x \in \mathbb{R},\ t > 0 \\ u_a(x,0) = g_a(x) & x \in \mathbb{R}, \end{cases} \tag{8}$$

where $g_a := \partial_a g$, $f_a(u, u_a) := f'(u)u_a + \partial_a f(u)$ and we dropped the time and space dependence in the equation for simplicity.

### 3.2 Analytical Solution of the Sensitivity Equation

The analytical solution of (8) can be computed in the regular zones by differentiating (2) with respect to $a$:

$$u_a(x,t) = g'(x - tf'(u(x,t)))(-tf''(u(x,t))u_a(x,t)) + g_a(x - tf'(u(x,t))),$$

therefore one can obtain an explicit expression for the sensitivity:

$$u_a(x,t) = \frac{g_a(x - tf'(u(x,t)))}{1 + tf''(u(x,t))g'(x - tf'(u(x,t)))} = \frac{g_a(x - tf'(u(x,t)))}{1 + th'(x - tf'(u(x,t)))}. \tag{9}$$

Let us observe that the denominator is zero if and only if there is an intersection between two characteristics, therefore only along the shock.

If the state $u$ is discontinuous, (9) is still valid on both sides of the shock. Along the shock the state $u$ is not differentiable in the classical sense; however, it admits a Dirac distribution as weak derivative. Therefore, by differentiating the compact expression (6) one obtains:

$$\begin{aligned} u_a(x,t) =& u_a^+(x,t)H(x - x_s(t)) + u_a^-(x,t)H(x_s(t) - x) \\ &+ (u^- - u^+)\partial_a x_s(t)\delta(x_s(t) - x), \end{aligned} \tag{10}$$

where $\delta$ is the Dirac delta function and $u_a^\pm$ are computed from (9) on either sides of the shock.

### 3.3 Correction Term

In previous works [1–4], a correction to the sensitivity systems was proposed in the case of discontinuous state, resulting in sensitivity solutions which do not exhibit spikes that mimic the Dirac distribution. To obtain such sensitivities, a source term is added, which should compensate the last term of (10). The corrected sensitivity equation is:

$$\partial_t u_a^c + \partial_x f_a(u, u_a^c) = S(u) \quad x \in \mathbb{R},\ t > 0, \tag{11}$$

where the source term has the following form:

$$S(u) = \partial_a v(u^+ - u^-)\delta(x_s(t) - x).$$

We use the superscript $c$ to indicate the corrected sensitivity. For more details on the derivation of the source term, see [3, 4] and Chap. 2 of [5] for the specific declination of the method for 1D scalar conservation laws. The analytical solution of (11) is:

$$u_a^c(x,t) = u_a^+(x,t)H(x - x_s(t)) + u_a^-(x,t)H(x_s(t) - x), \tag{12}$$

where $u_a^\pm$ are the same as in (10), since they are defined in the regular zones.

## 4 First Order Estimate of the Variance of the Shock Position

In this section we provide a first order estimate of the variance of the position of the shock $x_s(t)$. In this context, the parameter of interest $a$ is a random variable, whose expected value $\mu_a$ and variance $\sigma_a^2$ are known, and $x_s = x_s(t; a)$ is a function of this random variable. One can write a first order Taylor expansion of $x_s$ with respect to $a$, centred in $\mu_a$, obtaining:

$$x_s(t; a) = x_s(t; \mu_a) + (a - \mu_a)\partial_a x_s(t; \mu_a) + o((a - \mu_a)^2). \tag{13}$$

Then computing the average, one has:

$$\mathbb{E}[x_s(t; a)] \simeq x_s(t; \mu_a) + \mathbb{E}[a - \mu_a]\partial_a x_s(t; \mu_a) = x_s(t; \mu_a),$$

and for the variance:

$$\begin{aligned} \text{var}[x_s(t; a)] &\simeq \mathbb{E}[(x_s(t; a) - x_s(t; \mu_a))^2] \\ &= \mathbb{E}[(a - \mu_a)^2](\partial_a x_s(t; \mu_a))^2 = (\sigma_a \partial_a x_s(t; \mu_a))^2. \end{aligned}$$

Therefore, in order to estimate the variance of the position of the shock, we need to compute $\partial_a x_s(t)$. This can be obtained from $u$, $u_a$, and $u_a^c$ as follows:

$$\partial_a x_s(t) = \int_{\mathbb{R}} \frac{u_a(x,t) - u_a^c(x,t)}{u^- - u^+} dx. \tag{14}$$

The identity (14) can easily be verified by subtracting (12) from (10) and using the definition of a Dirac delta function.

## 5 Numerical Strategy

The aim of this section is to present the numerical strategy to compute (14), which can then be used to estimate the variance of the shock position.

From now on, we will consider the equations on an interval $(x_0, x_N)$, which is divided into $N$ cells, all of the same length $\Delta x$. The subscript $i$ indicates the $i$th cell, while $i \pm \frac{1}{2}$ refers to the interfaces. Concerning the time discretisation, we will use a variable time step $\Delta t$, chosen such that the CFL-condition is respected. The superscript $n$ indicates the $n$th time step.

Systems (1) and (8) are classical hyperbolic conservation laws, for which classical schemes can be used. However, for (11), a good discretisation of the source term is crucial.

The method chosen for the state equation is the approximate Riemann solver of Roe: the reason of this choice is that it provides the exact solution in the case of an isolated shock. For the sake of simplicity we consider the case $f'(u) > 0$. The scheme in the internal cells (i.e. $i \neq 0$) is the following:

$$u_i^{n+1} = u_i^n + \frac{\Delta t}{\Delta x}(f(u_{i-1}^n) - f(u_i^n)), \tag{15}$$

while for the first cell we impose $u_0^n = g(x_0)\ \forall n$.

For the sensitivity Eq. (8), i.e. the one without the source term, the same scheme used for the state can be applied:

$$u_{a,i}^{n+1} = u_{a,i}^n + \frac{\Delta t}{\Delta x}(f_a(u_{i-1}^n, u_{a,i-1}^n) - f_a(u_i^n, u_{a,i}^n)).$$

We remark that this scheme provides a sensitivity with spikes.

We here propose a second scheme, which comes from a quite natural discretisation of the source term:

$$u_{a,i}^{n+1} = u_{a,i}^n + \frac{\Delta t}{\Delta x}(f_a(u_{i-1}^n, u_{a,i-1}^n) - f_a(u_i^n, u_{a,i}^n)) + v_{a,i}^n(u_i^n - u_{i-1}^n)\delta_i^n, \tag{16}$$

where $\delta_i^n = 1$ if there is a shock in the $i$th cell at the $n$th time step, $\delta_i^n = 0$ otherwise. One of the main difficulties of this problem is to find an efficient shock detector. If one considers a Riemann problem, the simplest definition of a shock detector that works if $f$ is convex is:

$$\delta_i^n = \begin{cases} 0 & \text{if } u_{i-1}^n \leq u_i^n, \\ 1 & \text{otherwise.} \end{cases} \tag{17}$$

This of course does not work if the initial data is not a piecewise constant function. An option is the following shock detector based on the second derivative. Let $D$ be the maximum of the second derivative of the initial data and let $k$ be a constant to be tuned. Then

$$\delta_i^n = \begin{cases} 0 & \text{if } \dfrac{|u_{i-1}^n - 2u_i^n + u_{i+1}^n|}{\Delta x^2} \leq kD, \\ 1 & \text{otherwise.} \end{cases} \tag{18}$$

The shock detector is used to compute the source term for the corrected sensitivity system, but also to compute the jump of the state that appears in (14), which is computed as follows:

$$(u^+ - u^-)(t^n) = u_{k_1}^n - u_{k_2}^n,$$

where $k_1$ is such that $\delta_{k_1-1}^n = 1$ and $\delta_{k_1}^n = 0$, and $k_2$ is such that $\delta_{k_2}^n = 0$ and $\delta_{k_2+1}^n = 1$.

## 6 Numerical Results

In this section, we show some numerical results for different test cases: we consider three different initial conditions for the state. For every test case, the interval (0, 1) is discretised using $N = 1000$ cells.

1. The first test case is a Riemann problem, so that the analytical solution is known:

$$g_1(x) = u_L H(x_c - x) + u_R H(x - x_c), \tag{19}$$

with $u_L = 1$, $u_R = 0.1$, $x_c = 0.5$. For this problem, we consider two different uncertain parameters: the left value $u_L$, which gives a Riemann problem for the sensitivity as well, since $g_{1,u_L} = H(x_c - x)$; and the initial position of the discontinuity, which gives $g_{1,x_c}(x) = (u_L - u_R)\delta(x_c - x)$ as initial condition for the sensitivity.

2. A problem with a piecewise linear initial condition:

$$g_2(x) = \begin{cases} u_L & x < x_L, \\ \frac{u_R - u_L}{x_R - x_L} x + \frac{u_L x_R - u_R x_L}{x_R - x_L} & x_L \leq x \leq x_R, \\ u_R & x > x_R, \end{cases} \tag{20}$$

with $u_L = 2$, $u_R = 0.1$, $x_L = 0.25$, $x_R = 0.75$. The uncertain parameter for this case is $u_L$, which gives

$$g_{2,u_L}(x) = \begin{cases} 1 & x < x_L, \\ x\frac{1}{x_L - x_R} x + \frac{x_R}{x_R - x_L} & x_L \leq x \leq x_R, \\ 0 & x > x_R. \end{cases} \tag{21}$$

The interest of this case is that we are able to compute the analytical solution as for the Riemann problem, but the breaking time $t_s$ is strictly positive.

3. Finally, a problem with the following continuous initial condition:

$$g_3(x) = \begin{cases} A \sin^2\left(\frac{\pi}{L}(x - x_c) + \frac{\pi}{2}\right) & \text{if } -1 < 2\frac{x - x_c}{L} < 1 \\ 0 & \text{otherwise,} \end{cases} \tag{22}$$

where $A = 1$, $L = 0.6$, $x_c = 0.5$. The uncertain parameters considered here are $L$, which gives

$$g_{3,L}(x) = \begin{cases} A\frac{\pi}{L^2}(x - x_c) \sin\left(\frac{2\pi}{L}(x - x_c)\right) & \text{if } -1 < 2\frac{x - x_c}{L} < 1 \\ 0 & \text{otherwise,} \end{cases} \tag{23}$$

and $A$, which gives

$$g_{3,A}(x) = \begin{cases} \sin^2\left(\frac{\pi}{L}(x - x_c) + \frac{\pi}{2}\right) & \text{if } -1 < 2\frac{x-x_c}{L} < 1 \\ 0 & \text{otherwise.} \end{cases} \tag{24}$$

For each test case, the uncertain parameter has a variance of 5% of its average (i.e. $\sigma_a^2 = 0.05\mu_a$). In Fig. 1 we plot the confidence intervals for the position of the shock, which are computed as follows :

$$CI := \left[\mathbb{E}[x_s] - \sqrt{\frac{\text{var}[x_s]}{\alpha}}, \mathbb{E}[x_s] + \sqrt{\frac{\text{var}[x_s]}{\alpha}}\right] \tag{25}$$

where $\alpha$ is the level of the confidence interval and we chose $\alpha = 0.95$. The form (25) is a straightforward consequence of the Chebyshev inequality [6], for more details see for instance [7]. For the test cases for which an analytical solution is available, we observe perfect concordance with the numerical results. We remark that one drawback of this method is the inability to estimate the variance of the breaking time.

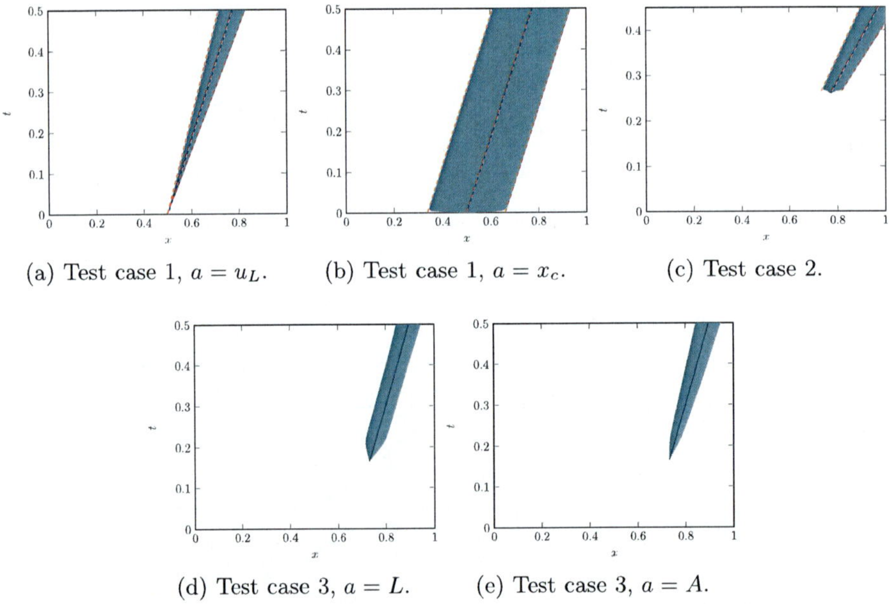


**Fig. 1** Confidence intervals for the position of the shock. In blue, the numerical solution. In red, the analytical solution when available

## 7 Conclusion and Perspectives

In this work, we use two different definitions of the sensitivity to provide a first order estimate of the variance of the position of discontinuities, present in hyperbolic conservation laws. This can then be used to compute confidence intervals for the position of the shock. This method can be easily extended to system of conservation laws, provided that anti-diffusive numerical schemes (such as, for instance, the Glimm method) are used to solve both the state and the sensitivity. The extension to higher space dimension is more challenging, mainly due to the fact that anti-diffusive numerical schemes are needed in order to correctly solve the sensitivity. Numerical schemes based on neural network, such as [8], or shock tracking techniques [9] could be used for this.

## References

1. Guinot, V.: Upwind finite volume solution of sensitivity equations for hyperbolic systems of conservation laws with discontinuous solutions. Comput. & Fluids **9**, 1697–1709 (2009)
2. Guinot, V., Delenne, C., Cappelaere, B.: An approximate Riemann solver for sensitivity equations with discontinuous solutions. Adv. Water Resour. **32**, 61–77 (2009)
3. Chalons, C., Duvigneau, R., Fiorini, C.: Sensitivity analysis and numerical diffusion effects for hyperbolic PDE systems with discontinuous solutions. The case of barotropic Euler equations in Lagrangian coordinates. SIAM J. Sci. Comput. **40**, A3955–A3981 (2018)
4. Fiorini, C., Chalons, C., Duvigneau, R.: A modified sensitivity equation method for Euler equations in presence of shocks. Numer. Methods Part. Diff. Eq. **36**, 839–867 (2020)
5. Fiorini, C.: Sensitivity analysis for nonlinear hyperbolic systems of conservation laws. Ph.D. Thesis, Université Paris Saclay (2018)
6. Jacod, J., Protter, P.: Probability essentials (2012)
7. Fiorini, C., Després, B., Puscas, M.A.: Sensitivity equation method for the Navier-Stokes equations applied to uncertainty propagation. Int. J. Numer. Methods Fluids **93**(1), 71–92 (2021)
8. Després, B., Jourdren, H.: Machine learning design of volume of fluid schemes for compressible flows. J. Comput. Phys. **408**, 109275 (2020)
9. Ciallella, M., Ricchiuto, M., Paciorri, R., Bonfiglioli, A.: Extrapolated Shock Tracking: bridging shock-fitting and embedded boundary methods. J. Comput. Phys. **412**, 109440 (2020)
10. Schwab, C., Tokareva, S.: High order approximation of probabilistic shock profiles in hyperbolic conservation laws with uncertain initial data. ESAIM: Math. Modell. Numer. Anal. **47**(3), 807–835 (2013)

# A MUSCL-Scheme on Staggered Grids for the Euler Equations on Unstructured Meshes

**Charbel Ghosn, Thierry Goudon, and Sebastian Minjeaud**

**Abstract** We set up a MUSCL approach for the $2D$ DDFV-like scheme introduced in [3] for the full Euler equations, discretized with unstructured meshes. It is a finite volume staggered discretization framework where numerical densities, energies and velocities are stored on different locations. For the MUSCL scheme, a combination of several approaches is needed for the reconstruction of the densities and the velocities, in order to reach the second order accuracy. This work focuses on the reconstruction of the mass density and internal energy. The improved accuracy is already sensitive on problems containing contact discontinuities. We provide a set of numerical simulations to illustrate the enhanced accuracy.

**Keywords** Euler model · Staggered grids · DDFV · MUSCL-scheme

## 1 Introduction

We consider in this work the Euler system of gas dynamics

$$\begin{cases} \partial_t \rho + \nabla \cdot (\rho \mathbf{u}) = 0, \\ \partial_t (\rho \mathbf{u}) + \nabla \cdot (\rho \mathbf{u} \otimes \mathbf{u}) + \nabla p = 0, \\ \partial_t (\rho E) + \nabla \cdot (\rho E \mathbf{u}) + \nabla \cdot (p \mathbf{u}) = 0, \end{cases} \tag{1}$$

C. Ghosn · T. Goudon · S. Minjeaud (✉)
LJAD, CNRS, Inria, Université Côte d'Azur, Nice, France
e-mail: minjeaud@unice.fr

C. Ghosn
e-mail: cghosn@unice.fr

T. Goudon
e-mail: thierry.goudon@inria.fr

Parc Valrose, 06108 Nice, France



E. Franck et al. (eds.), *Finite Volumes for Complex Applications X—Volume 2, Hyperbolic and Related Problems*, Springer Proceedings in Mathematics & Statistics 433,
https://doi.org/10.1007/978-3-031-40860-1_15

where $\rho$, $\mathbf{u}$, $E$ and $p$ stand for the mass density, the velocity field, the total energy and the pressure, respectively. These unknowns depend on the time and space variables $(t, x) \in [0, \infty) \times \Omega$ with $\Omega \subset \mathbb{R}^2$ a polygonal bounded domain. System (1) is supplemented by the following equation of state

$$E = \frac{\|\mathbf{u}\|^2}{2} + e \quad \text{and} \quad p = (\gamma - 1)\rho e,$$

where $e$ is the internal energy and $\gamma > 1$.

We aim at building a second order scheme on general staggered grids. Our approach is based on the first order scheme presented in [3] and a multislope MUSCL reconstruction of the variables (see [2] for a related but different approach). Since we work on staggered grids, the densities and velocities are stored at different locations of the mesh. The densities are cell-based variables and the velocities are defined on vertices. Thus, the reconstruction procedure will not be the same for these variables. We present here the MUSCL reconstruction of the mass density $\rho$ and the density of internal energy $\rho e$, the reconstruction for the velocity being an ongoing work. We use a multislope method to reconstruct the densities [4], which will be used to define the mass and internal energy fluxes. This is not enough to obtain a second order convergence but we show with the numerical simulations of 1D contact discontinuities (where the velocity is constant) that it already provides improved results.

## 2 Notation: Meshes, Unknowns

We construct three partitions of the domain $\Omega$: the primal mesh, the dual mesh and the diamond mesh. The steps of the construction are illustrated in Fig. 1 (see details in [3]).

The primal mesh $\mathfrak{M}$, in blue in Fig. 1, is a partition of $\Omega$ by polygonal convex subsets $K$ called "primal cells". The centers of these cells are labeled $x_K$. The dual mesh $\mathfrak{M}^*$, in red in Fig. 1, consists of cells "built around the vertices $x_{K^*}$" of the primal mesh. The dual cell $K^*$ is formed by joining the centers $x_K$ of all cells having $K^*$ as a vertex. The diamond mesh $\mathfrak{D}$, in green in Fig. 1, is made of quadrilateral cells $D_{\sigma,\sigma^*}$ obtained by joining the endpoints of the edge $\sigma = [x_{K^*}, x_{L^*}]$ of the primal mesh to the centers $x_K$ and $x_L$ of the primal cells that share this edge. The segment $\sigma^* = [x_K, x_L]$ is an edge of the dual mesh.

We do not pay attention in this article to the description of the meshes near the boundary $\partial\Omega$ and, more generally, to boundary conditions. We refer the reader to [3] for further details. As in [3], we use the following notation: we denote $\mathfrak{s} = D_{\sigma,\sigma^*}|D_{\sigma',\sigma^{*\prime}}$ the face separating two diamond cells $D_{\sigma,\sigma^*}$ and $D_{\sigma',\sigma^{*\prime}}$; for $K \in \mathfrak{M}$, we denote $\mathfrak{D}_K = \{D_{\sigma,\sigma^*} \in \mathfrak{D}, \sigma \in \partial K\}$; for $K^* \in \mathfrak{M}^*$, we similarly denote $\mathfrak{D}_{K^*} = \{D_{\sigma,\sigma^*} \in \mathfrak{D}, \sigma^* \in \partial K^*\}$; for a cell $X$ of $\mathfrak{M}$, $\mathfrak{M}^*$ or $\mathfrak{D}$ and for $\mathfrak{x} \in \partial X$, we define a unit vector $\mathbf{n}_{X,\mathfrak{x}}$ normal to the face $\mathfrak{x}$ of the cell $X$ and pointing outwards: $\mathbf{n}_{K,\sigma}$

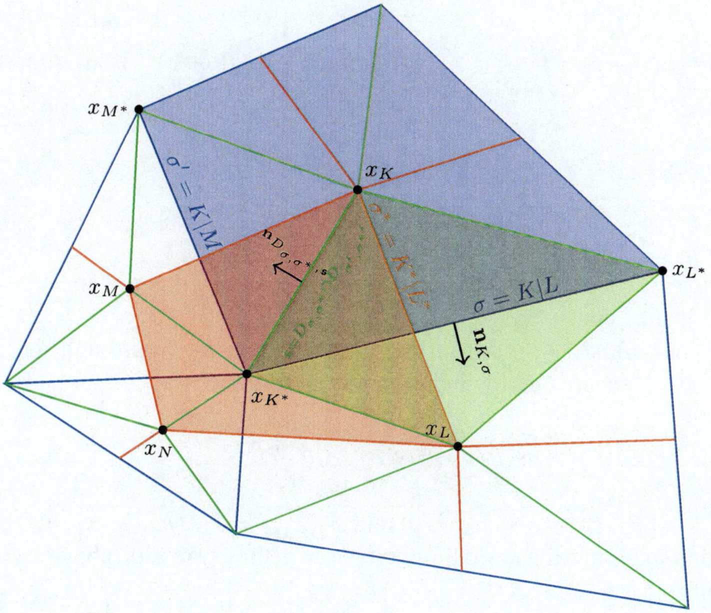


**Fig. 1** Meshes and associated notations

(with $\sigma \in \partial K$ for $K \in \mathfrak{M}$), $\mathbf{n}_{K^*,\sigma^*}$ (with $\sigma^* \in \partial K^*$ for $K^* \in \mathfrak{M}^*$), and $\mathbf{n}_{D_{\sigma,\sigma^*},\mathfrak{s}}$ (with $\mathfrak{s} \in \partial D_{\sigma,\sigma^*}$ for $D_{\sigma,\sigma^*} \in \mathfrak{D}$).

The unknowns of the mesh are defined as piecewise constants over the cells:

1. The mass density $(\rho_{\sigma,\sigma^*})_{D_{\sigma,\sigma^*} \in \mathfrak{D}}$ and the internal energy $(e_{\sigma,\sigma^*})_{D_{\sigma,\sigma^*} \in \mathfrak{D}}$ are piecewise constants over the diamond cells. We set $p_{\sigma,\sigma^*} = (\gamma - 1)\rho_{\sigma,\sigma^*} e_{\sigma,\sigma^*}$.
2. The numerical velocity fields $(\mathbf{u}_K)_{K \in \mathfrak{M}}$ and $(\mathbf{u}_{K^*})_{K^* \in \mathfrak{M}^*}$ are piecewise constants over the primal and dual cells, respectively.

## 3 A MUSCL Reconstruction on Staggered Grids

Let us first recall how the first order scheme is constructed in [3], the MUSCL scheme being described in a second part.

**The first order scheme**. The time discretization is explicit. We denote by $\delta t$ the time step and $\bar{q}$ the update at time $t + \delta t$ of any quantity $q$ at time $t$. The space discretization is a finite volume scheme based on a splitting of the mass flux inspired from the kinetic framework which involves the sound speed $c(e) = \sqrt{\gamma(\gamma - 1)e}$ of system (1). Let

$$\mathcal{F}^+(\rho, c, u) = \int_{\xi>0} \xi \mathcal{M}_{[\rho,c,u]}(\xi) d\xi = \begin{cases} 0 & \text{if } u \leqslant -c, \\ \dfrac{\rho}{4c}(u+c)^2 & \text{if } |u| \leqslant c, \\ \rho u & \text{if } u \geqslant c, \end{cases}$$

where

$$\xi \in \mathbb{R} \longmapsto \mathcal{M}_{[\rho,c,u]}(\xi) = \frac{\rho}{2c} \mathbb{1}_{|\xi-u|\leqslant c}.$$

The function $\mathcal{F}^-$ is defined by $\mathcal{F}^-(\rho, c, u) = -\mathcal{F}^+(\rho, c, -u)$, so that the functions $\mathcal{F}^\pm$ satisfies consistency property $\mathcal{F}^+(\rho, c, u) + \mathcal{F}^-(\rho, c, u) = \rho u$. We thus define the mass flux $\mathcal{F}_{D_{\sigma,\sigma^*},\mathfrak{s}}$ from the diamond cell $D_{\sigma,\sigma^*}$ through the interface $\mathfrak{s} = D_{\sigma,\sigma^*}|D_{\sigma',\sigma^{*\prime}}$ using the upwind principle as follows

$$\mathcal{F}_{D_{\sigma,\sigma^*},\mathfrak{s}} = \mathcal{F}^+_{D_{\sigma,\sigma^*},\mathfrak{s}} + \mathcal{F}^-_{D_{\sigma,\sigma^*},\mathfrak{s}}$$

with $\mathcal{F}^+_{D_{\sigma,\sigma^*},\mathfrak{s}} = \mathcal{F}^+(\rho_{\sigma,\sigma^*}, c_\mathfrak{s}, u_{D_{\sigma,\sigma^*},\mathfrak{s}})$ and $\mathcal{F}^-_{D_{\sigma,\sigma^*},\mathfrak{s}} = \mathcal{F}^-(\rho_{\sigma',\sigma^{*\prime}}, c_\mathfrak{s}, u_{D_{\sigma,\sigma^*},\mathfrak{s}})$.

The sound speed and the normal velocity used in the expression above are averaged values defined for $\mathfrak{s} = D_{\sigma,\sigma^*}|D_{\sigma',\sigma^{*\prime}} = [x_K, x_{K^*}]$ by:

$$u_{D_{\sigma,\sigma^*},\mathfrak{s}} = \frac{\mathbf{u}_K + \mathbf{u}_{K^*}}{2} \cdot \mathbf{n}_{D_{\sigma,\sigma^*},\mathfrak{s}} \quad \text{and} \quad c_\mathfrak{s} = c\left(\frac{e_{\sigma,\sigma^*} + e_{\sigma',\sigma^{*\prime}}}{2}\right).$$

The discrete mass equation on a cell $D_{\sigma,\sigma^*} \in \mathfrak{D}$ is given by

$$\frac{\overline{\rho}_{\sigma,\sigma^*} - \rho_{\sigma,\sigma^*}}{\delta t} + \frac{1}{|D_{\sigma,\sigma^*}|} \sum_{\mathfrak{s}\in\partial D_{\sigma,\sigma^*}} |\mathfrak{s}| \mathcal{F}_{D_{\sigma,\sigma^*},\mathfrak{s}} = 0.$$

The averaged density on a cell $K$ of the primal mesh is defined by

$$\rho_K = \sum_{D_{\sigma,\sigma^*}\in\mathfrak{D}_K} \frac{|D_{\sigma,\sigma^*} \cap K|}{|K|} \rho_{\sigma,\sigma^*} \text{ for } K \in \mathfrak{M}$$

and on a cell $K^*$ of the dual mesh, we set

$$\rho_{K^*} = \sum_{D_{\sigma,\sigma^*}\in\mathfrak{D}_{K^*}} \frac{|D_{\sigma,\sigma^*} \cap K^*|}{|K^*|} \rho_{\sigma,\sigma^*} \text{ for } K^* \in \mathfrak{M}^*.$$

To these densities we associate averaged mass fluxes $\mathcal{F}_{K,\sigma}$ outgoing from a primal cell $K$ and $\mathcal{F}_{K^*,\sigma^*}$ outgoing from a dual cell $K^*$, with

$$\mathcal{F}^\pm_{K,\sigma} = \frac{|D_{\sigma,\sigma^*} \cap K|}{|D_{\sigma,\sigma^*}|} \sum_{\substack{\mathfrak{s}\in\partial D_{\sigma,\sigma^*} \\ \mathfrak{s}\subset L}} \frac{|\mathfrak{s}|}{|\sigma|} \mathcal{F}^\pm_{D_{\sigma,\sigma^*},\mathfrak{s}} - \frac{|D_{\sigma,\sigma^*} \cap L|}{|D_{\sigma,\sigma^*}|} \sum_{\substack{\mathfrak{s}\in\partial D_{\sigma,\sigma^*} \\ \mathfrak{s}\subset K}} \frac{|\mathfrak{s}|}{|\sigma|} \mathcal{F}^\mp_{D_{\sigma,\sigma^*},\mathfrak{s}}.$$

$$\mathcal{F}^{\pm}_{K^*,\sigma^*} = \frac{|D_{\sigma,\sigma^*} \cap K^*|}{|D_{\sigma,\sigma^*}|} \sum_{\substack{\mathfrak{s}\in\partial D_{\sigma,\sigma^*}\\ \mathfrak{s}\subset L^*}} \frac{|\mathfrak{s}|}{|\sigma^*|} \mathcal{F}^{\pm}_{D_{\sigma,\sigma^*},\mathfrak{s}} - \frac{|D_{\sigma,\sigma^*} \cap L^*|}{|D_{\sigma,\sigma^*}|} \sum_{\substack{\mathfrak{s}\in\partial D_{\sigma,\sigma^*}\\ \mathfrak{s}\subset K^*}} \frac{|\mathfrak{s}|}{|\sigma^*|} \mathcal{F}^{\mp}_{D_{\sigma,\sigma^*},\mathfrak{s}}.$$

Using these fluxes, we can give the definition to the momentum fluxes $\mathcal{G}_{K,\sigma}$ for the primal cells and $\mathcal{G}_{K^*,\sigma^*}$ for the dual cells

$$\mathcal{G}_{K,\sigma} = \mathcal{F}^{+}_{K,\sigma}\mathbf{u}_K + \mathcal{F}^{-}_{K,\sigma}\mathbf{u}_L \quad \text{and} \quad \mathcal{G}_{K^*,\sigma^*} = \mathcal{F}^{+}_{K^*,\sigma^*}\mathbf{u}_{K^*} + \mathcal{F}^{-}_{K^*,\sigma^*}\mathbf{u}_{L^*}.$$

The discrete momentum equation now reads

$$\frac{\overline{\rho}_K\overline{\mathbf{u}}_K - \rho_K\mathbf{u}_K}{\delta t} + \frac{1}{|K|}\sum_{D_{\sigma,\sigma^*}\in\mathfrak{D}_K} |\sigma|\mathcal{G}_{K,\sigma} + (\nabla_d p)_K = 0,$$

$$\frac{\overline{\rho}_{K^*}\overline{\mathbf{u}}_{K^*} - \rho_{K^*}\mathbf{u}_{K^*}}{\delta t} + \frac{1}{|K^*|}\sum_{D_{\sigma,\sigma^*}\in\mathfrak{D}_{K^*}} |\sigma^*|\mathcal{G}_{K^*,\sigma^*} + (\nabla_d p)_{K^*} = 0,$$

with a suitable definition of the pressure gradients [3].

Finally, for the discretization of the internal energy equation, we define the following numerical fluxes, for all $D_{\sigma,\sigma^*} \in \mathfrak{D}$ and $\mathfrak{s} = D_{\sigma,\sigma^*}|D_{\sigma',\sigma^{*\prime}}$,

$$\mathcal{E}_{D_{\sigma,\sigma^*},\mathfrak{s}} = e_{\sigma,\sigma^*}\mathcal{F}^{+}_{D_{\sigma,\sigma^*},\mathfrak{s}} + e_{\sigma',\sigma^{*\prime}}\mathcal{F}^{-}_{D_{\sigma,\sigma^*},\mathfrak{s}}.$$

The discrete internal energy equation is given by

$$\frac{\overline{\rho}_{\sigma,\sigma^*}\overline{e}_{\sigma,\sigma^*} - \rho_{\sigma,\sigma^*}e_{\sigma,\sigma^*}}{\delta t} + \frac{1}{|D_{\sigma,\sigma^*}|}\sum_{\mathfrak{s}\in\partial D_{\sigma,\sigma^*}} |\mathfrak{s}|\mathcal{E}_{D_{\sigma,\sigma^*},\mathfrak{s}}$$
$$+ p_{\sigma,\sigma^*}(\nabla_d \cdot \overline{\mathbf{u}})_{\sigma,\sigma^*} = \mathcal{R}_{\sigma,\sigma^*}, \quad \forall D_{\sigma,\sigma^*} \in \mathfrak{D}$$

where $\nabla_d \cdot \overline{\mathbf{u}}$ is the discrete divergence operator, see [3], and $\mathcal{R}_{\sigma,\sigma^*}$ the correction term which should keep track of possible discontinuities. This remainder is defined so that it exactly balances the kinetic energy contributions that appear when summing the internal energy equation and the kinetic energy equation. This reconstruction allows us to derive a conservative discrete equation for an averaged total energy [3]. This construction shares many similarities with the DDFV framework since it works on dual grids and the discretization aims at preserving the duality of div-grad operators, which is crucial in order to preserve local conservations.

**The MUSCL-scheme**. We discuss how we adapt the MUSCL principles to the staggered framework. We first reconstruct second order quantities at edges of the cells (primal, dual or diamond) depending on the domain where the variables are defined. Then, concerning the discretization of the mass flux, we keep unchanged the velocity defined at the interface $\mathfrak{s}$ and we shall replace the UpWind value $\rho_{\sigma,\sigma^*}$

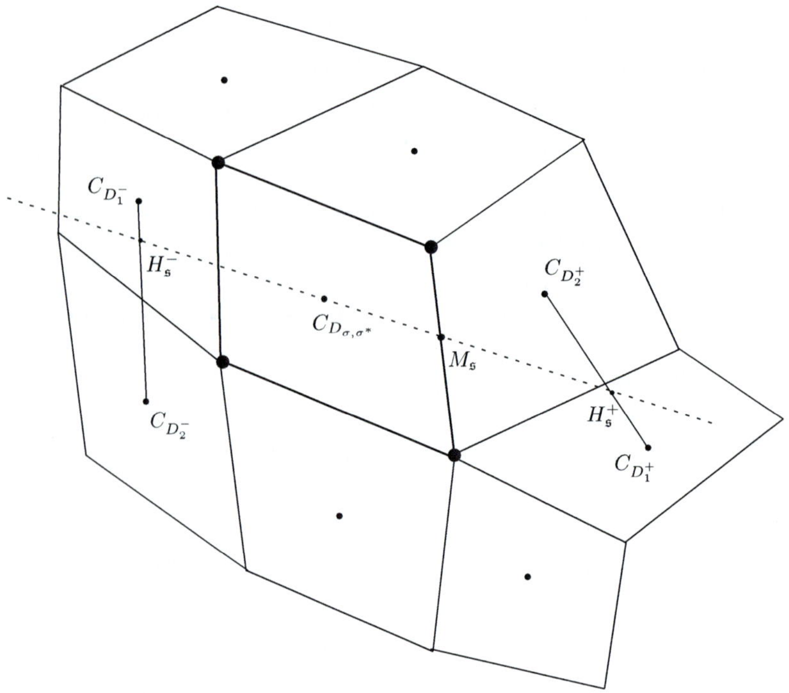


**Fig. 2** Forward and backward points $\mathbf{H}_{\mathfrak{s}}^+$ and $\mathbf{H}_{\mathfrak{s}}^-$

by a MUSCL reconstruction $\rho^{ML}_{D_{\sigma,\sigma^*},\mathfrak{s}}$ of the density: it defines the upgraded mass flux $\mathcal{F}^{ML}_{D_{\sigma,\sigma^*},\mathfrak{s}}$. For the internal energy, we combine the upgraded mass fluxes $\mathcal{F}^{ML}_{D_{\sigma,\sigma^*},\mathfrak{s}}$ with a MUSCL reconstruction of the internal energy defined from the ratio $\frac{(\rho e)^{ML}_{D_{\sigma,\sigma^*},\mathfrak{s}}}{\rho^{ML}_{D_{\sigma,\sigma^*},\mathfrak{s}}}$, where $(\rho e)^{ML}_{D_{\sigma,\sigma^*},\mathfrak{s}}$ is the MUSCL reconstruction of $\rho_{\sigma,\sigma^*} e_{\sigma,\sigma^*}$ at the interface $\mathfrak{s}$. We follow the multislope method introduced in [4]. We compute the reconstructed values at the centers of the interfaces $\mathbf{M}_{\mathfrak{s}}$ (see Fig. 2). As in the original MUSCL method, both a backward and a forward scalar slopes, respectively denoted $s^-_{\sigma,\mathfrak{s}}$ and $s^+_{\sigma,\mathfrak{s}}$, are computed for each interface $\mathfrak{s}$ of a given diamond cell $D_{\sigma,\sigma^*}$. In a classical way, we use a limiter function $\phi(s^-_{\sigma,\mathfrak{s}}, s^-_{\sigma,\mathfrak{s}})$ to ensure that no unphysical oscillation is introduced. Therefore, the reconstructed values read as follows:

$$\rho^{ML}_{D_{\sigma,\sigma^*},\mathfrak{s}} = \rho_{\sigma,\sigma^*} + \phi(s^-_{\sigma,\mathfrak{s}}, s^+_{\sigma,\mathfrak{s}}) \cdot ||\mathbf{C}_{D_{\sigma,\sigma^*}}\mathbf{M}_{\mathfrak{s}}||,$$

where $\mathbf{C}_{D_{\sigma,\sigma^*}}$ is the center of the diamond cell $D_{\sigma,\sigma^*}$.

We discuss next the building of the slopes. For that, we denote by $\mathcal{W}_{\sigma,\sigma^*}$ the set of the diamond cells sharing at least a vertex with $D_{\sigma,\sigma^*}$. The idea is to determine two points $\mathbf{H}^+_{\mathfrak{s}}$ and $\mathbf{H}^-_{\mathfrak{s}}$ both located on the axis $(\mathbf{C}_{D_{\sigma,\sigma^*}}\mathbf{M}_{\mathfrak{s}})$, respectively backward and forward the point $\mathbf{C}_{D_{\sigma,\sigma^*}}$ (see Fig. 2). These points are a priori neither vertices of the grid, nor element centers. However, these points are located by construction on a line joining two element centers so that densities $\rho_{\mathbf{H}_{\mathfrak{s}}^-}$ and $\rho_{\mathbf{H}_{\mathfrak{s}}^+}$ at these points can be obtained by a linear interpolation. The backward and forward slopes are then computed as follows:

$$s_{\sigma,\mathfrak{s}}^- = \frac{\rho_{\sigma,\sigma^*} - \rho_{\mathbf{H}_{\mathfrak{s}}^-}}{||\mathbf{C}_{D_{\sigma,\sigma^*}}\mathbf{H}_{\mathfrak{s}}^-||}, \quad s_{\sigma,\mathfrak{s}}^+ = \frac{\rho_{\mathbf{H}_{\mathfrak{s}}^+} - \rho_{\sigma,\sigma^*}}{||\mathbf{C}_{D_{\sigma,\sigma^*}}\mathbf{H}_{\mathfrak{s}}^+||}.$$

We present next the process to determine the points $\mathbf{H}_{\mathfrak{s}}^+$ and $\mathbf{H}_{\mathfrak{s}}^-$. Let $D_1^- \in \mathcal{W}_{\sigma,\sigma^*}$ be the most backward neighboring diamond cell of $D_{\sigma,\sigma^*}$ with respect to the direction $(\mathbf{C}_{D_{\sigma,\sigma^*}}\mathbf{M}_{\mathfrak{s}})$ and $\mathbf{C}_{D_1^-}$ its center, in the sense that

$$\cos\left(\mathbf{C}_{D_1^-}\mathbf{C}_{D_{\sigma,\sigma^*}}, \mathbf{C}_{D_{\sigma,\sigma^*}}\mathbf{M}_{\mathfrak{s}}\right) = \max_{D\in\mathcal{W}_{\sigma,\sigma^*}} \cos\left(\mathbf{C}_D\mathbf{C}_{D_{\sigma,\sigma^*}}, \mathbf{C}_{D_{\sigma,\sigma^*}}\mathbf{M}_{\mathfrak{s}}\right). \tag{2}$$

Let now $\mathbf{C}_{D_2^-}$ be the center of the next most backward diamond cell, provided that it is located on the other side of the axis $(\mathbf{C}_{D_{\sigma,\sigma^*}}\mathbf{M}_{\mathfrak{s}})$, that is

$$\cos\left(\mathbf{C}_{D_2^-}\mathbf{C}_{D_{\sigma,\sigma^*}}, \mathbf{C}_{D_{\sigma,\sigma^*}}\mathbf{M}_{\mathfrak{s}}\right) = \max_{D\in\overline{\mathcal{W}}_{\sigma,\sigma^*}} \cos\left(\mathbf{C}_D\mathbf{C}_{D_{\sigma,\sigma^*}}, \mathbf{C}_{D_{\sigma,\sigma^*}}\mathbf{M}_{\mathfrak{s}}\right), \tag{3}$$

where $\overline{\mathcal{W}}_{\sigma,\sigma^*}$ is the set of diamonds $D \in \mathcal{W}_{\sigma,\sigma^*}$ different from $D_1^-$ such that

$$\sin\left(\mathbf{C}_D\mathbf{C}_{D_{\sigma,\sigma^*}}, \mathbf{C}_{D_{\sigma,\sigma^*}}\mathbf{M}_{\mathfrak{s}}\right) \cdot \sin\left(\mathbf{C}_{D_1^-}\mathbf{C}_{D_{\sigma,\sigma^*}}, \mathbf{C}_{D_{\sigma,\sigma^*}}\mathbf{M}_{\mathfrak{s}}\right) \leq 0.$$

Next, we define the point $\mathbf{H}_{\mathfrak{s}}^-$ to be the intersection between the axis $(\mathbf{C}_{D_{\sigma,\sigma^*}}\mathbf{M}_{\mathfrak{s}})$ and the line $(\mathbf{C}_{D_1^-}\mathbf{C}_{D_2^-})$. The point $\mathbf{H}_{\mathfrak{s}}^-$ lies in the segment $(\mathbf{C}_{D_1^-}\mathbf{C}_{D_2^-})$. Therefore, we let $(\alpha_1^-, \alpha_2^-)$ to be the barycentric coordinates of $\mathbf{H}_{\mathfrak{s}}^-$ with respect of $(\mathbf{C}_{D_1^-}\mathbf{C}_{D_2^-})$, that is

$$\alpha_1^- = \frac{||\mathbf{C}_{D_2^-}\mathbf{H}_{\mathfrak{s}}^-||}{||\mathbf{C}_{D_1^-}\mathbf{C}_{D_2^-}||} \geq 0, \quad \alpha_2^- = \frac{||\mathbf{C}_{D_1^-}\mathbf{H}_{\mathfrak{s}}^-||}{||\mathbf{C}_{D_1^-}\mathbf{C}_{D_2^-}||} \geq 0, \quad \alpha_1^- + \alpha_2^- = 1.$$

In a symmetric way, we determine the points $\mathbf{C}_{D_1^+}$ and $\mathbf{C}_{D_2^+}$ for the forward direction (we take the minimum instead of the maximum in (2) and (3)), then $\mathbf{H}_{\mathfrak{s}}^+$ will be the intersection between $(\mathbf{C}_{D_{\sigma,\sigma^*}}\mathbf{M}_{\mathfrak{s}})$ and $(\mathbf{C}_{D_1^+}\mathbf{C}_{D_2^+})$, and $(\alpha_1^+, \alpha_2^+)$ its barycentric coordinates such that

$$\alpha_1^+ = \frac{||\mathbf{C}_{D_2^+}\mathbf{H}_{\mathfrak{s}}^+||}{||\mathbf{C}_{D_1^+}\mathbf{C}_{D_2^+}||} \geq 0, \quad \alpha_2^+ = \frac{||\mathbf{C}_{D_1^+}\mathbf{H}_{\mathfrak{s}}^+||}{||\mathbf{C}_{D_1^+}\mathbf{C}_{D_2^+}||} \geq 0, \quad \alpha_1^+ + \alpha_2^+ = 1.$$

Note that the computation of the coefficients $\alpha_1^{\pm}$ and $\alpha_2^{\pm}$ depends only on the mesh and it is done once for all. Finally we compute the densities at points $\mathbf{H}_{\mathfrak{s}}^-$ and $\mathbf{H}_{\mathfrak{s}}^+$ according to simple weighted means:

$$\rho_{\mathbf{H}_{\mathfrak{s}}^-} = \alpha_1^- \rho_{D_1^-} + \alpha_2^- \rho_{D_2^-}, \quad \rho_{\mathbf{H}_{\mathfrak{s}}^+} = \alpha_1^+ \rho_{D_1^+} + \alpha_2^+ \rho_{D_2^+}.$$

Similarly, we reconstruct the density of internal energy $\rho e$ denoted by $(\rho e)^{ML}_{D_{\sigma,\sigma^*},\mathfrak{s}}$. The reconstructed internal energy $e^{ML}_{D_{\sigma,\sigma^*},\mathfrak{s}}$ is then defined by $(\rho e)^{ML}_{D_{\sigma,\sigma^*},\mathfrak{s}} / \rho^{ML}_{D_{\sigma,\sigma^*},\mathfrak{s}}$.

Finally, the second order mass flux is defined by

$$\mathcal{F}^{ML}_{D_{\sigma,\sigma^*},\mathfrak{s}} = \mathcal{F}^{ML,+}_{D_{\sigma,\sigma^*},\mathfrak{s}} + \mathcal{F}^{ML,-}_{D_{\sigma,\sigma^*},\mathfrak{s}}$$

with

$$F^{ML,+}_{D_{\sigma,\sigma^*},\mathfrak{s}} = \mathcal{F}^{+}(\rho^{ML}_{D_{\sigma,\sigma^*},\mathfrak{s}}, c_{\mathfrak{s}}, u_{D_{\sigma,\sigma^*},\mathfrak{s}}) \text{ and } \mathcal{F}^{ML,-}_{D_{\sigma,\sigma^*},\mathfrak{s}} = \mathcal{F}^{-}(\rho^{ML}_{D_{\sigma',\sigma^{*\prime}},\mathfrak{s}}, c_{\mathfrak{s}}, u_{D_{\sigma,\sigma^*},\mathfrak{s}}),$$

and the internal energy flux is given by

$$\mathcal{E}^{ML}_{D_{\sigma,\sigma^*},\mathfrak{s}} = e^{ML}_{D_{\sigma,\sigma^*},\mathfrak{s}} \mathcal{F}^{ML,+}_{D_{\sigma,\sigma^*},\mathfrak{s}} + e^{ML}_{D_{\sigma',\sigma^{*\prime}},\mathfrak{s}} \mathcal{F}^{ML,-}_{D_{\sigma,\sigma^*},\mathfrak{s}}.$$

## 4 Numerical Test on a Riemann Problem

We illustrate the method with a comparison between the first order scheme and the MUSCL scheme presented above. We take $\gamma = 1.4$. We consider two 1D Riemann problems. The initial data $\rho$, $u$ and $p$ are piecewise constant functions with a discontinuity located at $x_0 = 0.5$. The initial constants at left and right of $x_0 = 0.5$ and the final time $T$ are given in Table 1. In the two cases, the solution consists in a left rarefaction, a contact discontinuity and a right shock. The tests are preformed with the 2D code on the square $[0, 1] \times [0, 0.1]$, with a triangular mesh with approximatively 256 grid points in the direction $x$. Figure 3 presents the results obtained at time $T$. The MUSCL approach does not reach the 2nd order accuracy since the velocity is not reconstructed but we can already see that the numerical diffusion at the contact discontinuity in the density and internal energy profiles is significantly reduced by the MUSCL approximation (the velocity being constant).

**Table 1** Initial data for the Riemann problems

| | $\rho_l$ | $\rho_r$ | $u_l$ | $u_r$ | $p_l$ | $p_r$ | T |
|---|---|---|---|---|---|---|---|
| Test 1 | 1 | 0.125 | 0 | 0 | 1 | 0.1 | 0.25 |
| Test 2 | 1 | 1 | 0 | 0 | 1000 | 0.1 | 0.012 |

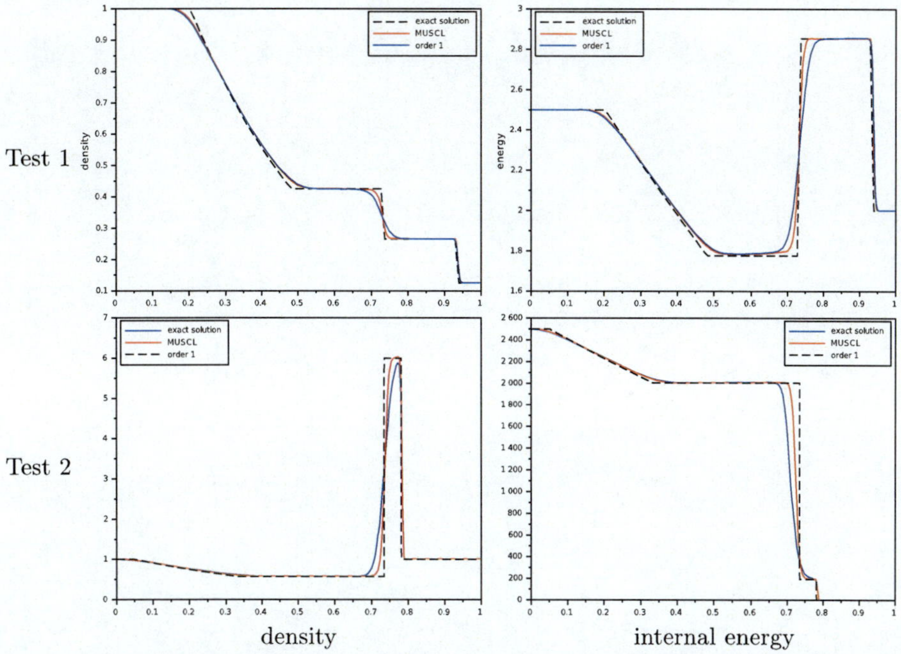


**Fig. 3** Numerical results (cutline along $x$-axis): density and internal energy at time $T$

# 5 Conclusion

We present a MUSCL approach based on the first order "DDFV-like" scheme introduced in [3]. We focus here on the reconstruction of the mass density and the internal energy, inspired from [4]. This approach gives improved results for 1D contact discontinuity problems. To reach 2nd order, it is needed to equally reconstruct the velocity. The strategy, to be detailed elsewhere, is based on similar reconstructions but the point $\mathbf{H}_{\mathfrak{s}}^{\pm}$ can be located on the edge of (primal/dual) cell and the interpolation becomes easier since discrete values are also available at cells vertices (see [1]).

# References

1. Calgaro, C., Creusé, E., Goudon, T.: An hybrid finite volume-finite element method for variable density incompressible flows. J. Comput. Phys. **227**(9), 4671–4696 (2008)
2. Gastaldo, L., Herbin, R., Latché, J.-C., Therme, N.: A muscl-type segregated-explicit staggered scheme for the euler equations. Comput. Fluids **175**, 91–110 (2018)
3. Goudon, T., Llobell, J., Minjeaud, S.: An explicit finite volume scheme on staggered grids for the euler equations: Unstructured meshes, stability analysis, and energy conservation. Int. J. Numer. Methods Fluids **94**(1), 76–119 (2022)
4. Le Touze, C., Murrone, A., Guillard, H.: Multislope muscl method for general unstructured meshes. J. Comput. Phys. **284**, 389–418 (2015)

# GEA: A New Finite Volume-Based Open Source Code for the Numerical Simulation of Atmospheric and Ocean Flows

**Michele Girfoglio, Annalisa Quaini, and Gianluigi Rozza**

**Abstract** We introduce GEA (Geophysical and Environmental Applications), a new open-source atmosphere and ocean modeling framework within the finite volume C++ library OpenFOAM®. Here, we present the development of a non-hydrostatic atmospheric model consisting of a pressure-based solver for the Euler equations written in conservative form using density, momentum, and total energy as variables. We validate the solver for two idealized test cases involving buoyancy driven flows: smooth and non-smooth rising thermal bubble. Through qualitative and quantitative comparisons against numerical data available in the literature, we show that our approach is accurate.

**Keywords** Compressible flow · Low Mach number · Stratified flow · Non-hydrostatic atmospheric flows · Finite volume approximation · Large eddy simulation

## 1 Introduction

Fast and accurate weather/climate forecasts need state-of-the-art numerical and computational methodologies. Open source software packages for weather and climate simulations, i.e. the Climate Machine [1] and WRF [2], are very useful tools for realistic simulations, but testing and assessing new numerical approaches within them is non-trivial. This paper is a follow-up of our work presented in [6] and it is meant to further lay the foundation for a new open source package, called GEA (Geophysical

M. Girfoglio (✉) · G. Rozza
mathLab, Mathematics Area, SISSA, via Bonomea 265, I-34136 Trieste, Italy
e-mail: michele.girfoglio@sissa.it

G. Rozza
e-mail: gianluigi.rozza@sissa.it

A. Quaini
Department of Mathematics, University of Houston, Houston, TX 77204, USA
e-mail: quaini@math.uh.edu

 
E. Franck et al. (eds.), *Finite Volumes for Complex Applications X—Volume 2, Hyperbolic and Related Problems*, Springer Proceedings in Mathematics & Statistics 433,
https://doi.org/10.1007/978-3-031-40860-1_16

and Environmental Applications) [3]. GEA is specifically created for the assessment of new computational approaches for the simulation of mesoscale atmospheric flows and ocean flows [4–6]. To maximize the reach and the impact, as well as to facilitate sharing it with the scientific community, we choose to build our software package on OpenFOAM® [7], an open source, freely available C++ finite volume library that has become widely used in Computational Fluid Dynamics (CFD).

As the core of our open source package, we present a solver for the Euler equations for non-hydrostatic mesoscale atmospheric modeling, and assess it through two well-known test cases involving a smooth and non-smooth rising thermal bubble. We consider the Euler equations written in conservative form using density, momentum, and total energy as variables.

By large, software for weather forecast relies on density-based approaches, which were historically developed for high-speed compressible flows (obviously not the case for atmospheric flows). In this paper, we opt for a pressure-based solver and show that it is able to provide results comparable with data in the literature.

The rest of the paper is organized as follows. In Sect. 2, we briefly describe the formulation of the Euler equations under consideration. Section 3 presents our pressure-based approach and provides the details of space and time discretization. Numerical results for the two benchmark tests are discussed in Sect. 4. Conclusions are drawn in Sect. 5.

## 2 Problem Definition

### 2.1 The Compressible Euler Equations

We consider the dynamics of the dry atmosphere (i.e., no moisture) in a spatial domain of interest $\Omega$ by neglecting the effects of solar radiation and heat flux from the ground. We assume that dry air behaves like an ideal gas.

Let $\rho$ be the air density, $\mathbf{u} = (u, v, w)$ the wind velocity, and $e$ the total energy density. Note that $e = c_v T + |\mathbf{u}|^2/2 + gz$, where $c_v$ the specific heat capacity at constant volume, $T$ is the absolute temperature, $g$ is the gravitational constant, and $z$ is the vertical coordinate. The equations stating conservation of mass, momentum, and energy for the dry atmosphere written in terms of $\rho$, $\mathbf{u}$, and $e$ over a time interval of interest $(0, t_f]$ read:

$$\frac{\partial \rho}{\partial t} + \nabla \cdot (\rho \mathbf{u}) = 0 \qquad \text{in } \Omega \times (0, t_f], \tag{1}$$

$$\frac{\partial (\rho \mathbf{u})}{\partial t} + \nabla \cdot (\rho \mathbf{u} \otimes \mathbf{u}) + \nabla p + \rho g \widehat{\mathbf{k}} = \mathbf{0} \qquad \text{in } \Omega \times (0, t_f], \tag{2}$$

$$\frac{\partial (\rho e)}{\partial t} + \nabla \cdot (\rho \mathbf{u} e) + \nabla \cdot (p \mathbf{u}) = 0 \qquad \text{in } \Omega \times (0, t_f], \tag{3}$$

where $\widehat{\mathbf{k}}$ is the unit vector aligned with the vertical axis $z$ and $p$ is pressure. To close system (1)–(3), we need a thermodynamics equation of state for $p$. Following the assumption that dry air behaves like an ideal gas, we have:

$$p = \rho R T, \tag{4}$$

where $R$ is the specific gas constant of dry air.

Let us write the pressure as the sum of a fluctuation $p'$ with respect to a background state

$$p = p' + \rho g z. \tag{5}$$

By plugging (5) into (2), we obtain:

$$\frac{\partial(\rho \mathbf{u})}{\partial t} + \nabla \cdot (\rho \mathbf{u} \otimes \mathbf{u}) + \nabla p' + g z \nabla \rho = \mathbf{0} \quad \text{in } \Omega \times (0, t_f]. \tag{6}$$

Let $c_p$ be the specific heat capacity at constant pressure for dry air and let

$$K = |\mathbf{u}|^2/2, \quad h = c_v T + p/\rho = c_p T, \tag{7}$$

be the kinetic energy density and the specific enthalpy, respectively. The total energy density can be written as $e = h - p/\rho + K + gz$. Then, Eq. (3) can be rewritten as:

$$\frac{\partial(\rho h)}{\partial t} + \nabla \cdot (\rho \mathbf{u} h) + \frac{\partial(\rho K)}{\partial t} + \nabla \cdot (\rho \mathbf{u} K) - \frac{\partial p}{\partial t} + \rho g \mathbf{u} \cdot \widehat{\mathbf{k}} = 0, \tag{8}$$

where we have used Eq. (1) for further simplification.

**Remark 1** To preserve the numerical stability, we add an artificial diffusion term to Eqs. (6) and (8):

$$\frac{\partial(\rho \mathbf{u})}{\partial t} + \nabla \cdot (\rho \mathbf{u} \otimes \mathbf{u}) + \nabla p' + g z \nabla \rho - \mu_a \Delta \mathbf{u} = \mathbf{0} \tag{9}$$

$$\frac{\partial(\rho h)}{\partial t} + \nabla \cdot (\rho \mathbf{u} h) + \frac{\partial(\rho K)}{\partial t} + \nabla \cdot (\rho \mathbf{u} K) - \frac{\partial p}{\partial t} + \rho g \mathbf{u} \cdot \widehat{\mathbf{k}} - \frac{\mu_a}{Pr} \Delta h = 0, \tag{10}$$

where $\mu_a$ is a constant diffusivity coefficient and $Pr$ is the Prandtl number.

**Remark 2** A quantity of interest for atmospheric problems is the potential temperature $\theta$ defined as

$$\theta = \frac{T}{\pi}, \quad \pi = \left(\frac{p}{p_0}\right)^{\frac{R}{c_p}}, \tag{11}$$

where $p_0 = 10^5$ Pa is the atmospheric pressure at the ground. Additionally, we define the potential temperature fluctuation $\theta'$:

$$\theta'(x, y, z, t) = \theta(x, y, z, t) - \theta_0(z), \tag{12}$$

where $\theta_0$ is the mean hydrostatic value, which is a function of the vertical coordinate $z$ only.

We will devise a splitting approach for problem (4)–(5), (7), (9)–(10), because this formulation of the Euler equation facilitates the decoupling of all variables and it allows for an explicit treatment of the kinetic and potential energies.

## 3 Time and Space Discretization

This section briefly presents a space and time discretization for the model (1), (4)–(5), (7), (9)–(10). For the space discretization, we adopt a finite volume method. This requires to partition the computational domain $\Omega$ into cells or control volumes $\Omega_i$, with $i = 1, \ldots, N_c$, where $N_c$ is the total number of cells in the mesh. Let $\mathbf{A}_j$ be the surface vector of each face of the control volume, with $j = 1, \ldots, M$. With the subindex $i$ we will denote a variable average in control volume $\Omega_i$. Let $\Delta t \in R$, $t^n = n\Delta t$, with $n = 0, \ldots, N_f$ and $t_f = N_f \Delta t$. Moreover, we denote by $y^n$ the approximation of a generic quantity $y$ at the time $t^n$. For time discretization, we adopt a Backward Differentiation Formula of order 1 (BDF1). A monolithic approach for coupled problem (4)–(5), (7), (9)–(10) would lead to high computational costs. Thus, to save computational time we adopt a splitting approach consisting of three steps detailed below.

Problem (4)–(5), (7), (9)–(10) discretized in time and space reads: given $\rho^0, \mathbf{u}^0, h^0$, $p^0$, and $T^0$, set $K^0 = |\mathbf{u}^0|^2/2$ and for $n \geq 0$ find solution ($\rho_i^{n+1}$, $\mathbf{u}_i^{n+1}$, $h_i^{n+1}$, $K_i^{n+1}$, $p_i^{n+1}$, $p_i^{\prime,n+1}$, $T_i^{n+1}$) of system:

- *Step 1*: find first intermediate density $\rho_i^{n+\frac{1}{3}}$, intermediate velocity $\mathbf{u}_i^{n+\frac{1}{3}}$ and associated kinetic energy density $K_i^{n+\frac{1}{3}}$ such that

$$\frac{1}{\Delta t}\rho_i^{n+\frac{1}{3}} + \sum_j \varphi_j^n = b_{\rho,i}^n, \quad \varphi_j^n = (\rho^n \mathbf{u}^n)_{i,j} \cdot \mathbf{A}_j, \tag{13}$$

$$\frac{1}{\Delta t}\rho_i^{n+\frac{1}{3}} \mathbf{u}_i^{n+\frac{1}{3}} + \sum_j \varphi_j^n \mathbf{u}_{i,j}^{n+\frac{1}{3}} + \nabla p_i^n - \sum_j \mu \nabla (\mathbf{u}_i^{n+\frac{1}{3}})_j \cdot \mathbf{A}_j = \boldsymbol{b}_{\mathbf{u},i}^n, \tag{14}$$

$$K_i^{n+\frac{1}{3}} = \frac{|\mathbf{u}_i^{n+\frac{1}{3}}|^2}{2}, \tag{15}$$

where $\varphi_j^n$ denotes the convective flux through face $j$ of $\Omega_i$, $b_{\rho,i}^n = \rho_i^n/\Delta t$ and $\mathbf{b}_{\mathbf{u},i}^n = \rho_i^n \mathbf{u}_i^n / \Delta t$.

- *Step 2*: find average specific enthalpy $h_i^{n+1}$, temperature $T_i^{n+1}$, and second intermediate density $\rho_i^{n+\frac{2}{3}}$ in control volume $\Omega_i$ such that

$$
\begin{aligned}
&\frac{1}{\Delta t}\rho_i^{n+\frac{1}{3}} h_i^{n+1} + \sum_j \varphi_j^n h_{i,j}^{n+1} - \sum_j \frac{\mu}{Pr}(\nabla h_i^{n+1})_j \cdot \mathbf{A}_j = \tilde{b}_{e,i}^n \\
&\quad - \frac{1}{\Delta t}\rho_i^{n+\frac{1}{3}} K_i^{n+\frac{1}{3}} - \sum_j \varphi_j^n K_{i,j}^{n+\frac{1}{3}} + \frac{1}{\Delta t} p_i^n - \rho_i^{n+\frac{1}{3}} g \mathbf{u}_i^{n+\frac{1}{3}} \cdot \widehat{\mathbf{k}},
\end{aligned} \tag{16}
$$

$$
h_i^{n+1} - c_p T_i^{n+1} = h_i^n - c_p T_i^n, \tag{17}
$$

$$
\rho_i^{n+\frac{2}{3}} R T_i^{n+1} = p_i^n, \tag{18}
$$

where $\tilde{b}_e^n = (\rho^n h^n + \rho^n K^{n-1} - p^{n-1})/\Delta t$.

- *Step 3*: find end-of-step velocity $\mathbf{u}^{n+1}$ and associated kinetic energy density $K^{n+1}$, pressure $p^{n+1}$ and pressure fluctuation $p'^{,n+1}$, and end-of-step density $\rho^{n+1}$ such that

$$
\begin{aligned}
&\frac{1}{\Delta t}\rho_i^{n+\frac{1}{3}} \mathbf{u}_i^{n+1} + \sum_j \varphi_j^n \mathbf{u}_{i,j}^{n+1} + \nabla p_i'^{,n+1} + g z_i \nabla \rho_i^{n+\frac{2}{3}} \\
&\quad - \sum_j \mu (\nabla \mathbf{u}_i^{n+1})_j \cdot \mathbf{A}_j = \boldsymbol{b}_{\mathbf{u},i}^{n+1},
\end{aligned} \tag{19}
$$

$$
\begin{aligned}
&\sum_j \rho_j^{n+\frac{2}{3}} (\nabla p_i'^{,n+1})_j \cdot \mathbf{A}_j = \sum_j \frac{\rho_j^{n+\frac{2}{3}} \Delta t}{\rho_j^{n+\frac{1}{3}}} \left( \mathbf{H}(\mathbf{u}_i^{n+1})_j - g z_j (\nabla \rho_i^{n+\frac{2}{3}})_j \right) \cdot \mathbf{A}_j \\
&\quad - b_{\rho,i}^{n+1} + \frac{1}{\Delta t} \rho_i^{n+\frac{2}{3}},
\end{aligned} \tag{20}
$$

$$
p_i^{n+1} = p_i'^{,n+1} + \rho_i^{n+\frac{2}{3}} g z_i, \quad K_i^{n+1} = \frac{|\mathbf{u}_i^{n+1}|^2}{2}, \quad \rho_i^{n+1} = \frac{p_i^{n+1}}{R T_i^{n+1}}, \tag{21}
$$

where $\mathbf{H}(\mathbf{u}_i^{n+1}) = -\sum_j \varphi_j^n \mathbf{u}_{i,j}^{n+1} + \sum_j \mu (\nabla \mathbf{u}_i^{n+1})_j \cdot \mathbf{A}_j + \boldsymbol{b}_{\mathbf{u},i}^{n+1}$.

In order to decouple the computation of the pressure from the computation of the velocity, we use the PISO algorithm [7]. We choose a second-order accurate scheme for the Laplacian and gradient terms and a fourth-order accurate scheme for the divergence term. For more details, we refer the reader to [6].

## 4 Numerical Results

We validate our solver with respect to two classic benchmarks, the smooth and non-smooth rising thermal bubble. Both test cases involve a perturbation of a neutrally stratified atmosphere with uniform background potential temperature over a flat terrain. In both tests, the computational domain in the $xz$-plane is $\Omega = [0, 1000]^2$ m$^2$ and the time interval of interest is $(0, 600]$ s. Impenetrable, free-slip boundary conditions are imposed on all walls.

Since these benchmarks do not have an exact solution, one can only have a comparison with other numerical data available in the literature.

## 4.1 Smooth Rising Thermal Bubble

The initial density is given by

$$\rho^0 = \frac{p_g}{R\theta_0}\left(\frac{p}{p_g}\right)^{c_v/c_p}, \quad p = p_g\left(1 - \frac{gz}{c_p\theta^0}\right)^{c_p/R}, \tag{22}$$

with $c_p = R + c_v$, $c_v = 715.5$ J/(Kg K), $R = 287$ J/(Kg K). In (22), $\theta^0$ is the initial potential temperature, which is defined as:

$$\theta^0 = 300 + \frac{0.5}{2}\left[1 + \cos\left(\frac{\pi r}{r_c}\right)\right] \text{ if } r \le r_c, \quad \theta^0 = 300 \text{ otherwise}, \tag{23}$$

where $r = \sqrt{(x - x_c)^2 + (z - z_c)^2}$, with $r_c = 250$ m and $(x_c, z_c) = (500, 350)$ m. The initial velocity field is zero everywhere and the initial specific enthalpy is given by

$$h^0 = c_p\theta^0\left(\frac{p}{p_g}\right)^{\frac{R}{c_p}}. \tag{24}$$

Following [8], we use a mesh with uniform resolution $h = \Delta x = \Delta z = 5$ m. The time step is set to $\Delta t = 0.1$ s which allows to obtain $CFL_{max} \approx 0.1$. Furthermore, we set $\mu = 0.15$ and $Pr = 1$.

Figure 1 (left) shows the potential temperature perturbation computed at $t = 600$ s. This plot is in very good agreement with the corresponding figure in [8]. Figure 1 (right) depicts the profile of the potential temperature perturbation along $z = 700$ m together with the data from [8]. The reference curve has been extracted manually from the corresponding figure in [8], hence it is a piecewise reconstruction. We observe that the two curves are very close, except for the oscillations at $t \approx 200$ s and $t \approx 800$ s that in our solution are completely damped and the slightly lower maxima.

For further comparison, in Table 1 we report the extrema for the horizontal velocity $u$ and vertical velocity $w$, together with the values obtained in [8, 10]. The results from [8, 10] are obtained using a density-based approach developed from a Godunov-type scheme. Moreover, the authors use spectral elements [10] and Discontinuous Galerkin methods [8] for the space discretization. Notice that despite the fact that they use the same approach a different space discretization method leads to a difference up to 9% in the values from [8, 10]. Other differences with our methodology include the orders of space and time discretizations, the adoption of implicit or explicit time

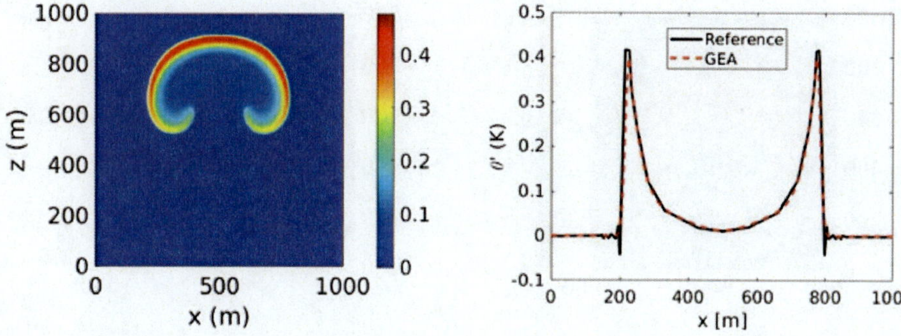


**Fig. 1** Smooth rising thermal bubble: perturbation of potential temperature computed at $t = 600$ s (left) and its profile at $z = 700$ m compared with data from [8] denoted as "Reference" (right)

**Table 1** Smooth rising thermal bubble: minimum and maximum horizontal velocity $u$ and vertical velocity $w$ at $t = 600$ s compared against results reported in [8, 10]

| Model | Res. (m) | $u_{min}$ (m/s) | $u_{max}$ (m/s) | $w_{min}$ (m/s) | $w_{max}$ (m/s) |
|---|---|---|---|---|---|
| GEA | 5 | −1.898 | 1.898 | 1.682 | 2.495 |
| Ref. [8] | 5 | −2.161 | 2.161 | 1.967 | 2.758 |
| Ref. [10] | 5 | −2.073 | 2.073 | 1.911 | 2.536 |

stepping techniques and the different treatment of the hydrostatic term. Given all these differences, we believe that our results are in line with the reference ones.

## 4.2 *Non-smooth Rising Thermal Bubble*

The second test case is analogous to the first one, except for a uniform thermal anomaly of 0.5 K:

$$\theta^0 = 300.5 \text{ if } r \leq r_c, \quad \theta^0 = 300 \text{ otherwise,} \tag{25}$$

and $(x_c, z_c) = (500, 260)$ m. The time step is set to $\Delta t = 0.1$, which gives $CFL_{max} \approx 0.1$. Furthermore, we set $\mu = 0.3$ and $Pr = 1$.

Figure 2 shows the spatial distribution of the potential temperature perturbation at $t = 420$ s and $t = 600$ s by using a mesh with uniform resolution $h = \Delta x = \Delta z = 5$ m. Qualitatively, these results are in very good agreement with those reported in [9]. For a more quantitative analysis, in Table 2 we report the extrema for $\theta'$ at $t = 420, 600$ s together with the values from [9]. We observe that the comparison is satisfactory considering that the method we use is very different from the approach in [9].

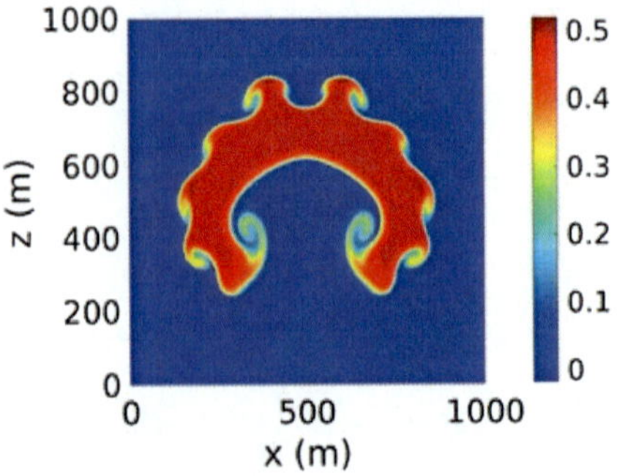


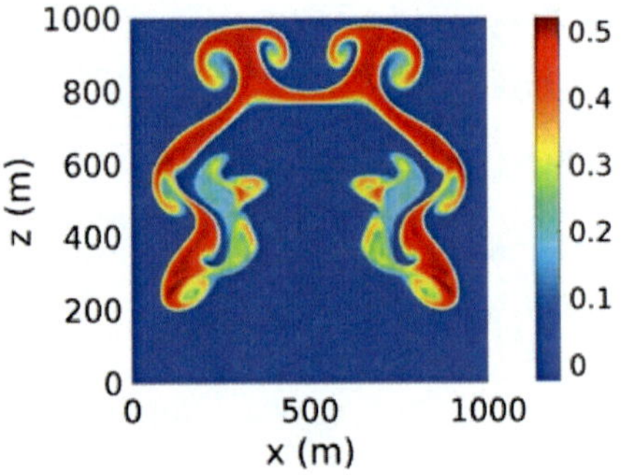


**Fig. 2** Non-smooth rising thermal bubble: perturbation of potential temperature computed at $t = 420$ s (left) and $t = 600$ s (right)

**Table 2** Non-smooth rising thermal bubble: minimum and maximum potential temperature perturbation $\theta'$ at $t = 420$ s and $t = 600$ s compared against results reported in [9]

| Model | Res. (m) | $\theta'_{min}$ (K) at 420 s | $\theta'_{min}$ (K) at 600 s | $\theta'_{max}$ (K) at 420 s | $\theta'_{max}$ (K) at 600 s |
|---|---|---|---|---|---|
| GEA | 5 | −0.018 | −0.023 | 0.521 | 0.522 |
| Ref. [9] | 5 | −0.020 | −0.016 | 0.522 | 0.527 |
| GEA | 10 | −0.047 | −0.062 | 0.577 | 0.563 |
| Ref. [9] | 10 | −0.057 | −0.053 | 0.550 | 0.561 |

## 5 Concluding Remarks

This paper discusses and further assessed GEA (Geophysical and Environmental Applications), a new C++ language simulation framework designed for the numerical simulation of atmospheric and ocean flows. We developed a pressure-based solver for the Euler equations written in conservative form using density, momentum, and total energy as variables. We validated the solver against numerical data available in the literature for two well-known benchmarks: the smooth and non-smooth thermal bubble. For both tests, we obtain good qualitative and quantitative comparisons. The code created for this paper is available on GitHub [3].

## References

1. Climate Modeling Alliance. https://github.com/CliMA
2. Weather Research and Forecasting. https://www.mmm.ucar.edu/models/wrf
3. GEA - Geophysical and Environmental Applications. https://github.com/GEA-Geophysical-and-Environmental-Apps/GEA
4. Girfoglio, M., Quaini, A., Rozza, G.: A novel large Eddy simulation model for the quasi-geostrophic equations in a finite volume setting. J. Comput. Appl. Math. **418**, 114656 (2023)

5. Girfoglio, M., Quaini, A., Rozza, G.: A linear filter regularization for POD-based reduced order models of the quasi-geostrophic equations. Comptes Rendus Mècanique, 1–21 (2023)
6. Girfoglio, M., Quaini, A., Rozza, G.: Validation of an OpenFOAM-based solver for the Euler equations with benchmarks for mesoscale atmospheric modeling. AIP Advances **13**, 055024 (2023)
7. Weller, H.G., Tabor, G., Jasak, H., Fureby, C.: A tensorial approach to computational continuum mechanics using object-oriented techniques. Comput. Phys. **12**, 620–631 (1998)
8. Restelli, M., Giraldo, F.X.: A conservative discontinuous Galerkin semi-implicit formulation for the Navier-Stokes equations in nonhydrostatic mesoscale modeling. SIAM J. Sci. Comput. **31** (2009)
9. Restelli, M.: Semi-Lagrangian and semi-implicit discontinuous Galerkin methods for atmospheric modeling applications. Ph.D. Thesis, Politecnico di Milano (2007)
10. Giraldo, F.X., Restelli, M.: A study of spectral element and discontinuous Galerkin methods for the Navier Stokes equations in nonhydrostatic mesoscale atmospheric modeling: equation sets and test cases. J. Comput. Phys. **227** (2008)

# Stable Second Order Boundary Conditions for Kinetic Approximations

**Romane Hélie and Philippe Helluy**

**Abstract** We introduce a general framework for applying boundary conditions in the vectorial kinetic scheme of Bouchut [2, 4]. The method ensures entropy stability of the resulting approximation but also high order accuracy.

**Keywords** Boundary conditions · Lattice boltzmann · Entropy stability

## 1 Introduction

The vectorial kinetic approach is a general method for approximating hyperbolic systems of conservation laws. It has been introduced by Bouchut [4] and Aregba and Natalini [2]. It consists in representing any system of conservation laws by a finite set of transport equations coupled through a stiff relaxation term. This representation has several advantages: it allows rigorous mathematical analysis [2, 4, 7]. Associated with time splitting techniques, it leads to the construction of very efficient schemes [3].

We first describe the one-dimensional framework, but it can be generalized to higher dimensions. We consider a system of $m$ conservation laws

$$\partial_t w + \partial_x q(w) = 0, \tag{1}$$

where the unknown conservative variable $w(x, t) \in \mathbb{R}^m$ depends on a space variable $x \in [L, R]$ and a real time variable $t \geq 0$. The system (1) is supplemented with an initial condition and boundary conditions at $x = L$ and $x = R$.

R. Hélie · P. Helluy (✉)
IRMA Université de Strasbourg and Inria Tonus, Strasbourg, France
e-mail: philippe.helluy@unistra.fr

R. Hélie
e-mail: romane.helie@math.unistra.fr

 
E. Franck et al. (eds.), *Finite Volumes for Complex Applications X—Volume 2, Hyperbolic and Related Problems*, Springer Proceedings in Mathematics & Statistics 433,
https://doi.org/10.1007/978-3-031-40860-1_17

We assume that (1) admits a Lax entropy $s : \mathbb{R}^m \to \mathbb{R}$, associated to an entropy flux $g : \mathbb{R}^m \to \mathbb{R}$. This means that $s$ is strictly convex and that $s'q' = g'$. A weak solution $w$ to (1) is admissible if the entropy is decreasing

$$\partial_t s(w) + \partial_x g(w) \leq 0.$$

## 2 Kinetic Approximation

The system of conservation laws is approximated by the following system of $2m$ kinetic equations

$$\partial_t f + \Lambda \partial_x f = \frac{1}{\varepsilon}(f^{eq}(w) - f), \tag{2}$$

where:

- the kinetic vector

$$f(x,t) = \begin{pmatrix} f_1(x,t) \\ f_2(x,t) \end{pmatrix} \in R^{2m}, \quad f_{1,2}(x,t) \in R^m,$$

- the matrix $\Lambda$ is defined by ($I$ is the identity matrix of size $m \times m$)

$$\lambda = \begin{pmatrix} \lambda I & 0 \\ 0 & -\lambda I \end{pmatrix}, \quad \lambda > 0,$$

- the parameter $\varepsilon > 0$. The equilibrium $f^{eq}$ is defined below,
- the conservative data are related to the kinetic data thanks to the $m \times 2m$ matrix

$$P = (I, I), \quad \text{and } w = Pf.$$

This kinetic model is consistent with the conservation laws (1) when $\varepsilon \to 0$ as soon as

$$Pf^{eq}(w) = w, \quad P\Lambda f^{eq}(w) = q(w). \tag{3}$$

The consistency conditions (3) lead to a unique possibility for the equilibrium data

$$f^{eq}(w) = \begin{pmatrix} \frac{w}{2} + \frac{q(w)}{2\lambda} \\ \frac{w}{2} - \frac{q(w)}{2\lambda} \end{pmatrix}.$$

For the numerical resolution of (2) we use a Lattice Boltzmann Method (LBM). For a given integer $N > 0$, we define the space step $h = (R - L)/N$, the grid points $x_i = L + ih + h/2$, $0 \leq i < N$ and the time step $\tau = h/\lambda$. The solution is approximated by

$$f_i^n \simeq f(x_i, n\tau).$$

For going from time step $n-1$ to time step $n$, we first solve the free transport equation $\partial_t f + \Lambda \partial_x f = 0$ exactly, which is possible by simple shift operations, thanks to the choice of the time step. We get

$$f_i^{n,-} = \begin{pmatrix} f_{1,i-1}^{n-1} \\ f_{2,i+1}^{n-1} \end{pmatrix}, \quad w_i^n = P f_i^{n,-}. \tag{4}$$

Then we apply a relaxation procedure for staying close to equilibrium

$$f_i^n = \omega f^{eq}(w_i^n) + (1-\omega) f_i^{n,-}. \tag{5}$$

The choice $\omega = 1$ consists in returning to equilibrium at the end of each step. The resulting time marching scheme is first order accurate. The over-relaxation choice $\omega = 2$ results in a second order accurate time scheme [5]. We observe that the LBM is particularly simple and efficient: it is made of independent shift operations followed by independent local relaxations. It is highly parallelizable and the second order extension has no additional cost.

It is clear that the shift algorithm (4) needs to be adapted at the boundaries, when $i = 0$ or $i = N-1$. At the left point, the value $f_{1,-1}^{n-1}$ is missing, and at the right point, the value $f_{2,N}^{n-1}$ is missing. The boundary conditions are given by two "ghost cell" functions $b_L$ and $b_R$ whose role is to reconstruct the missing values from the first and last cell values:

$$f_{1,-1}^{n-1} = b_L(f_{1,0}^{n-1}, f_{2,0}^{n-1}), \quad f_{2,N}^{n-1} = b_R(f_{1,N-1}^{n-1}, f_{2,N-1}^{n-1}).$$

We propose below a general strategy to apply the boundary conditions on the kinetic scheme. We shall test it on the very simple transport conservation law at constant velocity $v$. In this case, we thus have $m = 1$ and

$$q(w) = vw. \tag{6}$$

Let us explicit the kinetic model in this case. We find

$$P = (1, 1), \quad w = f_1 + f_2, \quad f^{eq}(w) = \begin{pmatrix} w(\frac{1}{2} + \frac{v}{2\lambda}) \\ w(\frac{1}{2} - \frac{v}{2\lambda}) \end{pmatrix}.$$

The chosen entropy and entropy flux are

$$s(w) = \frac{w^2}{2}, \quad g(w) = v\frac{w^2}{2}.$$

# 3 Entropy Stability and Boundary Conditions

## *3.1 Stability Analysis*

For constructing the kinetic model we follow [4, 7]. We assume that it is possible to find convex kinetic entropies $s_k$ such that

$$s(w) = \min_{w=f_1+f_2} s_1(f_1) + s_2(f_2) = s_1(f_1^{eq}) + s_2(f_2^{eq}). \tag{7}$$

For the simple transport Eq. (6) that we consider, we can take

$$s_1(f_1) = \frac{\lambda}{\lambda + v}(f_1)^2, \quad s_2(f_2) = \frac{\lambda}{\lambda - v}(f_2)^2.$$

These two kinetic entropies are convex under the sub-characteristic condition

$$|v| < \lambda.$$

For a general system of conservation laws with a Lax entropy, it is also possible to compute $s_k$. The practical calculations can be done thanks to the Legendre transform [7, 8].

Once the consistency (3) and the kinetic entropy property (7) are ensured, the lattice Boltzmann scheme (4) and (5) is entropy dissipative for $\omega \in [1, 2]$. Indeed, the shift step (4) preserves the kinetic entropy exactly and the relaxation step (5) makes it decrease, by design. This reasoning has to be adapted at the boundaries.

For establishing the entropy dissipative kinetic boundary condition, we assume that the initial condition is constant outside an interval $[a, b] \subset ]L, R[$

$$w(x, 0) = \overline{w}, \quad \text{if } x < a \text{ or } x > b.$$

Without loss of generality, we can assume that $s(\overline{w}) = 0$ and $g(\overline{w}) = 0$, because the properties of the entropy are not modified by the addition of affine functions. Then the boundary entropy condition simply states that, in the shift step (4), the kinetic entropy that enters the domain should be smaller than the kinetic entropy that leaves the domain. We obtain

$$s_1(b_L(f_1, f_2)) \le s_2(f_2), \quad s_2(b_R(f_1, f_2)) \le s_1(f_1). \tag{8}$$

## *3.2 First Order Stable Boundary Conditions*

We consider the simple transport case $q(w) = vw$. We assume that $\overline{w} = 0$ and $v > 0$. It is then mandatory to impose $w = 0$ at the left boundary. A condition is missing at the right boundary. As in [6] we apply a null boundary condition on the flux error $y$, defined by

$$y = \lambda f_1 - \lambda f_2 - q(f_1 + f_2) = (\lambda - v) f_1 - (\lambda + v) f_2.$$

This quantity vanishes when $f_k = f_k^{eq}$. The boundary conditions are thus

$$f_1 + f_2 = 0 \text{ at } x = L, \quad \lambda f_1 - \lambda f_2 - q(f_1 + f_2) = 0 \text{ at } x = R.$$

This allows to deducing the ghost cell functions $b_L$ and $b_R$

$$b_L(f_1, f_2) = -f_2, \quad b_R(f_1, f_2) = \frac{\lambda - v}{\lambda + v} f_1. \tag{9}$$

The kinetic entropy condition (8) is then satisfied, because $v > 0$.

The boundary conditions (9) are thus stable. However, we have observed that the right outgoing condition is only first order accurate, even when $\omega = 2$.

### 3.3 *Stabilized Second Order Boundary Conditions*

As proposed in [6] we replace the right Dirichlet boundary condition by a Neumann boundary condition, because it was experimentally observed a better second order accuracy with $w(L, t) = 0, \quad \partial_x y(R, t) = 0$. The spatial derivative is approximated by

$$\partial_x y(R, n\tau) \simeq \frac{y_{N-1}^n - y_{N-2}^n}{h} = 0.$$

But with this approximation, the right ghost cell boundary function now depends on the kinetic data in cell $N - 1$ and cell $N - 2$:

$$f_{2,N}^{n-1} = b_R(f_{1,N-1}^{n-1}, f_{2,N-1}^{n-1}, f_{1,N-2}^{n-1}, f_{2,N-2}^{n-1}). \tag{10}$$

It is then generally not possible to ensure directly (8). We thus propose the following nonlinear correction at the right boundary:

- Compute $f_{2,N}^{n-1}$ with formula (10).
- If the right entropy inequality is not satisfied, i.e. if $s_2(f_{2,N}^{n-1}) > s_1(f_{1,N-1}^{n-1})$, then replace $f_{2,N}^{n-1}$ by the closest value $\tilde{f}_{2,N}^{n-1}$ such that

$$s_2(\tilde{f}_{2,N}^{n-1}) = s_1(f_{1,N-1}^{n-1}).$$

This construction ensures the decrease of entropy, with a minimal adjustment. It is expected that it will remain a second order accurate approximation. Similar ideas are exposed for instance in [1] (see also included references) in the FV framework.

# 4 Extension to Higher Dimension

The above construction can be extended to higher dimensions. Let us comment a two-dimensional example. We solve the transport equation

$$\partial_t w + \nabla \cdot q(w) = 0, \quad q_i(w) = v_i w.$$

The unknown $w(x_1, x_2, t)$ depends on a two-dimensional variable $(x_1, x_2)$ in a square $\Omega \subset \mathbb{R}^2$. The velocity vector $v = (v_1, v_2)^T$. In this case, we can consider the D2Q4 model with four kinetic unknowns $f_k$, $1 \leq k \leq 4$. The kinetic velocities are ($\lambda$ is a positive parameter)

$$c_1 = \lambda \begin{pmatrix} 1 \\ 0 \end{pmatrix}, \quad c_2 = \lambda \begin{pmatrix} -1 \\ 0 \end{pmatrix}, \quad c_3 = \lambda \begin{pmatrix} 0 \\ 1 \end{pmatrix}, \quad c_4 = \lambda \begin{pmatrix} 0 \\ -1 \end{pmatrix}.$$

The kinetic equations read

$$\partial_t f_k + c_k \cdot \nabla f_k = \frac{1}{\varepsilon}(f_k^{eq} - f_k).$$

The equilibrium is given by

$$f_k^{eq}(w) = \frac{w}{4} + \frac{v \cdot c_k w}{2\lambda^2}, \quad w = \sum_{k=1}^{4} f_k.$$

We also define the flux error by

$$y = \begin{pmatrix} \lambda(f_1 - f_2) - v_1(f_1 + f_2) \\ \lambda(f_3 - f_4) - v_2(f_3 + f_4) \\ \lambda^2(f_1 + f_2 - f_3 - f_4) \end{pmatrix}.$$

As in the D1Q2 model, the flux error vanishes when $f = f^{eq}$. The kinetic entropy is given by

$$\Sigma(f) = \frac{1}{\lambda + 2v_1} f_1^2 + \frac{1}{\lambda - 2v_1} f_2^2 + \frac{1}{\lambda + 2v_2} f_3^2 + \frac{1}{\lambda - 2v_2} f_4^2.$$

Its convexity implies the sub-characteristic condition $2 \max(|v_1|, |v_2|) < \lambda$.

**Table 1** Two different boundary condition strategies for the D2Q4 model. Left: It is possible to prove that this boundary condition is entropy dissipative, but first order accurate. Right: second order boundary condition, but not stable for long time simulations. By projecting this second-order boundary condition on the space of the decreasing entropy boundary conditions, we obtain stability and second-order accuracy

| Boundary conditions | Entropy stable | Second order accurate |
|---|---|---|
| Inflow border | Exact solution on $w$<br>$y_3 = 0$ | Exact solution on $w$ |
| Outflow border | $y_1 n_1 + y_2 n_2 = 0$<br>$y_3 = 0$ | Neumann on $v_1 y_1 + v_2 y_2$ |
| Corner inflow/inflow | Exact solution on $w$<br>$y_3 = 0$ | Exact solution on $w$<br>$y_3 = 0$ |
| Corner inflow/outflow | Exact solution on $w$<br>$v_1 y_1 + v_2 y_2 = 0$<br>$y_3 = 0$ | Exact solution on $w$<br>Neumann on $v_1 y_1 + v_2 y_2$ |
| Corner outflow/outflow | $y_1 = 0$<br>$y_2 = 0$<br>$y_3 = 0$ | Neumann on $v_1 y_1 + v_2 y_2$<br>$y_3 = 0$ |

We summarize two boundary condition strategies in Table 1 ($(n_1, n_2)$ is the normal vector to the considered boundary).

# 5 Numerical Results

## *5.1 D1Q2 Model*

For our tests, the initial data is the compact support function

$$w(x, 0) = \begin{cases} 0 & \text{if } r(x) > 1, \\ (1 - r(x)^2)^5 & \text{otherwise,} \end{cases}$$

where $r(x) = \frac{|x - x_0|}{\sigma}$, with $\sigma = 0.2$. We consider several test cases corresponding to different initial positions $x_0$ of the peak: in test case 1, $x_0 = -1/2$; in test case 2, $x_0 = 0$ and in test case 3, $x_0 = 1/2$. In Fig. 1 we check the accuracy and the stability of the second order boundary condition.

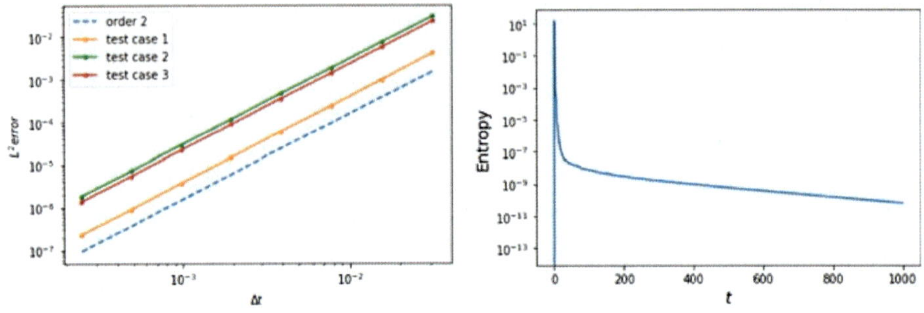


**Fig. 1** Second order boundary condition for the D1Q2 model (Dirichlet on $w$ at the left boundary and Neumann on $y$ at the right boundary). Left: error rates in the $L^2$ norm for the three test cases. Second order is achieved for all the tests. Right: entropy evolution for test case 1. Once the peak has entered the domain, the entropy decreases

## 5.2 *D2Q4 Model*

We choose a square geometry, aligned with the kinetic velocities: $\Omega = [0, 1] \times [0, 1]$. We initialize $w$ with a compact support function

$$w(x_1, x_2, t) = \begin{cases} 0 & \text{if } r(x_1, x_2) > 1, \\ (1 - r(x_1, x_2)^2)^5 & \text{otherwise.} \end{cases}$$

with $r(x_1, x_2) = \frac{\sqrt{(x_1 - x_{1,0})^2 + (x_2 - y_{2,0})^2}}{\sigma}$ and $\sigma = 0.4$. We consider several test cases in Table 2, corresponding to different initial positions of the peak and several velocities $(v_1, v_2)$.

We first check in Fig. 2 the second order accuracy and the instability of the boundary condition described in the second column of Table 1. Finally, we check in Fig. 2 the improvement given by the projection strategy (Fig. 3).

**Table 2** Parameters of the test cases for the $D2Q4$ model

| Test case | $x_{1,0}$ | $x_{2,0}$ | $v_1$ | $v_2$ |
|---|---|---|---|---|
| 1 | 0.5 | 0.5 | 0.8 | 0 |
| 2 | −0.4 | 0.5 | 0.8 | 0 |
| 3 | 0.4 | 0.5 | −0.8 | 0 |
| 4 | 0 | 0.5 | 0.8 | 0 |
| 5 | $-\sqrt{2}/4$ | $-\sqrt{2}/4$ | $\sqrt{2}/2$ | $\sqrt{2}/2$ |
| 6 | $\sqrt{2}/4$ | $\sqrt{2}/4$ | $-\sqrt{2}/2$ | $-\sqrt{2}/2$ |
| 7 | 0 | 0 | $\sqrt{2}/2$ | $\sqrt{2}/2$ |

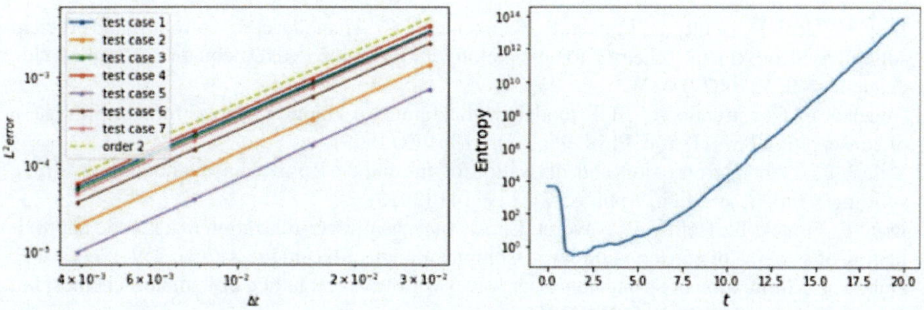


**Fig. 2** Second order boundary condition for the D2Q4 model, without entropy limiting. Left: error rates in the $L^2$ norm for the test cases of Table 2. Second order is achieved for all the tests for short times. Right: entropy evolution for test case 1. The scheme is unstable on the long time

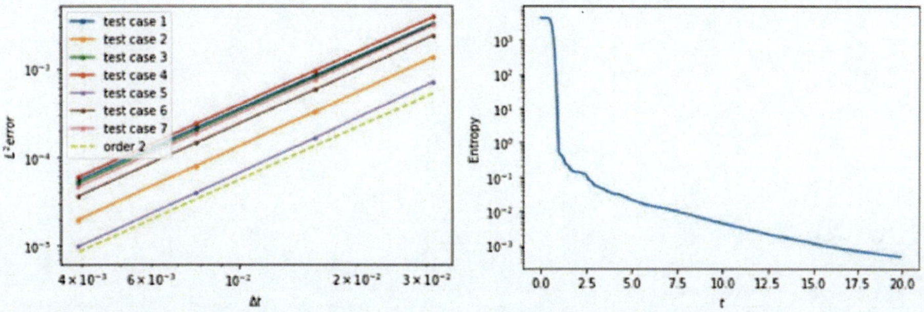


**Fig. 3** Second order boundary condition for the D2Q4 model, with entropy limiting. Left: error rates in the $L^2$ norm for the test cases of Table 2. Second order is achieved for all the tests. Right: entropy evolution for test case 1. The scheme is now stable thanks to the entropy limiting procedure

# 6 Conclusion

In this short paper, we have described a general strategy for imposing boundary conditions on a vectorial kinetic approximation. The difficulty is that the number of necessary boundary conditions is different between the kinetic model and the approximated conservation laws. With our method, it is possible to rigorously stabilize boundary schemes that are accurate but slightly unstable. In future works, we will extend the approach to nonlinear systems of conservation laws and more complex boundary geometries.

# References

1. Aregba-Driollet, D., Milišić, V.: Kinetic approximation of a boundary value problem for conservation laws. Numerische Mathematik **97**, 595–633 (2004)
2. Aregba-Driollet, D., Natalini, R.: Discrete kinetic schemes for multidimensional systems of conservation laws. SIAM J. Numer. Anal. **37**(6), 1973–2004 (2000)

3. Baty, H., Drui, F., Helluy, P., Franck, E., Klingenberg, C., Thanhäuser, L.: A robust and efficient solver based on kinetic schemes for magnetohydrodynamics (MHD) equations. Appl. Math. Comput. **440**, 127667 (2023)
4. Bouchut, F.: Construction of BGK models with a family of kinetic entropies for a given system of conservation laws. J. Stat. Phys. **95**(1–2), 113–170 (1999)
5. Dellar, P.J.: An interpretation and derivation of the lattice Boltzmann method using Strang splitting. Comput. & Math. Appl. **65**(2), 129–141 (2013)
6. Drui, F., Franck, E., Helluy, P., Navoret, L.: An analysis of over-relaxation in a kinetic approximation of systems of conservation laws. Comptes Rendus Mécanique **347**(3), 259–269 (2019)
7. Dubois, F.: Simulation of strong nonlinear waves with vectorial lattice Boltzmann schemes. Int. J. Mod. Phys. C. **25**(12), 1441014 (2014)
8. Guillon, K., Hélie, R., Helluy, P.: Stability analysis of the vectorial Lattice-Boltzmann Method (2023). HAL preprint hal-03986533

# Two-Dimensional Linear Implicit Relaxed Scheme for Hyperbolic Conservation Laws

**Angelo Iollo, Gabriella Puppo, and Andrea Thomann**

**Abstract** We present a two-dimensional extension to the linear implicit all-speed finite volume scheme for hyperbolic conservation laws based on Jin-Xin relaxation recently forwarded in [6]. It is based on stiffly accurate SDIRK methods in time and a convex combination of Rusanov and centered fluxes in space making it asymptotically consistent in the low Mach number regime and allows an accurate capturing of material waves under large time steps. The scheme is numerically tested on the Euler equations and a non-linear model for elasticity in the compressible and low Mach number regime.

**Keywords** All-speed scheme · Relaxation method · Eulerian elasticity · Linearly implicit schemes

## 1 Introduction

The importance of implicit schemes in the numerical integration of evolutionary equations typically lies in the possibility of circumventing the limitations imposed by the amplification of numerical instabilities from which explicit schemes suffer.

G. P. acknowledges the support of PRIN2017 and Sapienza, Progetto di Ateneo RM120172 B41DBF3A.

A. Iollo (✉)
IMB, Université Bordeaux and Equipe-projet Memphis, Inria Bordeaux-Sud Ouest, Bordeaux, France
e-mail: angelo.iollo@inria.fr

G. Puppo
Institute of Mathematics, La Sapienza Università di Roma, Roma, Italy
e-mail: gabriella.puppo@uniroma1.it

A. Thomann
Université de Strasbourg, CNRS, Inria, IRMA, F-67000 Strasbourg, France
e-mail: andrea.thomann@inria.fr

 
E. Franck et al. (eds.), *Finite Volumes for Complex Applications X—Volume 2, Hyperbolic and Related Problems*, Springer Proceedings in Mathematics & Statistics 433,
https://doi.org/10.1007/978-3-031-40860-1_18

Added to this motivation is, in the case of hyperbolic models, the need to preserve the consistency and accuracy of the integration scheme when the propagation phenomena described by the model have largely incomparable velocity scales. Nevertheless, an implicit scheme involves the inversion of a large often non linear system and, what is worse, the calculation of the Jacobian of the scheme residuals. The latter operation can be (i) cumbersome—when the numerical flux is non linear and the constitutive law complex; (ii) ill-posed—in presence of discontinuous solutions; (iii) limiting—if one wishes to devise a scheme independent of the constitutive laws of the model.

This work is in the wake of previous contributions in which the hyperbolic model considered is equivalent [2, 8], in the limit of an ad hoc relaxation, to a model in which the differential operator is linear and where the nonlinearity is transferred into the source term. This allows to resort to an implicit integration approach in which part of the difficulties enunciated above (the linear system is still there) can be overcome. Unlike other approaches, however, what is proposed below does not depend on relaxation parameters since it is an implicit relaxed scheme. Moreover, unlike previously, see [1, 7], the proposed implicit relaxed scheme does not involve additional relaxation variables due to a prediction-correction type technique projecting the relaxation source terms on the zero relaxation limit.

In what follows, the main contribution with respect to the work of [6] is the extension of the scheme to two-dimensional problems for compressible hyperelastic material flows in all-Mach regimes.

## 2 The Jin-Xin Relaxation Method

We consider the following hyperbolic conservation law on the two-dimensional domain $\Omega$ given by

$$\partial_t \boldsymbol{\psi} + \partial_x \boldsymbol{f}^x(\boldsymbol{\psi}) + \partial_y \boldsymbol{f}^y(\boldsymbol{\psi}) = 0, \tag{1}$$

with the state vector $\boldsymbol{\psi} \in \mathbb{R}^k$, directional fluxes $\boldsymbol{f}^x, \boldsymbol{f}^y : \mathbb{R}^k \to \mathbb{R}^k$. Following [5], we introduce a vector of relaxation variables $\boldsymbol{v} \in \mathbb{R}^k$ and consider the following relaxation system given by

$$\partial_t \boldsymbol{\psi} + \partial_x \boldsymbol{v} + \partial_y \boldsymbol{w} = 0, \tag{2a}$$

$$\partial_t \boldsymbol{v} + \boldsymbol{A}_x^2 \partial_x \boldsymbol{\psi} = -\frac{1}{\eta}\left(\boldsymbol{v} - \boldsymbol{f}^x(\boldsymbol{\psi})\right), \tag{2b}$$

$$\partial_t \boldsymbol{w} + \boldsymbol{A}_y^2 \partial_x \boldsymbol{\psi} = -\frac{1}{\eta}\left(\boldsymbol{w} - \boldsymbol{f}^y(\boldsymbol{\psi})\right), \tag{2c}$$

where $\eta > 0$ denotes the relaxation rate and $\boldsymbol{A}_x^2$, $\boldsymbol{A}_y^2$ are diagonal matrices with positive entries given by

$$\boldsymbol{A}_x^2 = \mathrm{diag}(a_x^2, \cdots, a_x^2), \quad \boldsymbol{A}_y^2 = \mathrm{diag}(a_y^2, \cdots, a_y^2). \tag{3}$$

The fluxes on the relaxation model are linear whereas the relaxation source terms on the right hand side of (2a) are non-linear and stiff for small $\eta$. Rewriting Eqs. (2b), (2c), using a Chapman-Enskog expansion of the relaxation variables for small $\eta$, we find

$$\boldsymbol{v} = \boldsymbol{f}^x(\boldsymbol{\psi}) - \eta(\nabla_{\psi}\boldsymbol{f}^x(\boldsymbol{\psi})^2 - \boldsymbol{A}_x^2))\partial_x\boldsymbol{\psi}, \tag{4a}$$

$$\boldsymbol{w} = \boldsymbol{f}^y(\boldsymbol{\psi}) - \eta(\nabla_{\psi}\boldsymbol{f}^y(\boldsymbol{\psi})^2 - \boldsymbol{A}_y^2))\partial_y\boldsymbol{\psi}, \tag{4b}$$

where $\nabla_{\psi}\boldsymbol{f}^x(\boldsymbol{\psi})$, $\nabla_{\psi}\boldsymbol{f}^x(\boldsymbol{\psi})$ denote the Jacobians of the flux functions $\boldsymbol{f}^x$, $\boldsymbol{f}^y$ respectively. We have neglected the $\mathcal{O}(\eta^2)$ terms in the expansion (4a), since we are only interested in the first order diffusion terms in $\eta$. Inserting (4a) into (2a), we have

$$\partial_t\boldsymbol{\psi} + \partial_x\boldsymbol{f}(\boldsymbol{\psi}) = \eta\left(\partial_x\left(\left(\boldsymbol{A}_x^2 - \nabla_{\psi}\boldsymbol{f}^x(\boldsymbol{\psi})\right)\partial_x\boldsymbol{\psi}\right) + \partial_y\left(\left(\boldsymbol{A}_y^2 - \nabla_{\psi}\boldsymbol{f}^y(\boldsymbol{\psi})^2\right)\partial_y\boldsymbol{\psi}\right)\right).$$

To obtain a diffusive approximation of the original system of Eq. (1), it has to be ensured that the diffusion term on the right hand side is non-negative. This yields the so called sub-characteristic condition, that for all $\boldsymbol{\psi}$ it has to hold $\boldsymbol{A}_x^2 - \nabla_{\psi}\boldsymbol{f}^x(\boldsymbol{\psi}) \geq 0$ and $\boldsymbol{A}_y^2 - \nabla_{\psi}\boldsymbol{f}^y(\boldsymbol{\psi})^2 \geq 0$. This positive semi-definite restriction is fulfilled for the choice (3) with

$$a_x = \max_{\boldsymbol{x}\in\Omega}\max_{j=1,\ldots,k}|\lambda_j^x(\boldsymbol{\psi}(\boldsymbol{x},t))|, \quad a_y = \max_{\boldsymbol{x}\in\Omega}\max_{j=1,\ldots,k}|\lambda_j^y(\boldsymbol{\psi}(\boldsymbol{x},t))|, \tag{5}$$

where $\lambda_j^x$, $\lambda_j^y$, $j = 1, \ldots, k$ are the characteristic speeds of the original Eq. (1) given by the eigenvalues of the directional Jacobians $\nabla_{\psi}\boldsymbol{f}^x(\boldsymbol{\psi})$, $\nabla_{\psi}\boldsymbol{f}^y(\boldsymbol{\psi})$. Thus from (4a) we recover at leading order the original system, namely

$$\boldsymbol{v} = \boldsymbol{f}^x(\boldsymbol{\psi}), \; \boldsymbol{w} = \boldsymbol{f}^y(\boldsymbol{\psi}), \quad \partial_t\boldsymbol{\psi} + \partial_x\boldsymbol{f}^x(\boldsymbol{\psi}) + \partial_y\boldsymbol{f}^y(\boldsymbol{\psi}) = 0, \tag{6}$$

also referred to as the relaxation limit.

## 3 The Numerical Scheme

Following the lines of construction of the one-dimensional scheme introduced in [6], we use stiffly accurate singular diagonally implicit Runge-Kutta (SA-SDIRK) methods in a finite volume framework. To exploit the linear fluxes of the relaxation system (2), we apply an operator splitting to separate the non-linear relaxation source terms from the linear right hand side. For each stage $k = 1, \ldots, s$ of an $s$-stages Runge-Kutta method we have

$$\begin{cases} \boldsymbol{\psi}^\star &= \boldsymbol{\psi}^n \\ \boldsymbol{v}^\star &= \boldsymbol{v}^n - \frac{\Delta t}{\eta}\left(\boldsymbol{v}^\star - \boldsymbol{f}^x(\boldsymbol{\psi}^\star)\right), \\ \boldsymbol{w}^\star &= \boldsymbol{w}^n - \frac{\Delta t}{\eta}\left(\boldsymbol{w}^\star - \boldsymbol{f}^y(\boldsymbol{\psi}^\star)\right), \end{cases} \tag{7}$$

where $\boldsymbol{\psi}^\star$, $\boldsymbol{v}^\star$, $\boldsymbol{w}^\star$ denote the states after the relaxation process. Taking the limit $\eta \to 0$, we obtain $\boldsymbol{v}^\star = \boldsymbol{f}^x(\boldsymbol{\psi}^\star)$, $\boldsymbol{w}^\star = \boldsymbol{f}^y(\boldsymbol{\psi}^\star)$ which is consistent with the relaxation equilibrium solution (6). On the homogeneous left hand side of (2) we apply a SDIRK method with weights $\alpha_{jl}$, $\beta_j$, $j = 1, \ldots, s$, $l = 1, \ldots j-1$, yielding

$$\begin{cases} \boldsymbol{\psi}^{(j)} &= \boldsymbol{\psi}^n - \Delta t \sum_{l=1}^{j} \alpha_{jl}\left(\partial_x \boldsymbol{v}^{(l)} + \partial_y \boldsymbol{w}^{(l)}\right), \\ \boldsymbol{v}^{(j)} &= \boldsymbol{v}^n - \Delta t \sum_{l=1}^{j} \alpha_{jl} \boldsymbol{A}_x^2 \partial_x \boldsymbol{\psi}^{(l)}, \\ \boldsymbol{w}^{(j)} &= \boldsymbol{w}^n - \Delta t \sum_{l=1}^{j} \alpha_{jl} \boldsymbol{A}_y^2 \partial_y \boldsymbol{\psi}^{(l)}, \end{cases} \tag{8a}$$

$$\begin{cases} \boldsymbol{\psi}^{n+1} &= \boldsymbol{\psi}^n - \Delta t \sum_{j=1}^{s} \beta_j\left(\partial_x \boldsymbol{v}^{(j)} + \partial_y \boldsymbol{w}^{(j)}\right), \\ \boldsymbol{v}^{n+1} &= \boldsymbol{v}^n - \Delta t \sum_{j=1}^{s} \beta_j \boldsymbol{A}_x^2 \partial_x \boldsymbol{\psi}^{(j)}, \\ \boldsymbol{w}^{n+1} &= \boldsymbol{w}^n - \Delta t \sum_{j=1}^{s} \beta_j \boldsymbol{A}_y^2 \partial_y \boldsymbol{\psi}^{(j)}. \end{cases} \tag{8b}$$

Note that this system has thrice as many variables per stage and update as the original system (1). To reduce the computational overhead associated with these additional auxiliary variables, we replace the stages $\boldsymbol{v}^{(l)}$, $\boldsymbol{w}^{(l)}$ in the update for $\boldsymbol{\psi}^{(j)}$ in (8a) and obtain the time-semi discrete stages

$$\begin{aligned} \boldsymbol{\psi}^{(j)} &- \Delta t^2 \alpha_{jj}^2\left(\boldsymbol{A}_x^2 \partial_x^2 \boldsymbol{\psi}^{(j)} + \boldsymbol{A}_y^2 \partial_y^2 \boldsymbol{\psi}^{(j)}\right) = \\ \boldsymbol{\psi}^n &- \Delta t \alpha_{jj} \partial_x \boldsymbol{v}^n - \Delta t \sum_{l=1}^{j-1} \alpha_{jl} \partial_x \boldsymbol{v}^{(l)} + \Delta t^2 \alpha_{jj} \sum_{l=1}^{j-1} \alpha_{jl} \boldsymbol{A}_x^2 \partial_x^2 \boldsymbol{\psi}^{(l)} \\ &- \Delta t \alpha_{jj} \partial_y \boldsymbol{w}^n - \Delta t \sum_{l=1}^{j-1} \alpha_{jl} \partial_y \boldsymbol{w}^{(l)} + \Delta t^2 \alpha_{jj} \sum_{l=1}^{j-1} \alpha_{jl} \boldsymbol{A}_y^2 \partial_y^2 \boldsymbol{\psi}^{(l)}. \end{aligned} \tag{9}$$

Using the fact, that from the relaxation source terms we have $\boldsymbol{v}^n = \boldsymbol{f}^x(\boldsymbol{\psi}^n)$, $\boldsymbol{v}^{(l)} = \boldsymbol{f}^x(\boldsymbol{\psi}^{(l)})$, $\boldsymbol{w}^n = \boldsymbol{f}^y(\boldsymbol{\psi}^n)$, $\boldsymbol{w}^{(l)} = \boldsymbol{f}^y(\boldsymbol{\psi}^{(l)})$ for all previous stages $l = 1, \ldots, j-1$, it follows

$$\begin{aligned} \boldsymbol{\psi}^{(j)} &- \Delta t^2 \alpha_{jj}^2\left(\boldsymbol{A}_x^2 \partial_x^2 \boldsymbol{\psi}^{(j)} + \boldsymbol{A}_y^2 \partial_y^2 \boldsymbol{\psi}^{(j)}\right) = \\ \boldsymbol{\psi}^n &- \Delta t \alpha_{jj} \partial_x \boldsymbol{f}^x(\boldsymbol{\psi}^n) - \Delta t \sum_{l=1}^{j-1} \alpha_{jl} \partial_x \boldsymbol{f}^x(\boldsymbol{\psi}^{(l)}) + \Delta t^2 \alpha_{jj} \sum_{l=1}^{j-1} \alpha_{jl} \boldsymbol{A}_x^2 \partial_x^2 \boldsymbol{\psi}^{(l)} \\ &- \Delta t \alpha_{jj} \partial_y \boldsymbol{f}^y(\boldsymbol{\psi}^n) - \Delta t \sum_{l=1}^{j-1} \alpha_{jl} \partial_y \boldsymbol{f}^y(\boldsymbol{\psi}^{(l)}) + \Delta t^2 \alpha_{jj} \sum_{l=1}^{j-1} \alpha_{jl} \boldsymbol{A}_y^2 \partial_y^2 \boldsymbol{\psi}^{(l)}, \end{aligned}$$

thus only depending on the state variable $\boldsymbol{\psi}$ and from the relaxation source term we obtain $\boldsymbol{v}^{(j)} = \boldsymbol{f}^x(\boldsymbol{\psi}^{(j)})$, $\boldsymbol{w}^{(j)} = \boldsymbol{f}^y(\boldsymbol{\psi}^{(j)})$. Note, that the implicit system (10) decouples into $k$ implicit subsystems for each state variable in $\boldsymbol{\psi}$, since $\boldsymbol{A}_x$ and $\boldsymbol{A}_y$ are diagonal matrices. Moreover, the coefficient matrices for each implicit subsystems are identical $M$-matrices since we use SDIRK methods. This drastically reduces the computational overhead compared to coupled implicit systems based on linearizations of Jacobians, see e.g. [1, 7]. Finally we obtain for the update

$$\boldsymbol{\psi}^{n+1} = \boldsymbol{\psi}^n - \Delta t \sum_{j=1}^{s} \beta_j \left(\partial_x \boldsymbol{f}^x(\boldsymbol{\psi}^{(j)}) + \partial_y \boldsymbol{f}^y(\boldsymbol{\psi}^{(j)})\right). \tag{10}$$

In space we consider the two-dimensional domain $\Omega \subset \mathbb{R}^2$ paved with rectangular cells $\Omega_I = [x_{i-/2}, x_{i+1/2}] \times [y_{j-1/2}, y_{j-1/2}]$ of length $\Delta x$, $\Delta y$ in x-and y-direction respectively with barycenter $(x_i, y_j) = (i\Delta x, j\Delta y)$. According to the finite volume framework, we evolve cell averages defined by

$$\boldsymbol{\psi}_I^n = \frac{1}{|\Omega_I|} \int_{\Omega_I} \boldsymbol{\psi}(x, y, t^n) d\Omega. \tag{11}$$

The finite volume scheme is then given by

$$\begin{aligned}
\boldsymbol{\psi}_I^{(j)} &- \Delta t^2 \alpha_{jj}^2 \left(\boldsymbol{A}_x^2 \mathcal{D}_x^{(j)} + \boldsymbol{A}_y^2 \mathcal{D}_y^{(j)}\right) = \qquad (12)\\
\boldsymbol{\psi}_I^n &- \Delta t \alpha_{jj} \mathcal{F}_x^n - \Delta t \sum_{l=1}^{j-1} \alpha_{jl} \mathcal{F}_x^{(l)} + \Delta t^2 \alpha_{jj} \sum_{l=1}^{j-1} \alpha_{jl} \boldsymbol{A}_x^2 \mathcal{D}_x^{(l)} \\
&- \Delta t \alpha_{jj} \mathcal{F}_y^n - \Delta t \sum_{l=1}^{j-1} \alpha_{jl} \mathcal{F}_y^{(l)} + \Delta t^2 \alpha_{jj} \sum_{l=1}^{j-1} \alpha_{jl} \boldsymbol{A}_y^2 \mathcal{D}_y^{(l)}, \\
\boldsymbol{\psi}^{n+1} &= \boldsymbol{\psi}^n - \Delta t \sum_{j=1}^{s} \beta_j \left(\mathcal{F}_x^{(j)} + \mathcal{F}_y^{(j)}\right) \qquad (13)
\end{aligned}$$

with the discretized diffusion operators $\mathcal{D}$ and numerical fluxes $\mathcal{F}$, given by

$$\mathcal{D}_x = \frac{1}{\Delta x^2} \left(\boldsymbol{\psi}_{i+1,j} - 2\boldsymbol{\psi}_{i,j} + \boldsymbol{\psi}_{i-1,j}\right), \quad \mathcal{D}_y = \frac{1}{\Delta y^2} \left(\boldsymbol{\psi}_{i,j+1} - 2\boldsymbol{\psi}_{i,j} + \boldsymbol{\psi}_{i,j-1}\right),$$

$$\mathcal{F}_x = \frac{1}{\Delta x} \left(\boldsymbol{f}_{i+1/2,j}^x - \boldsymbol{f}_{i-1/2,j}^x\right), \quad \mathcal{F}_y = \frac{1}{\Delta y} \left(\boldsymbol{f}_{i,j+1/2}^y - \boldsymbol{f}_{i,j-1/2}^y\right).$$

The fluxes at the interface are approximated by a convex combination between centered and Rusanov fluxes based on the local Mach number $M_{\text{loc}}$ given by the ratio between local fluid and sound speed. They read e.g.

$$f^x_{i+1/2,j} = \frac{1}{2}\left(f^x(\psi_{i+1,j}) + f^x(\psi_{i,j})\right) - a_{i+1/2,j}\theta\left(M_{\text{loc}}\right)\left(\psi_{i+1,j} - \psi_{i,j}\right), \tag{14}$$

with the local maximal characteristic speed

$$a_{i+1/2,j} = \max_l(|\lambda_l(\psi(x_i, y_j, t^n))|, |\lambda_l(\psi(x_{i+1}, y_j, t^n))|)$$

and an indicator function, e.g.,

$$\theta\left(M_{\text{loc}}\right) = \sin\left(\frac{\pi M_{loc}}{2}\right) \text{ for } M_{loc} \in [0, 1] \text{ and } \theta\left(M_{\text{loc}}\right) = 1 \text{ for } M_{loc} > 1.$$

This choice ensures the correct numerical diffusion for almost incompressible flows (centered) in the low Mach number limit and compressible flows (upwind scheme based on Rusanov fluxes with $M_{\text{loc}} \approx 1$). This, together with SA-SDIRK time integration, ensures an accurate description of low Mach flows, as well as compressible flows thus (12), (13) constitutes an *all-speed scheme*. In consequence, by construction, the numerical scheme is *asymptotic preserving* due to asymptotic stable time integrators and centered numerical fluxes in the low Mach number limit.

## 4 Numerical Results and Discussion

In the following, we use a first order scheme consisting the backward Euler scheme in time and a two stage second order SA-SDIRK method with $\alpha_{jj} = 1 - \sqrt{2}/2$ with a minmod reconstruction at the cell interfaces to ensure second order in space also for compressible flows. We test the numerical schemes on the Euler equations with an ideal gas law and a conservative model of non-linear elasticity given by

$$\psi = \begin{pmatrix} \rho \\ \rho\boldsymbol{u} \\ [\nabla Y] \\ E \end{pmatrix}, \quad f(\psi) = \begin{pmatrix} \rho\boldsymbol{u} \\ \rho\boldsymbol{u} \otimes \boldsymbol{u} - \sigma \\ \boldsymbol{u} \cdot [\nabla Y]\mathbb{I} \\ \boldsymbol{u}(E + p) \end{pmatrix} \tag{15}$$

from which the directional fluxes $f^x, f^y$ in (2) can be deduced. It consists of conservation of mass $\rho$, momentum $\rho\boldsymbol{u}$ and total energy $E = \rho e + 1/2\rho|\boldsymbol{u}|^2$ with an additional evolution equation for the deformation gradient $[\nabla Y]$. We consider a neo-hookean equation of state $\rho e = (\gamma - 1)p + p_\infty + \chi(\text{tr}\bar{B} - 2)$ where $p_\infty, \chi$ are material constants and $\bar{B}$ is the normalized Cauchy stress tensor $\bar{B} = \det([\nabla Y])([\nabla Y]^{-1}[\nabla Y]^{-T})$ and the stress tensor $\sigma$ is given by

$$\sigma = -p + 2\chi\det([\nabla Y])\left(\bar{B} - \frac{\text{tr}\bar{B}}{2}\mathbb{I}\right). \tag{16}$$

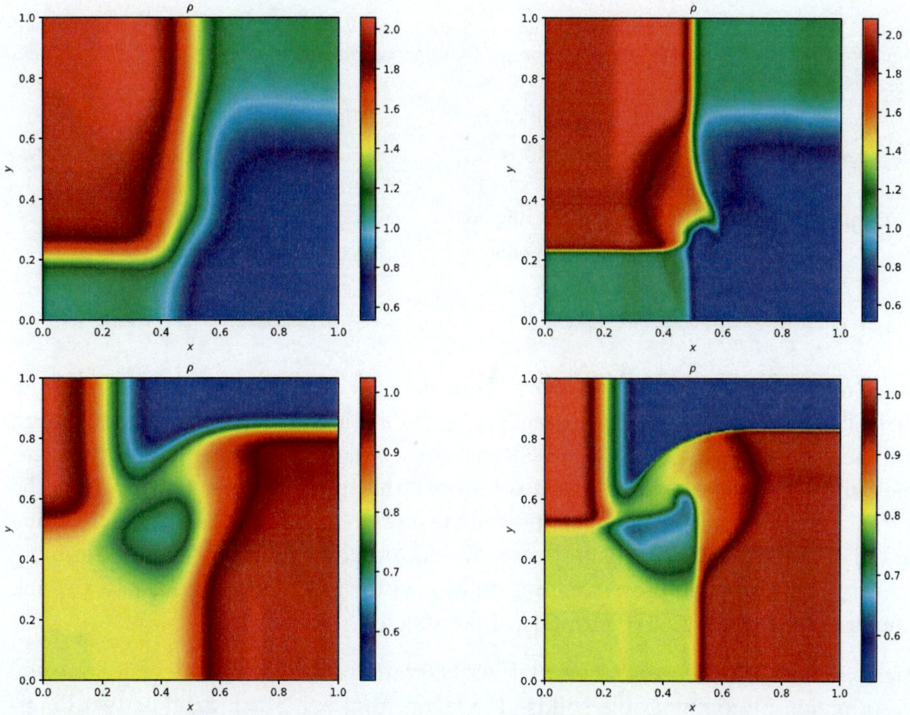


**Fig. 1** Riemann Problems for Euler equations in the compressible regime from [4] on a grid of $256 \times 256$ cells: Top panel: density for RP CRCS at time $T_f = 0.3$. Bottom panel: density for RP RCCS at time $T_f = 0.2$. Left side: first order scheme. Right side: Second order scheme

For more details on the model and its derivation see [2, 8]. Note that for $p_\infty = 0$ and $\chi = 0$. we recover the standard Euler equations with an ideal gas law.

*Compressible Riemann Problems* We consider two 2D Riemann Problems (RPs) for the Euler equations with an ideal gas law and $\gamma = 1.4$ taken from [4] containing contact discontinuities. Since both cases are in the compressible regime, we apply an acoustic CFL condition with CFL number 0.9 to *follow all waves* in the model. The first test consists of a contact, shock, contact and rafaction wave (RP CSCR) in counterclockwise rotation starting from the top right quarter. The second one consists of a rarefaction, two contacts and a shock wave (RP RCCS), both displayed in Fig. 1. Since the scheme contains the full upwind diffusion, the first order scheme is quite diffusive on all waves, whereas the second order scheme captures all waves accurately.

*Gresho Vortex* Next, we consider the Gresho vortex, which is a solution to the incompressible Euler equations with a divergence free velocity field and constant initial density and is a stationary exact solution to the Euler equations. For the set-up see e.g. [3]. We run it for three different Mach regimes on a gird of $40 \times 40$ on the domain $[0, 1]^2$ with $\Delta t = 2.2 \cdot 10^{-2}$. It is especially suited to verify the AP property,

**Table 1** Gresho vortex: $L^1$ error of the density at final time $T = 1$ with respect to initial constant density $\rho(x, 0) = 1$ and ratio of kinetic energy over the computational domain $[0, 1]^2$ for different Mach number regimes

| | Scheme | $M_{\max} = 10^{-2}$ | $M_{\max} = 10^{-4}$ | $M_{\max} = 10^{-6}$ |
|---|---|---|---|---|
| $L^1$ error $\rho$ | First order | 9.91e-8 | 8.77e-12 | 1.43e-15 |
| | Second order | 6.88e-7 | 2.11e-11 | 3.64e-15 |
| $E^T_{\text{kin}}/E^0_{\text{kin}}$ | First order | 0.998128 | 0.999945 | 0.999946 |
| | Second order | 0.999941 | 0.999945 | 0.999952 |

since the kinetic energy is preserved. Thus the loss of kinetic energy indicates the diffusiveness of the numerical scheme reported in Table 1. As we see, for all Mach regimes, the loss due to diffusion is almost equal and the numerical solutions resemble quite accurately the initial exact solution of the vortex. Moreover we present the $L^1$ error of the density at the final time with respect to the initial density which is, due to the stationary aspect of the test case, even of order $M^3$ thus the solution formally converges to the incompressible limit solution as the Mach number decreases which numerically verifies the AP property of the scheme.

*Deformation of hyperelastic solids* Finally, we consider two RPs concerning the deformation of compressible solids (15) taken from [1] which exhibits two longitudinal, two shear waves and a material wave. We consider two materials, copper with $p_\infty = 3.42 \cdot 10^{10}$, $\gamma = 4.33$ and shear modulus $\chi = 5 \cdot 10^{10}$ which lies in the low acoustic and low shear Mach number regime, i.e. the longitudinal and shear waves are considerably faster than the material wave. The initial condition consists of a jump in pressure from $10^9$ to $10^5$ in the bottom right quarter of the domain under constant density of 8900 and a zero velocity field. The second material under consideration is rubber ($p_\infty = 6.8 \cdot 10^8$, $\gamma = 4.4$, $\chi = 8 \cdot 10^5$) which lies in the low acoustic Mach number regime where the longitudinal waves are significantly faster than the shear and material waves. The initial condition consists of a jump in pressure from $10^8$ to $9.8 \cdot 10^7$ and velocities from 0 to 20 in the bottom right quarter of the domain under constant density of 1000. The numerical results for our implicit first and second order scheme, as well as an explicit SSP-RK2 scheme, with $256^2$ grid cells on a $[0, 2]^2$ domain are presented in Fig. 2 with $\Delta t = 1.17 \cdot 10^{-5}$ for copper and $\Delta t = 2.08 \cdot 10^{-5}$ for rubber. Note, that the second order explicit upwind scheme requires a time step of $\Delta t = 3.3 \cdot 10^{-7}$ for copper and $1.8 \cdot 10^{-6}$ for rubber. Our implicit first and second order schemes capture the resulting contact wave in the density accurately, in contrast to the explicit method, while diffusing shear and longitudinal waves. Due to the higher order approximation, we observe small oscillations on the negligible fast waves in our second order scheme.

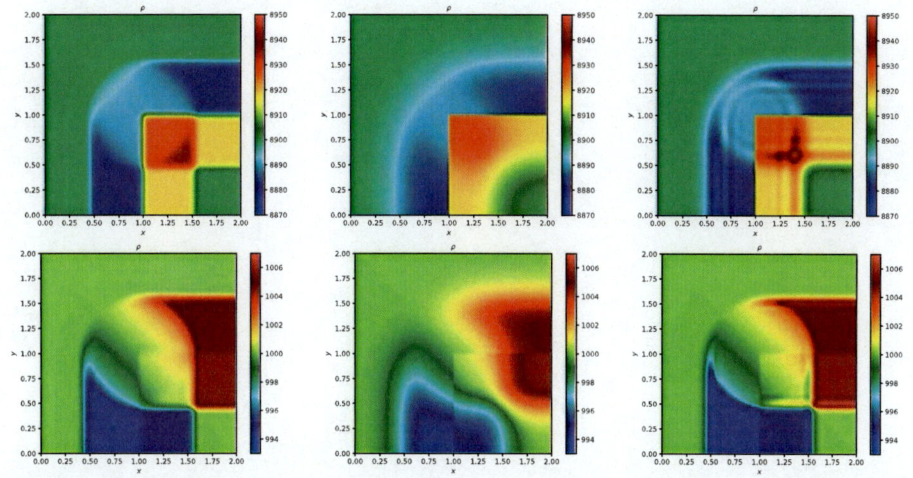


**Fig. 2** Riemann Problem elastic solids on the domain $[0, 2]^2$: Top panel: Copper density at time $T = 10^{-4}$. Bottom panel: Rubber density at time $T = 3 \cdot 10^{-4}$. Left: Explicit second order scheme. Middle: Our first order scheme. Right: Our second order scheme

# References


1. Abbate, E., Iollo, A., Puppo, G.: An asymptotic-preserving all-speed scheme for fluid dynamics and nonlinear elasticity. SIAM J. Sci. Comput. **41**, A2850–A2879 (2019)
2. de Brauer, A., Iollo, A., Milcent, T.: A Cartesian scheme for compressible multimaterial models in 3d. J. Comput. Phys. **313**, 121–143 (2016)
3. Miczek, F., Friedrich, K.R., Edelmann, P.V.F.: New numerical solver for flows at various Mach numbers. Astron. & Astrophys. **576**, A50 (2015)
4. Kurganov, A., Tadmor, E.: Solution of two-dimensional Riemann problems for gas dynamics without Riemann problem solvers. Numer. Methods Part. Differ. Equ. Int. J. **18**(5), 584–608 (2002)
5. Jin, S., Xin, Z.: The relaxation schemes for systems of conservation laws in arbitrary space dimensions. Commun. Pur. Appl. Math. **48**, 235–276 (1995)
6. Thomann, A., Iollo, A., Puppo, G.: Implicit relaxed all Mach number schemes for gases and compressible materials. SIAM J. Sci. Comput. (2021). arXiv.org/abs/2112.14126
7. Abbate, E., Iollo, A., Puppo, G.: An all-speed relaxation scheme for gases and compressible materials. J. Comput. Phys. **351**, 1–24 (2017)
8. Gorsse, Y., Iollo, A., Milcent, T., Telib, H.: A simple Cartesian scheme for compressible multimaterials. J. Comput. Phys. **272**, 772–798 (2014)

# Reducing Numerical Artifacts by Sacrificing Well-Balance for Rotating Shallow-Water Flow

**Håvard Heitlo Holm and Florian Beiser**

**Abstract** We consider the problem of rotational shallow-water flow for which non-trivial rotating steady-state solutions are of great importance. In particular, we investigate a high-resolution central-upwind scheme that is well-balanced for a subset of these stationary solutions and show that the well-balanced design is the source of numerical artifacts when applied to more general problems. We propose an alternative flux evaluation that sacrifices the well-balanced property and demonstrate that this gives qualitatively better results for relevant test cases and real-world oceanographic simulations.

**Keywords** High-resolution finite-volume schemes · Rotating shallow-water equations · Oceanography · Numerical artifacts

## 1 Introduction

Over the past few years, we have developed a GPU-accelerated simulation framework [1] for oceanographic applications and ensemble prediction systems. At the core of the framework is a simplified ocean model based on the shallow-water equations in a rotational frame of reference. This hyperbolic system is solved using a slightly modified version of the numerical scheme proposed by Chertock et al. [2] (abbreviated as CDKLM). This is an explicit high-resolution finite-volume scheme that is conservative, second order, and consistent, formulated on Cartesian grids. Its main advantage compared to similar schemes is that it is well-balanced with respect to some non-trivial steady-state solutions in geostrophic balance, namely geostrophic jets along the

H. H. Holm (✉) · F. Beiser
Mathematic and Cybernetics, SINTEF Digital, Oslo, Norway
e-mail: havard.heitlo.holm@sintef.no

F. Beiser
Department of Mathematical Sciences, Norwegian University of Science and Technology, Trondheim, Norway



E. Franck et al. (eds.), *Finite Volumes for Complex Applications X—Volume 2, Hyperbolic and Related Problems*, Springer Proceedings in Mathematics & Statistics 433,
https://doi.org/10.1007/978-3-031-40860-1_19

coordinate axes. Hereby, the rotational pull on the jets is equal to the gravity-induced potential energy from a non-flat ocean surface. The geostrophic jets are however not the only steady-state solutions based on geostrophic balance, and in real-world oceanographic applications, rotational steady-states are often of greater importance.

The CDKLM scheme is based on central-upwind schemes [3], which are attractive as "black-box" solvers as they do not require solutions to the Riemann problem. An alternative to CDKLM is the earlier scheme proposed by Kurganov and Petrova [4] (abbreviated as KP), which also uses central-upwind flux evaluations but is not well-balanced for any non-trivial rotational steady-state solutions. When considering alternative numerical schemes, one that is well-balanced for some geostrophic steady-state solutions and one that is not, it is easy to choose to implement the former also for more complex applications. However, as we will show in this paper, one should be more careful in this selection.

In this work, we demonstrate that the CDKLM scheme gives oscillatory artifacts and show that we can achieve solutions that are qualitatively better for rotational flow by using a non-well-balanced variation of CDKLM. Thereafter, we continue by investigating how to tune the resulting finite-volume scheme when applied to real-world oceanographic problems.

## 2 Rotational Shallow Water Equations

The shallow water equations in a rotational frame of reference are given by

$$\begin{bmatrix} \eta \\ hu \\ hv \end{bmatrix}_t + \begin{bmatrix} hu \\ hu^2 + \frac{1}{2}gh^2 \\ huv \end{bmatrix}_x + \begin{bmatrix} hv \\ huv \\ hv^2 + \frac{1}{2}gh^2 \end{bmatrix}_y = \begin{bmatrix} 0 \\ fhv \\ -fhu \end{bmatrix} + \begin{bmatrix} 0 \\ ghH_x \\ ghH_y \end{bmatrix} \tag{1}$$

and conserve mass and momentum. The variable $\eta$ represents the water surface as a deviation from the mean equilibrium depth $H$, so that $h = H + \eta$ becomes the total water depth. Momentum is given by $hu$ and $hv$ in $x$-and $y$-direction, respectively. Furthermore, $g$ is the gravitational constant, and $f$ is the Coriolis parameter, which is a function of latitude. We define the vector $\mathbf{q} = [\eta, hu, hv]^T$ to contain the conserved variables, so that we can write (1) as

$$\mathbf{q}_t + \mathcal{F}(\mathbf{q})_x + \mathcal{G}(\mathbf{q})_y = \mathbf{S}(\mathbf{q}) \ . \tag{2}$$

Rotating steady-states solutions of (1) are given by geostrophic balances, where

$$u_x + v_y = 0 \ , \qquad g\eta_x = fv \ , \quad \text{and} \quad g\eta_y = -fu \ . \tag{3}$$

In [2], the last two expressions are rewritten in terms of potential energies as

$$K(x, y) := g\left(\eta - \int_{-\infty}^{x} \frac{f}{g} v \, dx'\right), \quad L(x, y) := g\left(\eta - \int_{-\infty}^{y} \frac{f}{g} u \, dy'\right), \tag{4}$$

so that they can be written as $K_x = 0$ and $L_y = 0$. In particular, the CDKLM scheme is well-balanced with respect to the geostrophic jets

$$u \equiv 0, \quad v_y \equiv 0, \quad \eta_y \equiv 0, \quad H_y \equiv 0, \quad K \equiv \text{const}, \tag{5}$$

$$v \equiv 0, \quad u_x \equiv 0, \quad \eta_x \equiv 0, \quad H_x \equiv 0, \quad L \equiv \text{const}, \tag{6}$$

which are special cases of (3).

## 3 Finite Volume Schemes

In the following, we summarize only the relevant parts of the CDKLM scheme; the reader can find all details in [2]. We start by discretizing the domain in a regular Cartesian grid with cells of size $(\Delta x, \Delta y)$, and define $\mathbf{Q}_{j,k}$ to be the average of $\mathbf{q}$ within the cell with center at $\left((j + \frac{1}{2})\Delta x,\ (k + \frac{1}{2})\Delta y\right)$. Equation (2) can then be written in semi-discrete form as

$$\frac{\partial \mathbf{Q}_{j,k}}{\partial t} = -\frac{\mathbf{F}_{j+1/2,k} - \mathbf{F}_{j-1/2,k}}{\Delta x} - \frac{\mathbf{G}_{j,k+1/2} - \mathbf{G}_{j,k-1/2}}{\Delta y} + \mathbf{S}(\mathbf{Q}_{j,k}). \tag{7}$$

The system is solved in time with the second-order total-variation-diminishing Runge-Kutta scheme [5] with time steps satisfying the CFL condition. In the rest of the paper, we focus on the spatial reconstruction and fluxes.

### *3.1 Well-Balanced Scheme*

To design a well-balanced finite-volume scheme for (1) with respect to (5) and (6), Chertock et al. [2] propose to reconstruct the face values of $\mathbf{q}$ by computing the slopes of the equilibrium variables $\mathbf{p} = [u, v, K, L]^T$ rather than the slopes of $\mathbf{q}$ directly. Values for $K_{j,k}$ and $L_{j,k}$ are obtained by evaluating the integrals in (4) by a recursive sum starting from zero at the boundary. Note that the recursive sums for $K$ and $L$ mostly cancel in the reconstruction and are reduced to local operations [1, Sect. 3.1]. The slopes are then found by the generalized minmod limiter (c.f. [6]), which in the $x$-direction is

$$(\mathbf{p}_x)_{j,k} = \text{minmod}\left(\theta \frac{\mathbf{p}_{j+1,k} - \mathbf{p}_{j,k}}{\Delta x}, \frac{\mathbf{p}_{j+1,k} - \mathbf{p}_{j-1,k}}{2\Delta x}, \theta \frac{\mathbf{p}_{j,k} - \mathbf{p}_{j-1,k}}{\Delta x}\right), \tag{8}$$

where $\theta \in [1, 2]$ controls numerical dissipation.

Using $(\mathbf{p}_x)_{j,k}$ and $(\mathbf{p}_x)_{j+1,k}$, we can reconstruct one-sided point values for $\mathbf{q}$ at the face between the two cells as $\mathbf{Q}^r_{j+1/2,k}$ and $\mathbf{Q}^l_{j+1/2,k}$ from the right and left, respectively, using the discretized version of (4) to reconstruct values for $\eta$. The flux terms in (7) are computed using the central-upwind (CU) scheme [3]

$$\mathbf{F}^{(i)}_{j+1/2,k} = \frac{a^+\mathcal{F}\left(\mathbf{Q}^l\right) - a^-\mathcal{F}\left(\mathbf{Q}^r\right)}{a^+ - a^-} + \frac{a^+a^-}{a^+ - a^-}\left(\mathbf{Q}^r - \mathbf{Q}^l\right) \tag{9}$$

for $i = 1, 2$, where we have omitted the subscript $(\cdot)_{j+1/2,k}$ for all $\mathbf{Q}$ (including its entries) and $a$ variables for readability. The variables $a^+$ and $a^-$ represent the largest positive and negative wave speeds at the cell interface, respectively,

$$a^{\pm} = \begin{matrix}\max\\ \min\end{matrix}\left\{u^l \pm \sqrt{gh^l},\quad u^r \pm \sqrt{gh^r},\quad 0\right\}. \tag{10}$$

For the scheme to be well-balanced with respect to (6), however, Chertock et al. [2] propose to use a standard-upwind (SU) scheme for $\mathbf{F}^{(3)}_{j+1/2,k}$, given by

$$\mathbf{F}^{(3)}_{j+1/2,k} = \begin{cases} h^r u^r v^r, & \text{if } u^r + u^l > 0, \\ h^l u^l v^l, & \text{otherwise.} \end{cases} \tag{11}$$

The scheme is defined analogously in the $y$-direction, where the CU flux in (9) is used for $\mathbf{G}^{(i)}_{j,k+1/2}$, $i = 1, 3$, and the SU in (11) is used for $\mathbf{G}^{(2)}_{j,k+1/2}$.

### 3.2 Unbalanced Alternatives

As we will see in Sect. 4, the use of the SU flux in (11) produces artifacts when applying CDKLM to rotational and realistic problems. To investigate the problem, we consider a convex combination of the two fluxes

$$\mathbf{F}^{(3)}_{j+1/2,k} = \phi(\mathbf{F}_{SU})^{(3)}_{j+1/2,k} + (1-\phi)(\mathbf{F}_{CU})^{(3)}_{j+1/2,k}, \tag{12}$$

in which $\mathbf{F}_{CU}$ comes from (9), $\mathbf{F}_{SU}$ comes from (11), and $\phi \in [0, 1]$.

In contrast to CDKLM, KP reconstructs the slopes of $\mathbf{q}$ using the conserved variables directly and evaluates all flux components using the central-upwind scheme. We implement (12) also for KP and compare it with CDKLM to better assess the effects of different reconstructions and flux terms in isolation.

## 4 Numerical Experiments

We consider five experiments to evaluate the pros and cons of using the reconstruction and flux that make the scheme well-balanced for geostrophic jets.

***Case A: Radial wave without rotation.*** First, we consider a simple radial wave in a non-rotating domain (meaning $f = 0$), initialized by

$$\eta(x, y, 0) = \exp^{-10^{-5}(x^2+y^2)}, \quad \text{and} \quad hu(x, y, 0) = hv(x, y, 0) = 0. \tag{13}$$

The domain consists of $400 \times 400$ cells with $\Delta x = \Delta y = 100$ m, centered at the origin. The equilibrium depth is $H(x, y) = 60$ m.

Figure 1 shows $hv$ at $t = 1000$ s in the central part of the domain after most of the energy has traveled outwards. In the plots of the $y$-momentum shown in the two leftmost plots for the CDKLM scheme with SU and CU fluxes, we clearly see an oscillating artifact when using SU. This is even more pronounced in the right-hand plot of Fig. 1, which shows a cross section at $y = -1$ km. The CU flux gives numerical artifacts along the grid diagonals, but these are smooth in contrast to the SU artifacts. Both solutions are symmetric in $hv$ and $hu$, and the artifacts are observed for any spatiotemporal resolution satisfying the CFL conditions with Courant number less or equal to 0.5.

Note that with $f = 0$, the CDKLM reconstruction reduces to that of the the KP scheme. We have implemented the KP scheme independently, and get exactly the same results as for CDKLM using both types of fluxes. This excludes the possibility that the artifacts are caused by a programming bug.

***Case B: Rossby adjustment.*** We continue by considering Rossby adjustment, in which a rotational steady state is generated when running Case A with $f = 0.0012s^{-1}$. The presence of Coriolis forces will establish a rotating bump in the center of the domain and create a steady-state that is not covered by the jets described in (5) and (6).

In Fig. 2, we consider the flux in (12) and assess different values for $\phi$. The leftmost plot uses $\phi = 1.0$, which corresponds to a pure SU flux as in the original CDKLM scheme, whereas $\phi = 0.0$ to the far right corresponds to the CU flux. Again, $hu$ and $hv$ are symmetric, so we only show results for $hv$. We see a similar nonphysical

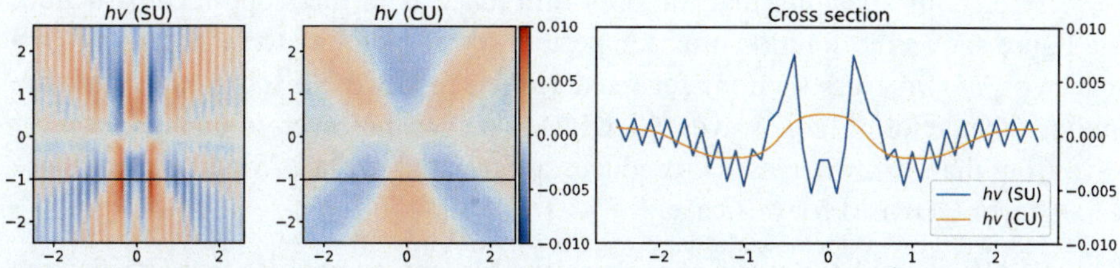


**Fig. 1** Artifacts in $hv$ from an initial bump using CDKLM with SU flux (left) and CU flux (center), and cross section of $hv$ at $y = -1$ km (right)

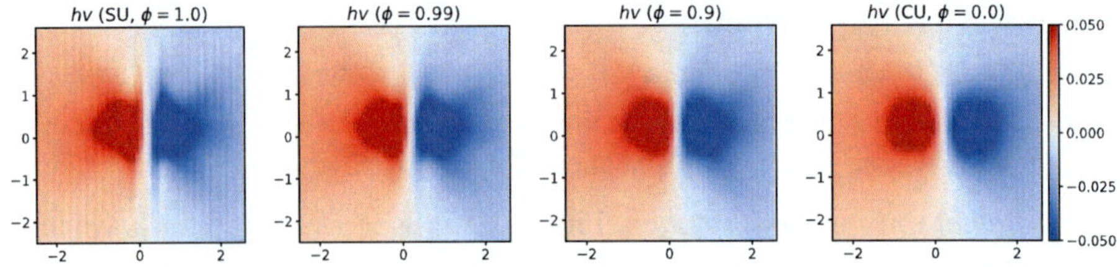


**Fig. 2** Numerical steady-states in $hv$ from an initial bump with different $\phi$-values in flux (12) in the presence of Coriolis forces

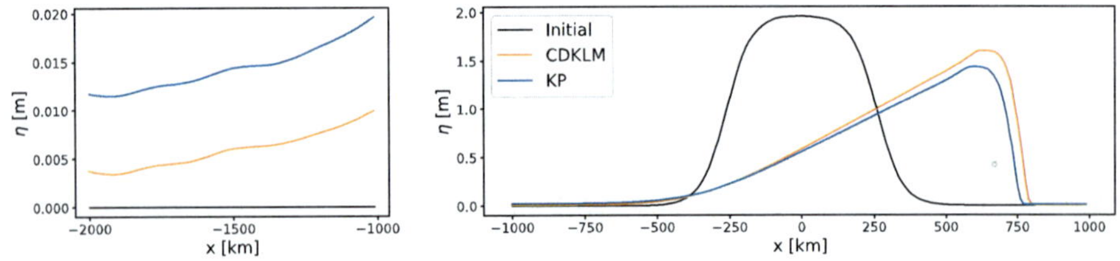


**Fig. 3** Cross-sectional plots of a Kelvin wave after ten periods for CDKLM and KP with CU flux and $\theta = 1.5$. Tail on the left and wave center on the right panel

oscillation pattern as in Case A when using the SU flux, and the rotating bump is dragged along the grid axes. In contrast, the CU flux results in a circular bump that captures (3) much better. It is interesting that it is sufficient to introduce just a small CU contribution to the SU flux to improve the numerical behavior, and the results are already significantly better with $\phi = 0.9$. The figures are visually the same when using KP.

***Case C: Kelvin wave.*** Kelvin waves are relatively fast travelling waves trapped along the coast by the Coriolis forces. They are travelling with the coast to their right on the Northern hemisphere and are important in realistic applications, as the tides often form them. Here, we consider a periodic domain in the east-west direction with wall boundary conditions in the north and south, and initialize a wave travelling eastwards along the southern wall. Sufficiently small waves follow linear physics and give periodic solutions, but the amplitudes used herein give nonlinear responses that develop a shock over time; see [7, Sect. 4.4] for details.

Figure 3 compares the CDKLM and KP schemes when using the CU flux and $\theta = 1.5$ for both, meaning that the only difference is in the slope reconstruction. The figure shows the solution after ten periods, corresponding to more than 60,000 time steps, at two cross sections for fixed $y$. We see how both schemes capture the nonlinear behavior of large waves. CDKLM manages, however, to maintain a taller wave front than KP, whereas KP produces a long nonzero tail. We argue that these results favor CDKLM for this case.

***Case D: Geostrophic jets.*** We consider a double geostrophic jet that satisfies the steady-state given by (6), with one jet travelling eastward in the northern part of the domain and an westward jet in the southern part. We use periodic boundary

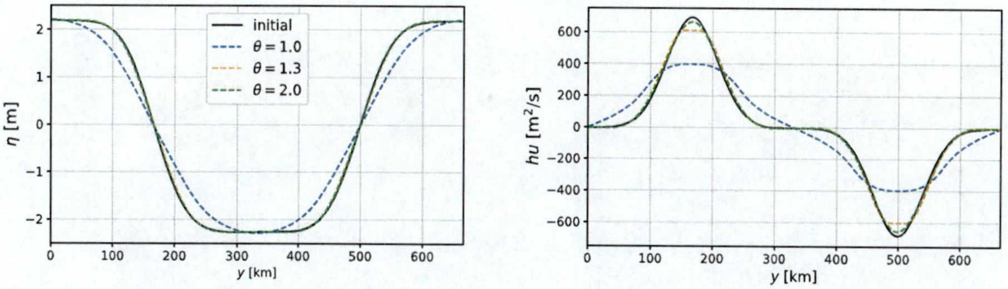


**Fig. 4** A double jet state after 9 simulation days with the CU flux: the plots show cross-sections of $\eta$ (left) and $hu$ (right)

conditions; the case is described in full in [8, Sect. 5]. This case exposes what we lose by not using the well-balanced SU flux for CDKLM, as this flux preserves this steady state for any slope limiter $\theta$.

Figure 4 shows a cross section along $y$ after running CDKLM for a long time using the CU flux with different values for $\theta$. With $\theta = 1.0$, the numerical dispersion smooths out the jet streams significantly, but a moderate increase in $\theta$ compensates for using a non-balanced flux, and the jets stay captured and only cap the extrema slightly. With the maximal possible $\theta = 2.0$, we lose only 4.5% of the momentum in the center of the jets.

To see how much of the CU flux we can incorporate without diminishing the preservation of geostrophic jets, we run the experiment for decreasing values of $\phi$ with fixed $\theta = 1.8$. For all $\phi \geq 0.52$, we conserve the maximum momentum within 4.5%.

***Case E: Real-world simulation.*** Finally, we study the consequence of using different flux terms and slope limiters when using the CDKLM scheme for simulating a real-world oceanographic scenario. Here, we utilize the full GPU-accelerated framework from [1], in which land mask, bathymetry, initial and boundary conditions, domain orientation, and wind forcing are taken from the operational model NorKyst800 [9], with 800 m horizontal resolution.

Figure 5 shows water velocity at 23 h for a few selected values of $\phi$ and $\theta$ for CDKLM. The first thing to notice is how the SU flux gives visible artifacts around the outermost islands at (100, 150). Secondly, the CU scheme gives visibly slower currents than SU, e.g., at (300, 150). Since the tidal Kelvin wave follows the coast from left towards right, the Lofoten islands trap the tidal wave, causing strong currents in between the islands. With CU, we do not resolve this, but setting $\phi = 0.8$ helps significantly. It is hard to see visual differences between setting $\theta = 1.3$ and $\theta = 1.8$, but examining cross sections (not shown) reveals that we indeed get sharper fronts but no oscillations with the latter, which is preferred.

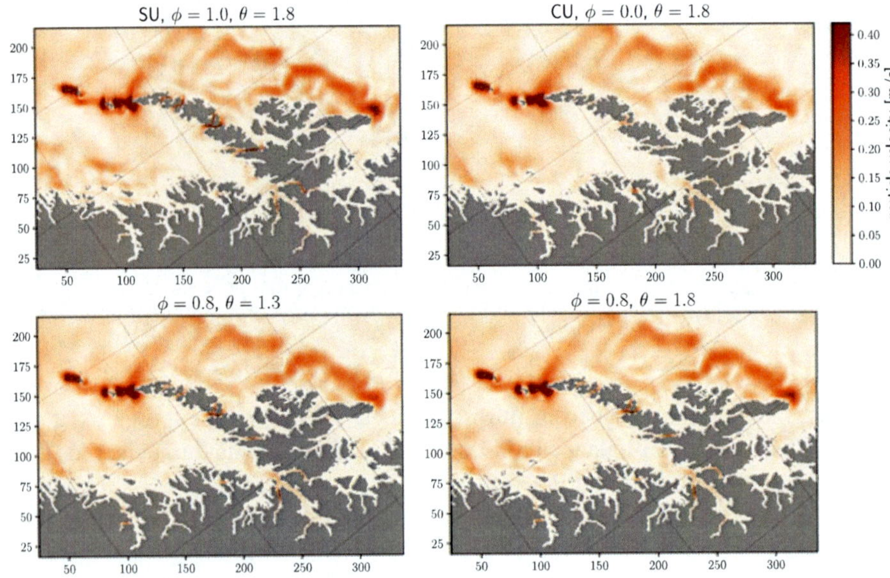


**Fig. 5** Real-world simulation for the Lofoten archipelago in Northern Norway. The axes show position within the computational domain in km

## 5 Conclusion

We have demonstrated numerical artifacts in the CDKLM scheme in cases that differ from geostrophic balanced jets along the coordinate axes. We identified the source of the oscillations to be the standard-upwind flux in the $huv$ fluxes. By combining the standard-upwind flux with the central-upwind flux with a factor $\phi$, we reduced unintended numerical behavior significantly and made the results more physical, even though we lose the well-balanced property for the jets.

We have illustrated the artifacts in Cases A and B, where Case B indicates that combining the SU flux with a small factor of CU is sufficient to smooth out these oscillations. In Case C, we see that the slope reconstruction in CDKLM still gives a small advantage over the more standard reconstruction used by the KP scheme. By using a large value $\theta$ in the slope limiter, we demonstrated that also pure CU can preserve the geostrophic jets from Case D within acceptable tolerances. Case E, however, shows that the pure CU flux is unable to capture strong realistic currents in narrow straits. The conclusion is then to use values close to $\phi = 0.8$ and $\theta = 1.8$ as flux parameter and slope limiter, respectively, for real-world oceanographic applications. This reduces the observed artifacts and gives a practical compromise between the well-balanced SU flux and the smoother behaving CU flux.

**Acknowledgements** This work is supported by the Research Council of Norway (RCN) through grant number 310515 (Havvarsel). The numerical examples are made with the GPU Ocean software [10], and can be reproduced by [11].

## References

1. Brodtkorb, A., Holm, H.: Coastal ocean forecasting on the GPU using a two-dimensional finite-volume scheme. Tellus A. **73**(1), 1–22 (2021)
2. Chertock, A., Dudzinski, M., Kurganov, A., Lukácová-Medvidová, M.: Well-balanced schemes for the shallow water equations with Coriolis forces. Numer. Math. **138**(4), 939–973 (2017)
3. Kurganov, A., Noelle, S., Petrova, G.: Semidiscrete central-upwind schemes for hyperbolic conservation laws and Hamilton-Jacobi equations. SIAM J. Sci. Comput. **23**(3), 707–740 (2001)
4. Kurganov, A., Petrova, G.: A second-order well-balanced positivity preserving central-upwind scheme for the Saint-Venant system. Commun. Math. Sci. **5**(1), 133–160, 03 (2007)
5. Gottlieb, S., Shu, C.-W.: Total variation diminishing Runge-Kutta schemes. Math. Comput. **67**(221), 73–85 (1998)
6. Lie, K.-A., Noelle, S.: On the artificial compression method for second-order nonoscillatory central difference schemes for systems of conservation laws. SIAM J. Sci. Comput. **24**(4), 1157–1174 (2003)
7. Holm, H., Brodtkorb, A., Broström, G., Christensen, K., Sætra, M.: Evaluation of selected finite-difference and finite-volume approaches to rotational shallow-water flow. Commun Comput. Phys **27**(4), 1234–1274 (2020)
8. Holm, H., Sætra, M., van Leeuwen, P.: Massively parallel implicit equal-weights particle filter for ocean drift trajectory forecasting. J. Comput. Phys. X. **6**(0314), 100053 (2020)
9. Albretsen, J., Sperrevik, A., Staalstrøm, A., Sandvik, A., Vikebø,F.: NorKyst-800 report no. 1: User manual and technical descriptions. Technical Report 2, Fisken og Havet, Institute of Marine Research (2011)
10. Brodtkorb, A., Holm, H., Sætra, M., Beiser, F.: GPU Ocean (2023). https://doi.org/10.5281/zenodo.7938844
11. Holm, H., Beiser, F.: Supplementary material for "Reducing numerical artifacts by sacrificing well-balance for rotating shallow-water flow" (2023). https://doi.org/10.5281/zenodo.7938887.

# Relaxation Process in an Immiscible Three-Phase Flow Model

**Jean-Marc Hérard and Guillaume Jomée**

**Abstract** This paper deals with the numerical modeling of immiscible three-phase flows. The main focus here is on the numerical treatment of the source terms of the model. A new scheme based on a more coupled approach than the preexisting fractional step strategy is presented. Properties of this scheme are given. Numerical applications highlight the benefits of this scheme in terms of both accuracy and stability.

**Keywords** Three-phase flow · Source term · Relaxation · Finite volume

## 1 Introduction

Some recent computations of immiscible three-phase flow models have highlighted some weaknesses of existing numerical schemes, for instance when representing vapor explosion [1], and thus have urged the building of robust implicit schemes. This has also motivated to investigate the behavior of solutions of rather complex sets of PDEs, such as those arising in [7], which are used in [1]. Moreover, it is widely admitted by the scientific community that the hyperbolic models at stake here contain inner relaxation processes that guarantee the return-to-equilibrium. The purpose of the current work precisely consists in better understanding the associated (at least expected) relaxation process first, and then taking advantage of the latter in order to stabilize approximate solutions obtained with help of rather classical Finite Volume procedures. In the sequel, we will focus on the model [7], and propose a two-step numerical strategy. This strategy accounts for the expected inner relaxation process

J.-M. Hérard · G. Jomée (✉)
EDF R and D, 6 Quai Watier, 78400 Chatou, France
e-mail: guillaume.jomee@edf.fr

J.-M. Hérard
e-mail: jean-marc.herard@edf.fr

I2M, 39 rue Frédéric Joliot-Curie, 13453 Marseille Cedex 13, France



E. Franck et al. (eds.), *Finite Volumes for Complex Applications X—Volume 2, Hyperbolic and Related Problems*, Springer Proceedings in Mathematics & Statistics 433,
https://doi.org/10.1007/978-3-031-40860-1_20

in an suitable implicit way, and also uses an explicit Riemann-type solver in order to cope with convective effects.

First, the model presented in [7] is recalled as well as its main properties. We emphasize that the strategy that has been used to derive model [7] is similar to the one that has been used in [2], in the two-phase flow framework. This guarantees a meaningful definition of non-conservative products for these models. Then, the new numerical strategy to compute approximate solutions of [7] is presented. Eventually, numerical cases depict the effectiveness of the new method. This paper is the sequel of a work that is dedicated to two-phase flow models [6].

## 2 The Immiscible Three-Phase Flow Model [7]

The system of PDEs governing the time-space evolution of the variables is:

$$\begin{cases} \frac{\partial \alpha_k}{\partial t} + \mathcal{V}_I(W) \cdot \nabla \alpha_k = S_k^{\alpha}(W) \, , \\ \frac{\partial m_k}{\partial t} + \nabla \cdot (m_k U_k) = S_k^m(W) \, , \\ \frac{\partial m_k U_k}{\partial t} + \nabla \cdot (m_k U_k \otimes U_k + \alpha_k P_k \mathcal{I}) + \sum\limits_{l=1, l \neq k}^{3} \Pi_{kl}(W) \nabla \alpha_l = S_k^U(W) \, , \\ \frac{\partial \alpha_k E_k}{\partial t} + \nabla \cdot (\alpha_k E_k U_k + \alpha_k P_k U_k) - \sum\limits_{l=1, l \neq k}^{3} \Pi_{kl}(W) \frac{\partial \alpha_l}{\partial t} = S_k^E(W) \, , \end{cases} \tag{1}$$

where $\forall k \in [\![1, 3]\!]$ , $\alpha_k \in ]0, 1[$ denote the statistical fractions which satisfy:

$$\alpha_1 + \alpha_2 + \alpha_3 = 1 \, . \tag{2}$$

$\rho_k$ stand for the densities, $m_k = \alpha_k \rho_k$ are the partial densities, $U_k$ the velocities, $P_k$ the pressures and $E_k$ the total energies which read $E_k = \rho_k(\epsilon_k + U_k^2/2)$. The specific internal energies $\epsilon_k$ are obtained through an Equation Of State (EOS): $\epsilon_k = \epsilon_k(\rho_k, P_k)$, and $\mathcal{I}$ stands for the identity. The velocity $\mathcal{V}_I(W)$ is chosen as:

$$\mathcal{V}_I = U_1 \, , \tag{3}$$

see [7], which leads to the following unique interfacial pressure definition, owing to the entropy inequality (see **Appendix** G of [7]):

$$\begin{cases} \Pi_{12} = \Pi_{21} = \Pi_{23} = P_2 \, , \\ \Pi_{13} = \Pi_{31} = \Pi_{32} = P_3 . \end{cases} \tag{4}$$

The state variable $W \in \mathbb{R}^{11}$ is defined as:

$$W = (\alpha_2, \alpha_3, m_1, m_1 U_1, \alpha_1 E_1, m_2, m_2 U_2, \alpha_2 E_2, m_3, m_3 U_3, \alpha_3 E_3)^{\mathsf{T}} \, . \tag{5}$$

In this work, the phase transitions are neglected, hence:

$$\forall k \in [\![1, 3]\!],\ l \neq k\ \ S_k^m(W) = 0\,. \tag{6}$$

In order to close the source terms, the total entropy $\eta(W)$ paired with its entropy-flux $\mathcal{F}_\eta(W)$, the temperature $T_k$ and the celerity $c_k$ are introduced:

$$\begin{cases} \eta = m_1 s_1(P_1, \rho_1) + m_2 s_2(P_2, \rho_2) + m_3 s_3(P_3, \rho_3)\,, \\ \mathcal{F}_\eta = m_1 U_1 s_1(P_1, \rho_1) + m_2 U_2 s_2(P_2, \rho_2) + m_3 U_3 s_3(P_3, \rho_3)\,. \end{cases} \tag{7}$$

$\frac{1}{T_k} = \frac{\partial_{P_k}(s_k(P_k,\rho_k))|_{\rho_k}}{\partial_{P_k}(\epsilon_k(P_k,\rho_k))|_{\rho_k}}$ and $c_k^2\ \partial_{P_k}(s_k(P_k, \rho_k))\big|_{\rho_k} + \partial_{\rho_k}(s_k(P_k, \rho_k))\big|_{P_k} = 0$. Functions $s_k$ denote specific entropies. Source terms are written, $\forall k \in [\![1, 3]\!]$, as:

$$\begin{aligned} S_k^\alpha &= \sum_{l=1, l\neq k}^{3} K_{kl}(W)(P_k - P_l)\ \ ;\ \ S_k^U = \sum_{l=1, l\neq k}^{3} d_{kl}(W)(U_l - U_k)\,; \\ S_k^E &= \sum_{l=1, l\neq k}^{3} q_{kl}(W)(T_l - T_k) + \sum_{l=1, l\neq k}^{3} d_{kl} V_{kl}(U_l - U_k)\,, \end{aligned} \tag{8}$$

(see [1, 7]), with the positive parameters $d_{kl}(W) = m_k m_l/(M\tau_{kl}^U(W))$, $K_{kl}(W) = \alpha_k\alpha_l/(\Pi_0\tau_{kl}^P(W))$ and $q_{kl}(W) = m_k m_l C_{v_k} C_{v_l}/(\tau_{kl}^T(W)(mC_v)_{kl})$. Quantities $C_{V_k}$ denote the specific heat capacities. $V_{kl}$ are phasic mean velocities: $V_{kl} = (U_k + U_l)/2$. $\Pi_0$ is a positive pressure of reference, $M = m_1 + m_2 + m_3$ and $(mC_v)_{kl} = m_k C_{v_k} + m_l C_{v_l}$. For each phasic connection $k - l$, $\tau_{kl}^P(W)$, $\tau_{kl}^T(W)$ and $\tau_{kl}^U(W)$ are the positive relaxation time scales related to the return to equilibrium of the associated thermodynamic quantity. Closure laws for the relaxation time scales can be found in the literature, see [4] for the pressure or [9] for the velocity and temperature. Model (1) with closures (3), (4), (6) and (8) is such that:

**Property 1** (Three-phase flow model in a 1D framework)

If $\forall k \in [\![1, 3]\!]$, $\alpha_k$ stay in $]0, 1[$ and $\|U_k - U_1\| \neq c_k$, then:

- The convective part of the model is hyperbolic and the associated eigenvalues are:

$$\begin{aligned} &\lambda_{1,2,3}(W) = U_1\ ;\ \lambda_4(W) = U_2\ ;\ \lambda_5(W) = U_3\ ; \\ &\lambda_{6,7}(W) = U_1 \pm c_1\ ;\ \lambda_{8,9}(W) = U_2 \pm c_2\ ;\ \lambda_{10,11}(W) = U_3 \pm c_3\,. \end{aligned} \tag{9}$$

- Fields associated with $\lambda_k$ $(k = 6 - 11)$ are **GNL**. Other fields are **LD**.
- Smooth solutions of the model comply with an entropy inequality:

$$\partial_t(\eta(W)) + \nabla \cdot \mathcal{F}_\eta(W) \geq 0\,. \tag{10}$$

- The model is symmetrizable.

***Proof*** The first three items can be found in [7] and the latter is detailed in [8].

## 3 Numerical Scheme

The overall scheme is quite similar to the one detailed in [1]. It is a Finite Volume scheme for unstructured meshes. It consists of two main steps:

- I: First, compute **explicitly** a solution between $t^n \overset{\Delta t}{\longrightarrow} t^*$ of the convective subsystem:

$$A(W)\partial_t (W) + \nabla \cdot F(W) + C(W)\nabla \cdot W = 0 \,, \tag{11}$$

  using a Rusanov scheme [10] adapted for non-conservative products. Equation (11) exactly corresponds to system (1), while setting all source terms to zero.
- II: Then, compute an approximate solution of the ODE:

$$A(W)\partial_t (W) = S(W) \,, \tag{12}$$

  between $t^* \overset{\Delta t}{\longrightarrow} t^{n+1}$. Equation (12) corresponds to system (1), when all spatial derivatives are set to zero. This step is solved with an **implicit** scheme.

The discretization of the time derivative is done with a first-order Euler scheme. The time step $\Delta t$ is determined by the Courant-Friedrich-Levy (CFL) condition of the first step. Details of the convective step are given in [1]. An extension of the scheme proposed in [11] for a barotropic three-phase flow model to the current three-phase flow model (which includes energy equations) would be useful. This would give higher accuracy than the scheme used in the present work. This strategy has already been used in the case of a two-phase flow model in [3].

We focus in the sequel on the numerical treatment of the source terms for this model. A fractional step algorithm, which fully decouples each relaxation effect, is employed in [1]. However, in some cases, this method is not robust enough to compute a solution (see Part 4.4 of [1]). Hereafter, a new method for a better handling of the source terms is presented. Setting for $k \in [\![1, 3]\!]$, $\Phi_k \in \{U_k, P_k, T_k\}$, $\Delta\Phi_{kl} = \Phi_k - \Phi_l$, system (12) reads:

$$\left\{\begin{array}{l}
\frac{\partial \alpha_k}{\partial t} = \sum\limits_{l=1, l\neq k}^{3} K_{kl}(W)\Delta P_{kl} \,, \\
\frac{\partial m_k}{\partial t} = 0 \,, \\
\frac{\partial m_k U_k}{\partial t} = -\sum\limits_{l=1, l\neq k}^{3} d_{kl}(W)\Delta U_{kl} \,, \\
\frac{\partial \alpha_k E_k}{\partial t} - \sum\limits_{l=1, l\neq k}^{3} \Pi_{kl}(W)\frac{\partial \alpha_l}{\partial t} = -\sum\limits_{l=1, l\neq k}^{3} q_{kl}(W)\Delta T_{kl} - \sum\limits_{l=1, l\neq k}^{3} d_{kl}(W)\frac{U_k^2 - U_l^2}{2} \,.
\end{array}\right. \tag{13}$$

Setting: $\Delta_S \in \mathbb{R}^6$ , $\Delta_S = (\Delta U_{12}, \Delta U_{13}, \Delta P_{12}, \Delta P_{13}, \Delta T_{12}, \Delta T_{13})^{\intercal} \in \mathbb{R}^6$, it can be derived from (13):

$$\partial_t (\Delta_S) = -R^{relax}(W)\Delta_S \,, \tag{14}$$

and the five stationary constraints:

$$\forall k \in [\![1,3]\!] \,,\ \partial_t m_k = 0 \,;\ \partial_t \sum_{k=1}^{3} m_k U_k = \partial_t \sum_{k=1}^{3} \alpha_k E_k = 0 \,, \tag{15}$$

with $R^{relax}$ a matrix in $\mathcal{M}_6(\mathbb{R})$ of the form:

$$R^{relax}(W) = \begin{pmatrix} R_{UU}(W) & 0 & 0 \\ R_{PU}(W) & R_{PP}(W) & R_{PT}(W) \\ R_{TU}(W) & R_{TP}(W) & R_{TT}(W) \end{pmatrix} \,, \tag{16}$$

where each sub-matrix $R_{IJ}(W),\ (I, J) \in \{U, P, T\}^2$ is in $\mathcal{M}_2(\mathbb{R})$. All coefficients of the sub-matrices are given in detail in [5] for a class of three-phase flow models. The counterpart for a similar class of two-phase flow models is given in [6]. From (14) and (16), conditions that guarantee the relaxation process over time can be exhibited, as it has been done in [6] for a two-phase flow model. Because of the block triangular structure of $R^{relax}$, the overall method to compute an approximate solution of (13) is:

(i) Estimate the evolution of $\Delta U = (\Delta U_{12}, \Delta U_{13})^{\mathsf{T}}$ between $t^* \xrightarrow{\Delta t} t^{n+1^-}$ according to:

$$\partial_t\,(\Delta U) = -R_{UU}(W)\Delta U \,, \tag{17}$$

with $R_{UU}(W) \in \mathcal{M}_2(\mathbb{R})$. Then update the state variable at time $t^{n+1^-}$ using the conservation law of the total momentum (see [1] Sect. 3.3.1).

(ii) Compute the evolution of $\Delta \in \mathbb{R}^4$ , $\Delta = (\Delta P_{12}, \Delta P_{13}, \Delta T_{12}, \Delta T_{13})^{\mathsf{T}}$ between $t^{n+1^-} \xrightarrow{\Delta t} t^{n+1}$, solution of:

$$\partial_t\,(\Delta) = -\mathcal{R}^{PT}(W)\Delta \,, \tag{18}$$

with $\mathcal{R}^{PT}(W) \in \mathcal{M}_4(\mathbb{R})$ defined as:

$$\mathcal{R}^{PT}(W) = \begin{pmatrix} R_{PP}(W) & R_{PT}(W) \\ R_{TP}(W) & R_{TT}(W) \end{pmatrix} . \tag{19}$$

Moreover, as the velocity relaxation is handled beforehand, the equation of evolution of the sum of the internal energies turns into a conservation law:

$$\partial_t \left( \sum_{k=1}^{3} m_k \epsilon_k \right) = 0 \,, \tag{20}$$

which enables to update $W$ at time $t^{n+1}$, together with constraint (2).

Hereafter, setting time $t^{n+1^-}$ as $t^n$ for clarity, the algorithm in stage (ii) is:

Step 1: Compute the solution of (18) at $t^{n+1}$ with a linear implicit Euler scheme and $\mathcal{R}^{PT}(W)$ frozen at time $t^n$:

$$\Delta^{n+1} = \left(\mathcal{I}_4 + \Delta t\, \mathcal{R}^{PT}(W^n)\right)^{-1} \Delta^n \,, \tag{21}$$

with $\mathcal{I}_4$ the identity matrix of size 4.

Step 2: Setting:

$$\begin{cases} \xi_n = m_1\epsilon_1(P_1^n, T_1^n) + m_2\epsilon_2(P_2^n, T_2^n) + m_3\epsilon_3(P_3^n, T_3^n) \,, \\ T_2^{n+1} = T_1^{n+1} - \Delta T_{12}^{n+1} \quad ; \quad T_3^{n+1} = T_1^{n+1} - \Delta T_{13}^{n+1} \,; \\ P_2^{n+1} = P_1^{n+1} - \Delta P_{12}^{n+1} \quad ; \quad P_3^{n+1} = P_1^{n+1} - \Delta P_{13}^{n+1} \,. \end{cases} \tag{22}$$

Compute $P_1^{n+1}$ and $T_1^{n+1}$, solutions of:

$$\begin{cases} m_1^n\epsilon_1(P_1^{n+1}, T_1^{n+1}) + m_2^n\epsilon_2(P_2^{n+1}, T_2^{n+1}) + m_3^n\epsilon_3(P_3^{n+1}, T_3^{n+1}) = \xi_n \,, \\ \frac{m_1^n}{\rho_1(P_1^{n+1}, T_1^{n+1})} + \frac{m_2^n}{\rho_2(P_2^{n+1}, T_2^{n+1})} + \frac{m_3^n}{\rho_3(P_3^{n+1}, T_3^{n+1})} = 1 \,. \end{cases} \tag{23}$$

System (23) comes from (20) and the immiscible constraint (2).

Step 3: Update $\alpha_2^{n+1}$ and $\alpha_3^{n+1}$:

$$\alpha_2^{n+1} = \frac{m_2^n}{\rho_2(P_2^{n+1}, T_2^{n+1})} \quad ; \quad \alpha_3^{n+1} = \frac{m_3^n}{\rho_3(P_3^{n+1}, T_3^{n+1})} \,, \tag{24}$$

and the total energies :

$$\begin{cases} (\alpha_1 E_1)^{n+1} = m_1^n\epsilon_1(P_1^{n+1}, T_1^{n+1}) + \frac{1}{2}m_1^n(U_1^n)^2 \,, \\ (\alpha_2 E_2)^{n+1} = m_2^n\epsilon_2(P_2^{n+1}, T_2^{n+1}) + \frac{1}{2}m_2^n(U_2^n)^2 \,, \\ (\alpha_3 E_3)^{n+1} = \alpha_1^n E_1^n + \alpha_2^n E_2^n + \alpha_3^n E_3^n - (\alpha_1 E_1)^{n+1} - (\alpha_2 E_2)^{n+1} \,. \end{cases} \tag{25}$$

Note that system (23) is non linear implicit. Hence we get:

**Property 2** (P-T coupled relaxation algorithm)

(a) For a mixture of ideal gases, the solution of (23) exists and is unique inside the relevant domain of definition and statistical fractions remain in [0, 1].

(b) If all of the principal minors of $\mathcal{R}^{PT}(W)$ are positive, then (21) ensures the discrete relaxation process over time, whatever the time step $\Delta t$ is.

*Sketch of proof:* (a) Thanks to the specific structure of the ideal gas EOS, the internal energies are functions of only one thermodynamic variable. Because of that, the conservation of the sum of the internal energies in (23) gives the temperature at time $t^{n+1}$. The complete proof is similar to the one given in [6] for a two-phase flow model. (b) The proof of the second property is also an extension of the one given in [6].

# 4 Numerical Results

In the sequel, for a quantity $\Psi \in \{P, T\}$, the relaxation time scales are supposed to be identical on each phasic connection: $\tau_{12}^{\Psi} = \tau_{13}^{\Psi} = \tau_{23}^{\Psi} = \tau^{\Psi}$. For simplicity, they are set as constants: $\tau^P = 10^{-5}s$ and $\tau^T = 10^{-3}s$. Besides, in all of the following numerical test cases, each phase $k$ is given a stiffened gas EOS: $\gamma_1 = 22.84\ (Jkg^{-1}K^{-1})$, $\Pi_1 = 1.884\ 10^9\ (Pa)$, $\gamma_2 = 1.615\ (Jkg^{-1}K^{-1})$, $\Pi_2 = 3.564\ 10^8\ (Pa)$, $\gamma_3 = 1.086\ (Jkg^{-1}K^{-1})$, $\Pi_3 = 0\ (Pa)$. It represents a mixture of hot liquid metal, liquid water and its vapour.

First, let us consider a homogeneous flow. Thus, for each phase $k$, $U_k = 0$ and for each quantity $\Psi_k$ we have: $\nabla_x \Psi_k = 0$. With this hypothesis, the set of PDEs associated with model (1) is identical to the one approximated by step (ii). At time $t = 0\ s$, the pressures are set equal: $P_1 = P_2 = P_3 = 1.0 \times 10^5\ Pa$ and the temperatures are set as: $T_1 = 2500K$, $T_2 = 363K$ and $T_3 = 1000K$. Figure 1 highlights the poor behaviour of the fractional step algorithm for the pressure $P_1$ with coarse time steps. As the time step is given by the convective part of the simulation under a CFL condition, the numerical method used to compute an approximate solution of (13) must be robust. The coupled algorithm obviously better captures the global behaviour of the solution on coarse time steps. Figure 2 confirms that the two schemes converge to the same solution with order 1.

The second case consists in a liquid water shock tube. For the numerical simulation, two meshes are considered: 50 000 cells and 100 000 cells. The CFL number is set to 1/2. At time $t = 0s$, the shock tube is set as in Table 1. All phasic velocities are set to zero at time $t = 0s$. When the numerical simulation starts, the tube contains three zones: a high pressure chamber filled with liquid water, a low pressure zone full of liquid water, and an interaction zone. In the high pressure chamber, all phasic pressures are set to $15 \times 10^5\ Pa$ at time $t = 0s$. Throughout the interaction zone, aluminium droplets of 500 $\mu$m diameter are introduced at time $t = 0s$. The droplet temperature is set to $T_1 = 2500K$. Thus, two Riemann problems have to be solved at $x_{I_{12}} = 2.0$ m and $x_{I_{23}} = 2.15$ m in a short transient regime, and then interact with one another. Figure 3 shows the evolution of the total pressure $P_{mix} = \alpha_1 P_1 + \alpha_2 P_2 + \alpha_3 P_3$ over time at two stations. Station 1 is located at $x_{S_1} = 1.4$ m and Station 2 is positioned at $x_{S_2} = 2.4$ m. Focusing first on Station 2, the minor decrease of the total pressure $P_{mix}$ at the beginning of the computation is the counterpart of what happens in the homogeneous case. After that, the small decrease of the total pressure at time $t = 0.4$ ms corresponds to the right-going wave associated with the initial interface located at $x_{I_{23}}$. At time $t = 0.6$ ms, the incident right-going shock wave of magnitude $8.5 \times 10^5\ Pa$, generated at time $t = 0s$ at $x_{I_{12}}$ arrives at Station 2. Then, the return-to-equilibrium process takes place till 3.8 ms. Afterwards the left-going rarefaction wave reflected from the right-side reaches station 2. Besides, as Station 1 is located inside the initial high pressure chamber, $P_{mix}$ plateaus at $15 \times 10^5\ Pa$ till $t = 0.9$ ms, which corresponds to the arrival of the left-going rarefaction wave generated at $x_{I_{12}}$.

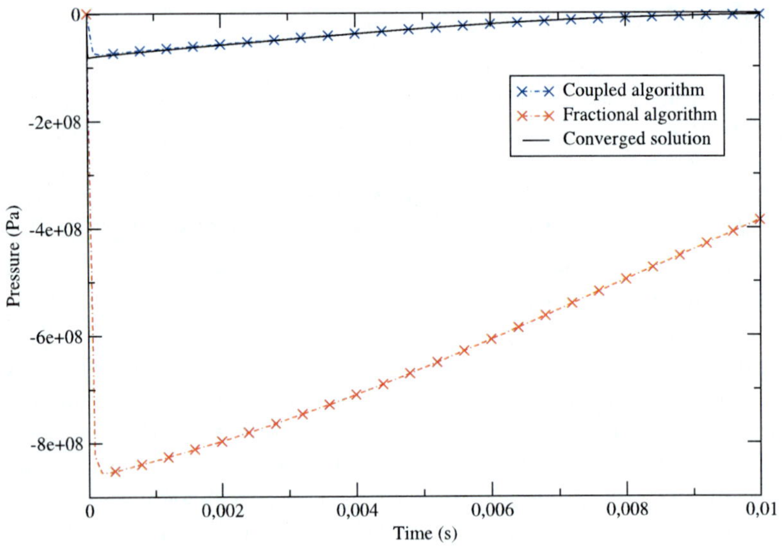


**Fig. 1** Evolution of pressure $P_1$ computed with the coupled and fractional schemes for $\Delta t = 10^{-4}s$. For the converged solution, a time step $\Delta t = 10^{-11}s$ is considered

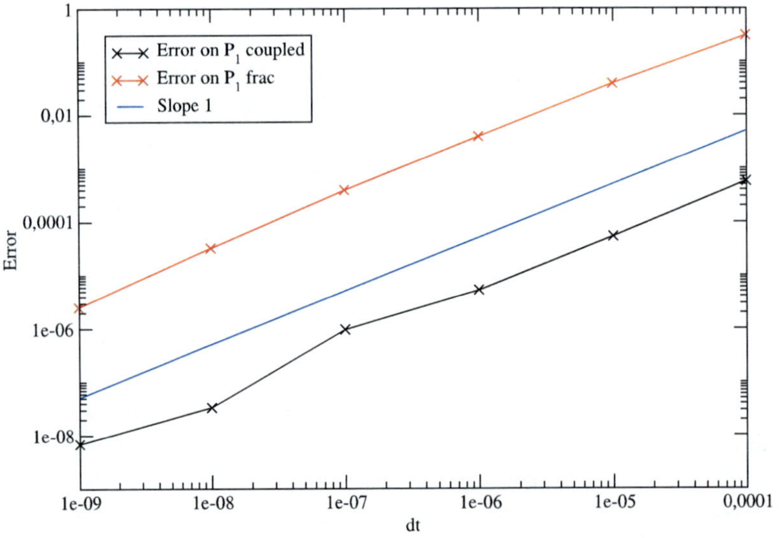


**Fig. 2** Convergence curve for the coupled and fractional algorithm; error on pressure $P_1$ in $L_1$ norm

**Table 1** Initial conditions for test case 2. All phasic velocities are set to zero

| Abscissa (m) | $x \in [0; 2.0]$ | $x \in ]2.0; 2.15[$ | $x \in [2.15; 3.75]$ |
|---|---|---|---|
| $\alpha_1$ | $\epsilon_{lim} = 10^{-8}$ | $\epsilon_{lim}$ | 0.026 |
| $\alpha_2$ | $1 - 2\epsilon_{lim}$ | $1 - 2\epsilon_{lim}$ | 0.884 |
| $P_1 = P_2 = P_3\ (Pa)$ | $1{,}5 \times 10^6$ | $1{,}0 \times 10^5$ | $1{,}0 \times 10^5$ |
| $T_1$ (K) | 1000 | 363 | 2500 |
| $T_2$ (K) | 1000 | 363 | 363 |
| $T_3$ (K) | 1000 | 363 | 1000 |

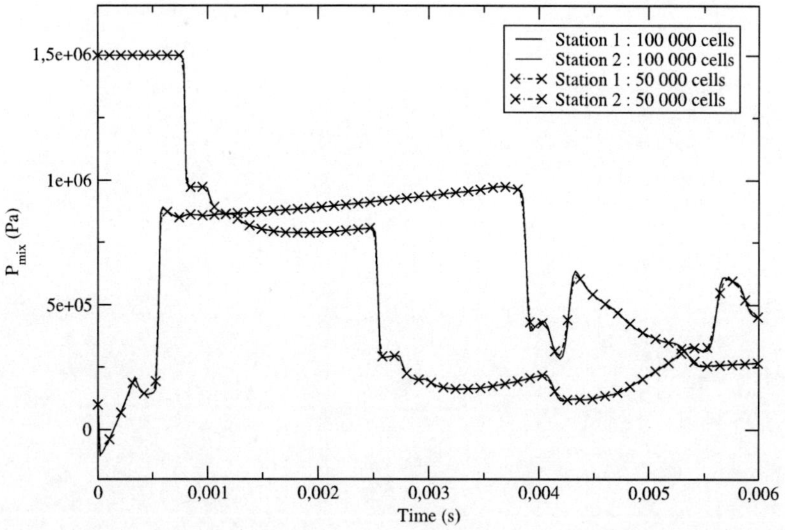


**Fig. 3** Evolution of the total pressure $P_{mix} = \alpha_1 P_1 + \alpha_2 P_2 + \alpha_3 P_3$ over time at two different stations and for two different meshes (50 000 cells / 100 000 cells)

# References


1. Boukili, H., Hérard, J.M.: Simulation and preliminary validation of a three-phase flow model with energy. Comput. Fluids. **221**, 104868 (2021)
2. Coquel, F., Gallouët, T., Hérard, J.M., Seguin, N.: Closure laws for a two-fluid two-pressure model. Comptes Rendus Mathematique **334**(10), 927–932 (2002)
3. Coquel, F., Hérard, J.M., Saleh, K.: A positive and entropy-satisfying finite volume scheme for the Baer-Nunziato model. J. Comput. Phys. **330**, 401–435 (2017)
4. Gavrilyuk, S.: The structure of pressure relaxation terms: the one-velocity case. EDF report H-I83-2014-0276-EN (2014)
5. Herard, J.M., Jomée, G.: A coupled algorithm to compute unsteady compressible immiscible three-phase flow models with stiff relaxation source terms. In prep
6. Herard, J.M., Jomée, G.: Two approaches to compute unsteady compressible two-phase flow models with stiff relaxation terms. EDF report (2022)
7. Hérard, J.M.: A three-phase flow model. Math. Comput. Model. **45**(5), 732–755 (2007)
8. Hérard, J.M.: A class of three-phase flow models with energy. EDF report 6125-3016-2020-01853-EN (2020)

9. Ishii, M.: Thermo-fluid dynamic theory of two-phase flow. Eyrolles-Collection de la Direction des Etudes et Recherches EDF (1975)
10. Rusanov, V.: The calculation of the interaction of non-stationary shock waves and obstacles. USSR Comput. Math. Math. Phys. **1**(2), 304–320 (1962)
11. Saleh, K.: A relaxation scheme for a hyperbolic multiphase flow model - Part I: Barotropic EOS. ESAIM Math. Model. Numer. Anal. **53**(5), 1763–1795 (2019)

# On the Convergence of the Godunov Scheme with a Centered Discretization of the Pressure Gradient

**Jonathan Jung, Ibtissem Lannabi, and Vincent Perrier**

**Abstract** This paper deals with the numerical resolution of a linear wave system using the Godunov scheme with a centered discretization of the pressure gradient. The interest in such schemes is motivated by the low Mach number accuracy problem. We prove that for both steady and unsteady flows, an oscillatory mode appears in the numerical solution. This can be explained by the loss of the Total Variation Diminishing property on the characteristic variables. Moreover, we illustrate numerically that the long-time numerical solution includes an oscillatory mode, which jeopardizes the convergence towards the expected solution.

**Keywords** Euler system · Low mach number flows · Finite volume methods · Oscillatory modes · Linear wave system

## Introduction

It is well-known that explicit Godunov-type schemes for solving the Euler equations fail, in general, to be accurate at low Mach number [7], in particular, on Cartesian grids. The most common argument to explain this failure is based on an asymptotic analysis of both the continuous equations and the numerical scheme. Initially, the low Mach number accuracy problem was explained via a single-scale asymptotic analysis that takes into account only the convective waves [7]. Afterwards, both acoustic and convective waves were taken into account by performing a two-scale asymptotic

J. Jung · I. Lannabi (✉) · V. Perrier
CNRS/Univ Pau and Pays Adour/E2S UPPA, Laboratoire de Mathématiques et de leurs Applications de Pau - Fédération IPRA, UMR5142, 64000 Pau, France
e-mail: ibtissem.lannabi@univ-pau.fr

J. Jung
e-mail: jonathan.jung@univ-pau.fr

V. Perrier
e-mail: vincent.perrier@inria.fr

Cagire team, Inria Bordeaux Sud-Ouest, France

 
E. Franck et al. (eds.), *Finite Volumes for Complex Applications X—Volume 2, Hyperbolic and Related Problems*, Springer Proceedings in Mathematics & Statistics 433,
https://doi.org/10.1007/978-3-031-40860-1_21

analysis [2, 8, 10], which led to a coupling of the zeroth-order momentum and the first-order pressure through a wave system.

A natural question that arises when considering a discretization of a Cauchy-problem for a wave system which has a long-time limit, is whether a discretization of this system also has a long-time limit and whether this long-time limit is consistent with that of the continuous system [9]. In [8], a link was elaborated between the low Mach number accuracy problem and the long-time consistency problem of a discrete wave system: a Godunov-type scheme is low Mach number accurate if and only if the discrete linear wave system derived from the two-scale asymptotic analysis is long-time limit consistent.

To ensure accuracy at low Mach number for the numerical resolution of the Euler equations on Cartesian grids, several fixes have been proposed. In this paper, we are interested in fixes of type [3, 11]. These schemes are also known to be two-scale asymptotic consistent with a discretization of the wave system, in which a centered discretization of the pressure gradient is used [2]. These schemes will be called *pressure-centered type schemes.* Since these fixes may introduce other issues, for instance, the appearance of an oscillatory mode (or checkerboard mode) in the numerical solution, it is interesting to address the study of the long-time limit of the wave system. Consider the linear wave system

$$\begin{cases} \partial_\tau \widehat{p} + \frac{1}{\widehat{\rho}} \nabla_{\mathbf{x}} \cdot \widehat{\mathbf{u}} = 0, \\ \partial_\tau \widehat{\mathbf{u}} + \widehat{\kappa} \nabla_{\mathbf{x}} \widehat{p} = 0, \end{cases} \tag{1}$$

solved on a domain $\Omega$, associated with the initial condition $\widehat{p}(\tau = 0, \mathbf{x}) = \widehat{p}^0(\mathbf{x})$ and $\widehat{\mathbf{u}}(\tau = 0, \mathbf{x})^T = \widehat{\mathbf{u}}^0(\mathbf{x})^T$, where $\widehat{p}$ denotes the pressure, $\widehat{\mathbf{u}}$ denotes the velocity field, $\widehat{\kappa}$, $\widehat{\rho} > 0$ and $\widehat{c} = \sqrt{\widehat{\kappa}/\widehat{\rho}}$ is the sound velocity. Let $\mathcal{T}_h = \bigcup_i \Omega_i$ denotes a mesh of the domain $\Omega$, on which a cell-centered finite volume discretization is applied. We denote by $|\Omega_i|$ the volume of the cell $\Omega_i$, by $|\Gamma_{ij}|$ the size of the side $\Gamma_{ij}$ linking the cell $\Omega_i$ to the cell $\Omega_j$, by $\mathbf{n}_{ij}$ the unit normal outgoing from the cell $\Omega_i$ and going to the cell $\Omega_j$, by $\mathcal{V}_{\text{int}}(i)$ the set of the cells sharing a face with the cell $\Omega_i$, by $\mathcal{V}_{\text{b}}(i)$ the set of the boundary sides of $\Omega_i$. We consider the numerical scheme

$$\begin{aligned} |\Omega_i| \left( \frac{\widehat{\mathbf{U}}_i^{n+1} - \widehat{\mathbf{U}}_i^n}{\delta\tau} \right) + \sum_{j \in \mathcal{V}_{\text{int}}(i)} |\Gamma_{ij}| \mathbf{F}^\theta \left( \widehat{\mathbf{U}}_i^n, \widehat{\mathbf{U}}_j^n, \mathbf{n}_{ij} \right) \\ + \sum_{j \in \mathcal{V}_{\text{b}}(i)} |\Gamma_{ij}| \mathbf{F}^{\text{b}} \left( \widehat{\mathbf{U}}_i^n, \widehat{\mathbf{U}}_b, \mathbf{n}_{ij} \right) = \mathbf{0}, \end{aligned} \tag{2}$$

where $\delta\tau$ denotes the time-step, $\widehat{\mathbf{U}}_i^n = \left( \widehat{p}_i^n, \widehat{\mathbf{u}}_i^n \right)^T$ is the vector of the discretized variables, $\widehat{\mathbf{U}}_b = (\widehat{p}_b, \widehat{\mathbf{u}}_b)^T$ is a state weakly imposed on the boundary of the domain $\Omega$, $\mathbf{F}^\theta \left( \widehat{\mathbf{U}}_i^n, \widehat{\mathbf{U}}_j^n, \mathbf{n}_{ij} \right)$ is the numerical flux inside the domain and is given by

$$\mathbf{F}^{\theta}\left(\widehat{\mathbf{U}}_i^n, \widehat{\mathbf{U}}_j^n, \mathbf{n}_{ij}\right) = \begin{pmatrix} \dfrac{1}{\widehat{\rho}}\dfrac{\widehat{\mathbf{u}}_i^n + \widehat{\mathbf{u}}_j^n}{2} \cdot \mathbf{n}_{ij} + \dfrac{\widehat{c}}{2}\left(\widehat{p}_i^n - \widehat{p}_j^n\right) \\ \widehat{\kappa}\dfrac{\widehat{p}_i^n + \widehat{p}_j^n}{2}\mathbf{n}_{ij} + \theta\dfrac{\widehat{c}}{2}\left(\left(\widehat{\mathbf{u}}_i^n - \widehat{\mathbf{u}}_j^n\right) \cdot \mathbf{n}_{ij}\right)\mathbf{n}_{ij} \end{pmatrix}, \tag{3}$$

and $\mathbf{F}^{\mathrm{b}}\left(\widehat{\mathbf{U}}_i^n, \widehat{\mathbf{U}}_b, \mathbf{n}_{ij}\right)$ is the numerical flux on the boundary of $\Omega$. The parameter $\theta \in \{0, 1\}$ is introduced in order to consider both the Godunov ($\theta = 1$) and the pressure-centered ($\theta = 0$) schemes. Note that (2) with (3) is $L^2$ stable for $\theta = 1$ with CFL 1 and for $\theta = 0$ with CFL 0.5, see [3]. In Sect. 1, periodic boundary conditions will be considered, so $\mathcal{V}_{\mathrm{b}}(i) = \varnothing$ for all $i$. In Sect. 2, wall and Steger-Warming boundary conditions will be considered. We denote by $\mathcal{V}_{\mathrm{wall}}(i)/\mathcal{V}_{\mathrm{SW}}(i)$ the set of the boundary sides of $\Omega_i$ on which wall/Steger-Warming boundary conditions are imposed: $\mathcal{V}_{\mathrm{b}}(i) = \mathcal{V}_{\mathrm{wall}}(i) \cup \mathcal{V}_{\mathrm{SW}}(i)$. The flux $\mathbf{F}^{\mathrm{b}}\left(\widehat{\mathbf{U}}_i, \widehat{\mathbf{U}}_b, \mathbf{n}_{ij}\right)$ is then given by

$$\mathbf{F}^{\mathrm{b}}\left(\widehat{\mathbf{U}}_i, \widehat{\mathbf{U}}_b, \mathbf{n}_{ij}\right) = \begin{cases} \begin{pmatrix} 0 \\ \widehat{\kappa}\widehat{p}_i\mathbf{n}_{ij} + \widehat{c}\left(\widehat{\mathbf{u}}_i \cdot \mathbf{n}_{ij}\right)\mathbf{n}_{ij} \end{pmatrix}, & \text{if } j \in \mathcal{V}_{\mathrm{wall}}(i), \\ \mathbf{F}^{\theta=1}\left(\widehat{\mathbf{U}}_i, \widehat{\mathbf{U}}_b, \mathbf{n}_{ij}\right), & \text{if } j \in \mathcal{V}_{\mathrm{SW}}(i). \end{cases} \tag{4}$$

In Sect. 1, we show that the Total Variation Diminishing (TVD) property is lost with the pressure-centered scheme. In Sect. 2, a problem in which the continuous wave system has a long-time limit is addressed. We show that the long-time limit of the numerical solution contains an oscillatory mode, which we are able to isolate. Lastly, we show that the long-time limit consistency is lost.

# 1 Loss of the TVD Property

In this section, we are interested in the one-dimensional first-order wave system (1), solved on an infinite domain $\Omega$, using a finite volume scheme of the form (2)–(3). Let us recall that the TV norm of $v = (v_i)_{i\in\mathbb{N}}$ is given by $\mathrm{TV}(v) = \sum_j |v_{j+1} - v_j|$. In what follows, the characteristic variables $\dfrac{\widehat{p}}{2} - \dfrac{\widehat{u}}{2\widehat{\rho c}}$ and $\dfrac{\widehat{p}}{2} + \dfrac{\widehat{u}}{2\widehat{\rho c}}$ are denoted by $\widehat{\mathcal{C}}_-$ and $\widehat{\mathcal{C}}_+$, respectively. The following proposition is presented.

**Proposition 1** (Total Variation Diminishing property)

- *The explicit Godunov scheme (2)–(3) with $\theta = 1$ is TVD on the characteristic variables $\widehat{\mathcal{C}}_-$ and $\widehat{\mathcal{C}}_+$ under the CFL condition $0 \le \widehat{c}\delta\tau/\delta x \le 1$, i.e*

$$\mathrm{TV}\left(\widehat{\mathcal{C}}_{\mp}^{n+1}\right) \le \mathrm{TV}\left(\widehat{\mathcal{C}}_{\mp}^{n}\right), \quad \forall\, \widehat{\mathcal{C}}_{\mp}^{n}.$$

- *The explicit pressure-centered scheme (2)–(3) with $\theta = 0$ is not TVD, i.e., for all $\delta\tau \in ]0, \delta x/(2\widehat{c})[$, there exists an initial condition $\left(\widehat{p}^0, \widehat{u}^0\right)$ for which*

$$\mathrm{TV}\left(\widehat{\mathcal{C}}_{\mp}^{1}\right) > \mathrm{TV}\left(\widehat{\mathcal{C}}_{\mp}^{0}\right).$$

***Proof*** Regarding the conservation of the TVD property for the explicit Godunov scheme, it follows immediately since the Godunov scheme written in terms of the characteristic variables is an upwind scheme. We refer to [4], for a detailed proof.

Regarding the loss of the TVD property for the explicit pressure-centered scheme, it is obtained by considering the following initial condition: $\widehat{p}^0/2 = \widehat{u}^0/(2\widehat{\rho c})$ where $\widehat{u}^0$ will be taken such that it has a discontinuity at some point. For example, we take the discrete initial condition defined by $\widehat{u}_j^0 = -1$ for $j > k$ and $\widehat{u}_j^0 = 1$ for $j \leq k$. We thus obtain

$$\begin{aligned} \mathrm{TV}\left(\widehat{C}_-^1\right) &= \frac{\widehat{c}\delta t}{2\delta x}\sum_j \left|\frac{\widehat{u}_{j+2}^0 - 3\widehat{u}_{j+1}^0 + 3\widehat{u}_j^0 - \widehat{u}_{j-1}^0}{2\widehat{\rho c}}\right| = \frac{\delta t}{\widehat{\rho}\delta x}\left|\widehat{u}_k^0 - \widehat{u}_{k+1}^0\right| \\ &= \frac{2\delta t}{\widehat{\rho}\delta x} > 0 = \mathrm{TV}\left(\widehat{C}_-^0\right), \end{aligned}$$

which ends the proof.

In order to numerically illustrate the loss of the TVD property we consider the one dimensional Riemann test case for unsteady flows, associated with the first-order wave system (1). The initial condition considered for the selected test case is the following

$$\widehat{p}_L^0 = \widehat{u}_L^0 = 1, \quad \widehat{p}_R^0 = \widehat{u}_R^0 = -1, \tag{5}$$

with final time equal to 0.1, $\widehat{c} = \widehat{\rho} = 1$ and a CFL number equal to 0.45.

With the chosen initial condition, we remark that the characteristic variable $\widehat{C}_-$ is initialized as zero. In Fig. 1, we observe that with the Godunov scheme, the uniformity of $\widehat{C}_-$ is preserved over time since the Godunov scheme is TVD on the characteristic variables. However, with the pressure-centered scheme, the uniformity of this characteristic variable is not preserved. Moreover, oscillations appear on the characteristic variables, as well as on the pressure and velocity. Note that the same phenomenon is observed when considering an initial uniform characteristic variable $\widehat{C}_+$. In Fig. 2, the plot of the TV norms is presented. It highlights that the TV norms of the two characteristic variables are preserved for the Godunov scheme but increase for the pressure-centered scheme. Similar results were obtained by considering an implicit time integration for both the Godunov and the pressure-centered schemes, but these results are not presented in this paper.

## 2 Loss of Mesh Convergence of the Numerical Solution in the Long Time Limit

In this section, we are interested in the linear wave system (1), solved on a bounded domain $\Omega$, with wall and Steger-Warming boundary conditions. In this framework, the long-time limit of the solution of both the continuous wave system (1) and the

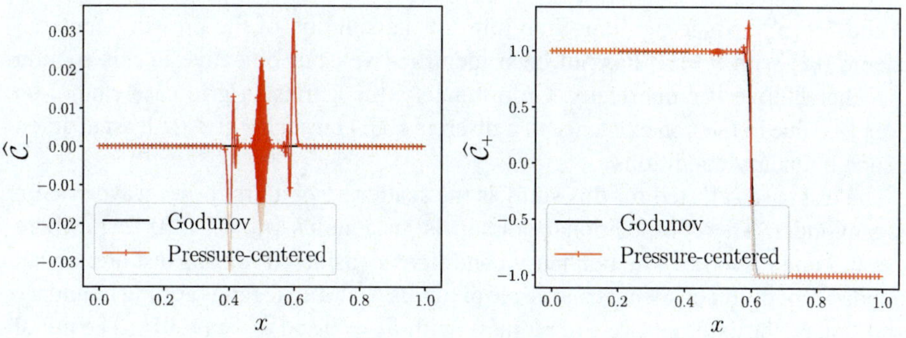


**Fig. 1** One-dimensional Riemann problem: characteristic variables at $t = 0.1$

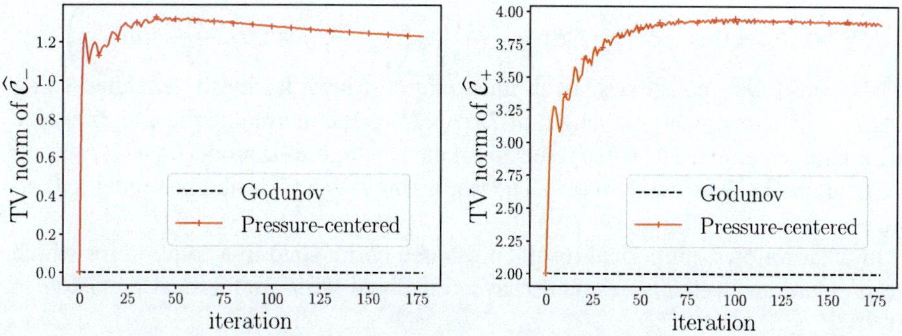


**Fig. 2** One-dimensional Riemann problem: TV norm of the characteristic variables as a function of time iterations

numerical schemes (2)–(4) with $\theta \in \{0, 1\}$ exists (see [9]). This limit can be identified thanks to a Hodge-Helmholtz Decomposition (HHD). Indeed, at the continuous level, by considering a vector field $\widehat{\mathbf{u}} \in L^2(\Omega)$ defined on a bounded domain $\Omega$ with a smooth boundary $\partial\Omega$, $\widehat{\mathbf{u}}$ can be uniquely decomposed into a sum of an irrotational component (which is a gradient of a scalar potential function), and a divergence free one (which is a curl of a vector potential function). The proof for this decomposition in the case where $\widehat{\mathbf{u}}$ satisfies wall/Steger-Warming boundary conditions can be found in [8] and is based on solving a Poisson equation with Neumann boundary conditions. At the discrete level, a similar exact decomposition is obtained on triangular grids in [8], see also [1, 4–6] for other boundary conditions. By considering a piecewise constant vector field $\widehat{\mathbf{u}}_h$, its discrete HHD is given by $\widehat{\mathbf{u}}_h = \mathcal{P}^{\psi}[\widehat{\mathbf{u}}_h] + \mathcal{P}^{\varphi}[\widehat{\mathbf{u}}_h]$, where $\mathcal{P}^{\psi}[\widehat{\mathbf{u}}_h]$ is the curl of a continuous $\mathbb{P}^1$ potential vector, $\mathcal{P}^{\varphi}[\widehat{\mathbf{u}}_h]$ is the gradient of a Crouzeix-Raviart scalar potential and $\mathcal{P}^{\psi}[\widehat{\mathbf{u}}_h] \cdot \mathbf{n} = \widehat{\mathbf{u}}_b \cdot \mathbf{n}$ on the boundary (see [8, Proposition 5], for details about the proof). Back to the identification of the long-time limit, at the continuous level, it is identified as the uniform pressure $\widehat{p}_b$ and the divergence free component of the HHD of the initial velocity $\widehat{\mathbf{u}}^0$. On triangular meshes, with the Godunov scheme ((2)–(4) with $\theta = 1$), the long-time limit of $\widehat{p}_h$ and $\widehat{\mathbf{u}}_h$ is consistent with the one obtained at the continuous level: it corresponds to

$\widehat{p}_b$ and $\mathcal{P}^{\psi}[\widehat{\mathbf{u}}_h^0]$. Since the long-time limit of the solution of the pressure-centered scheme ((2) with $\theta = 0$) has not been identified yet, our objective in this section is to characterize it numerically. Unfortunately, the Cartesian grid case cannot be addressed due to the non existence of a discrete HHD on such a grid (at least adapted to such boundary conditions).

The test case selected for this study is the scattering of a first-order wave system by a cylinder. The computational domain is an annulus $[r_0, r_1] \times [0, 2\pi[$, where $r_0 = 0.5$ and $r_1 = 5.5$. The boundary conditions considered for this test are: a wall boundary condition on the internal circle of radius $r_0$ and a Steger-Warming boundary condition on the external circle of radius $r_1$ with $\widehat{p}_b = 0$ and $\widehat{\mathbf{u}}_b = (1, 0)^T$. The initial condition is set to $\widehat{p}^0 = 0$ and $\widehat{\mathbf{u}}^0 = (0, 0)^T$.

It is worth emphasizing that the exact steady solution is known [9] and is given by $\widehat{p}^{\text{exact}}(r, \theta) = 0$ and $\widehat{\mathbf{u}}^{\text{exact}}(r, \theta) = \dfrac{r_1^2}{r_1^2 - r_0^2}\left(1 - \dfrac{r_0^2}{r^2}\cos(2\theta), -\dfrac{r_0^2}{r^2}\sin(2\theta)\right)^T$.

The mesh type considered is an unstructured triangular mesh generated using GMSH, with a characteristic length of $2\pi r_0/4N$ on the internal circle and $2\pi r_1/4N$ on the external circle. The mesh selected for the next figures is produced with $N = 40$ and contains 1332 triangular cells. The test is run with a CFL number equal to 0.45 and $\widehat{\rho} = \widehat{\kappa} = 1$.

In what follows, numerical results presented correspond to a solution for which the pressure and velocity residuals have converged with a numerical precision of $10^{-14}$.

In Fig. 3, the iso-contours of the $L^2$ norm of the long-time limit of the exact and numerical velocity fields are plotted. We observe that the numerical velocity field computed using the Godunov scheme is close to the exact velocity field, while the one obtained with the pressure-centered scheme is also close but polluted by spurious oscillations. Using the long-time consistency result proved in [8, 9] and recalled above, the long-time limit of the numerical velocity field $\widehat{\mathbf{u}}_h^{\infty}$ computed using the Godunov scheme is $\mathcal{P}[\widehat{\mathbf{u}}_h^0]$, the divergence free component of the discrete HHD of the initial velocity[1] $\widehat{\mathbf{u}}_h^0$. If $\widehat{\mathbf{u}}_h^{\infty}$ is the long-time limit of the numerical velocity field, we define

$$\widehat{\mathbf{u}}_h^{\infty,\text{Spurious}} = \widehat{\mathbf{u}}_h^{\infty} - \mathcal{P}^{\psi}[\widehat{\mathbf{u}}_h^0]. \tag{6}$$

The difference (6) is plotted in Fig. 4 for the Godunov and the pressure-centered schemes. With the Godunov scheme, the difference is zero, as expected, while with the pressure-centered scheme, (6) is a spurious oscillatory mode and its scale is of order $10^{-1}$, which is not negligible compared to the magnitude of the velocity field on the boundary which is of magnitude 1.

In order to study the long-time consistency of the scheme, a grid convergence study is performed. To do so, unstructured triangular meshes obtained with $N \in \{12, 20, 40, 80, 160, 320\}$ containing 96, 332, 1332, 5154, 20546, 83810 cells respectively, are considered. In Fig. 5, the $L^2$ norm of the error between the exact

[1] $\widehat{\mathbf{u}}_h^0$ is equal to $\mathbf{0}$ for this test case (note that $\mathcal{P}^{\psi}[\widehat{\mathbf{u}}_h^0]$ is not equal to $\mathbf{0}$ due to non homogeneous boundary conditions).

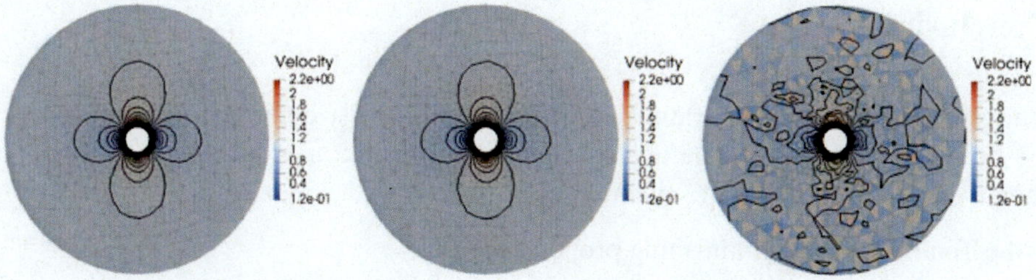


Fig. 3 Numerical velocity field computed using the Godunov and the pressure-centered schemes: Exact (left), Godunov (middle), Pressure-centered (right)

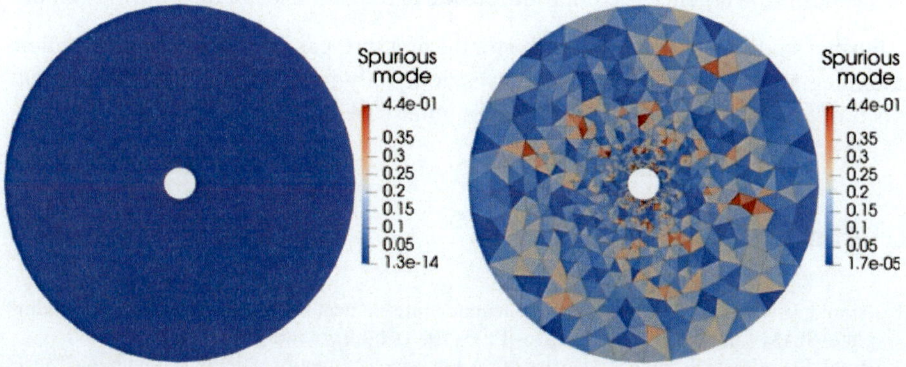


Fig. 4 Difference (6) computed using the Godunov and the pressure-centered schemes: Godunov (left), pressure-centered (right)

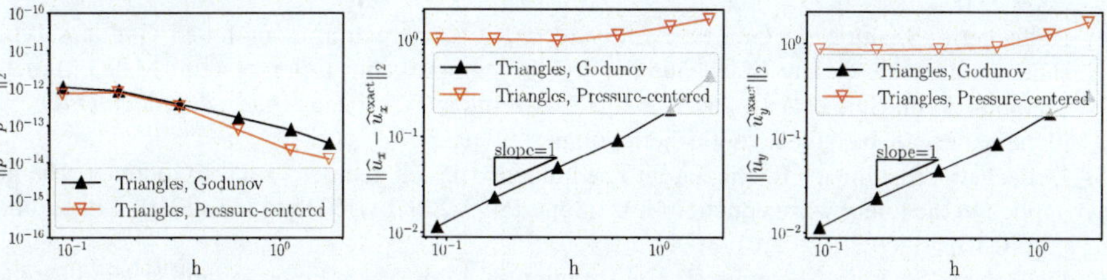


Fig. 5 $L^2$ norm of the error between the exact solution and the long-time numerical solution computed using the Godunov and the pressure-centered schemes as a function of the mesh step

solution and the long-time numerical solution computed using the Godunov and the pressure-centered schemes is plotted. We observe that with both schemes, the long-time limit of the pressure matches a uniform pressure, which is consistent with the long-time limit of the continuous pressure, as explained in [9]. However, regarding the long-time velocity field, the order of convergence is 1 with the Godunov scheme, while the pressure-centered scheme does not converge. This provides numerical evidence that the presence of the spurious mode jeopardizes the long-time limit consistency for the velocity.

## Conclusion

In this paper, the study of the numerical resolution of a first-order wave system using the pressure-centered scheme and explicit time integration is addressed. We have shown that, with this type of scheme,

- the Total Variation Diminishing property is lost.
- Oscillations appear in the unsteady numerical solution.
- The long-time numerical solution contains a spurious mode in the velocity field, which is an oscillatory mode.
- The mesh convergence is compromised due to the existence of this spurious mode.

Further study needs to be conducted for the pressure-centered scheme with implicit time integration. Moreover, the study was done only for triangular cell geometries, thus it needs to be extended to the quadrangular case.

## References

1. Arnold, D.N., Falk, R.S.: A uniformly accurate finite element method for the Reissner-Mindlin plate. SIAM J. Numer. Anal. **26**, 1276–1290 (1989). https://doi.org/10.1137/0726074
2. Bruel, P., Delmas, S., Jung, J., Perrier, V.: A low Mach correction able to deal with low Mach acoustics. J. Comput. Phys. **378**, 723–759 (2019). https://doi.org/10.1016/j.jcp.2018.11.020
3. Dellacherie, S.: Analysis of Godunov type schemes applied to the compressible Euler system at low Mach number. J. Comput. Phys. **229**(4), 978–1016 (2010). https://doi.org/10.1016/j.jcp.2009.09.044
4. Dellacherie, S., Jung, J., Omnes, P., Raviart, P.A.: Construction of modified Godunov-type schemes accurate at any Mach number for the compressible Euler system. Math. Models Methods Appl. Sci. **26**(13), 2525–2615 (2016). https://hal.science/file/index/docid/776629/filename/article_bas_mach_dellacherie_omnes_et_par.pdf
5. Dellacherie, S., Omnes, P., Rieper, F.: The influence of cell geometry on the Godunov scheme applied to the linear wave equation. J. Comput. Phys. **229**(14), 5315–5338 (2010). https://doi.org/10.1016/j.jcp.2010.03.012
6. Dellacherie S., Jung J., Omnes P.: Construction of a low Mach finite volume scheme for the isentropic Euler system with porosity. ESAIM: Math. Model. Numer. Anal. **1**;**55**(3), 1199–237 (2021). https://doi.org/10.1051/m2an/2021016
7. Guillard, H., Viozat, C.: On the behaviour of upwind schemes in the low Mach number limit. Comput. Fluids **28**(1), 63–86 (1999). https://doi.org/10.1016/S0045-7930(98)00017-6
8. Jung, J., Perrier, V.: Steady low Mach number flows: identification of the spurious mode and filtering method. J. Comput. Phys. **468**, 111462 (2022). https://doi.org/10.1016/j.jcp.2022.111462
9. Jung, J., Perrier, V.: Long time behavior of finite volume discretization of symmetrizable linear hyperbolic systems. IMA J. Numer. Anal. **43**(1), 326–356 (2023). https://doi.org/10.1093/imanum/drab092
10. Müller, B.: Low-Mach-number asymptotics of the Navier-Stokes equations. Floating, Flowing, Flying: Pieter J. Zandbergen's Life as Innovator, Inspirator and Instigator in Numerical Fluid Dynamics, pp. 97–109 (1998). https://doi.org/10.1007/978-94-017-1564-5_6
11. Rieper, F.: A low-Mach number fix for Roe"s approximate Riemann solver. J. Comput. Phys. **230**(13), 5263–5287 (2011). https://doi.org/10.1016/j.jcp.2011.03.025

# A Reinforcement Learning Based Slope Limiter for Two-Dimensional Finite Volume Schemes

**Jens Keim, Anna Schwarz, Simone Chiocchetti, Christian Rohde, and Andrea Beck**

**Abstract** Limiters in second-order finite volume schemes are generally too dissipative in smooth regions or require empirical parameter tuning. This paper aims to develop a new slope limiter for a two-dimensional finite volume scheme with reinforcement learning. The proposed limiter is based on two admissibility constraints: positivity of the solution and a relaxed discrete maximum principle. The new limiter is applied to selected well-known, two-dimensional Riemann problems, which illustrates its applicability and the potential of reinforcement learning in numerics.

**Keywords** Reinforcement learning · Finite volume · MUSCL

## 1 Introduction

Non-linear hyperbolic conservation laws are widely used in science and engineering. An intrinsic characteristic of such partial differential equations is the breakdown of classical solutions and the occurrence of discontinuities [1]. If one is interested in the solution of such systems, first-order finite volume (FV) schemes combined with a monotone numerical flux function turned out to be well suited for this task [2]. However, in practical applications, discontinuities occur only on sets of small measure while for the by far larger part of the domain the solution varies smoothly. Thus,

J. Keim (✉) · A. Schwarz · A. Beck
Institute of Aerodynamics and Gas Dynamics, University of Stuttgart, Stuttgart, Germany
e-mail: keim@iag.uni-stuttgart.de

A. Schwarz
e-mail: schwarz@iag.uni-stuttgart.de

S. Chiocchetti
Laboratory of Applied Mathematics, University of Trento, Trento, Italy

C. Rohde
Institute of Applied Analysis and Numerical Simulation, University of Stuttgart, Stuttgart, Germany



E. Franck et al. (eds.), *Finite Volumes for Complex Applications X—Volume 2, Hyperbolic and Related Problems*, Springer Proceedings in Mathematics & Statistics 433,
https://doi.org/10.1007/978-3-031-40860-1_22

first-order schemes require a massive grid resolution to be accurate. Especially for two- and three-dimensional problems, high-order methods are favorable in general.

Over the last decades, much effort has been dedicated to the development of higher-order schemes [3–7]. However, high-order methods are prone to oscillations at strong discontinuities. Therefore, a popular ansatz for such schemes is to introduce localized artificial viscosity, see for instance [8–11]. In the context of FV schemes, a non-linear reconstruction or limiting procedure is required if schemes of higher-order shall be constructed [12]. Hence, non-linear methods based on TVD limiting strategies [13, 14] or essentially non-oscillatory reconstruction schemes of (W)ENO-type [6, 15] have been proposed. Second-order TVD schemes of MUSCL-type [5, 16] are popular due to their algorithmic simplicity and robustness. However, a major drawback of these schemes is the optimal choice of the limiter function with respect to stability and accuracy. While the *minmod* limiter provides a robust scheme, other less dissipative limiters, such as the *superbee* limiter, may fail at strong discontinuities and local extrema. To resolve this issue, more sophisticated limiters have been proposed that allow the limiter to be tuned to the particular test case and, in general, to strike a balance between robustness and accuracy of the results [13, 14]. However, these tuning parameters are strongly test case dependent and a priori unknown. An alternative strategy to the discussed a priori limiters is the MOOD approach [7, 17], where the validity of the solution is examined in an a posteriori fashion based on some predefined admissibility criteria. If any of the defined criteria is violated, the solution of the corresponding elements has to be recomputed with a more robust scheme. This procedure enables a nearly oscillation-free treatment of discontinuities. However, depending on the actual implementation, this procedure can become computationally expensive. Therefore, the present work aims to design an a priori limiter for a two-dimensional MUSCL-Hancock scheme [16] which has the MOOD properties of an a posteriori limiter. This task is tackled by means of data-driven methods from machine learning.

Several researchers have proposed data-driven shock detection [18, 19] or capturing methods [20, 21]. However, all of them have in common that they were trained in a supervised learning (SL) fashion, i.e., none of the models actually considered the response of the numerical scheme to their prediction. In the approaches available, either known shock capturing schemes were replaced by a data-driven model [20, 21], or models were trained to distinguish between a numerically sound and a troubled solution [18, 19]. In the present work, methods from reinforcement learning (RL) are applied to train a data-driven a priori shock-capturing scheme.

Early successes of reinforcement learning have been achieved in the construction of artificial intelligences capable of playing complex games [22, 23]. In contrast to SL, the training process is carried out by allowing the trained model itself, called the agent, to interact with an environment. Besides games, this approach is well suited for many types of control processes, e.g., in computational fluid dynamics [24–26]. In this paper, the environment is a two-dimensional MUSCL-Hancock scheme and the agent is a neural network. The inputs to the neural network are the integral mean values of the primitive variables, density and pressure, of a nine-point stencil, which extends the five-point stencil generally employed for a two-dimensional second-order reconstruction scheme on a Cartesian mesh. In the present work, the framework in [27] for the Euler equations and an ideal gas equation of state is extended to two-dimensions using a slightly modified reward function and convolutional neural networks.

The remainder of the paper is structured as follows. First, in Sect. 2, we briefly outline the applied RL algorithm. In Sect. 3, the one-dimensional framework in [27] is revisited and its extension to two dimensions is discussed. Turning to numerical experiments, the applicability and predictive performance of the developed limiter on selected well-known standard test cases is discussed in Sect. 4. We conclude the paper with a short summary and an outlook in Sect. 5.

## 2 Reinforcement Learning

The main idea of RL is to learn a time-dependent problem through an agent which receives a feedback for its interaction with a given environment. For this, the agent has to be endowed with all essential information to transit to the next state. The problem is described as a Markov decision process (MDP), where at each discrete time level $t^n \in \mathbb{R}_+$ with $n \in \mathbb{N}$, the agent selects an action $a \in \mathcal{A}$ ($\mathcal{A}$: action space) given by the policy $\pi$ and based on the state $s^n \in \mathcal{S}$ ($\mathcal{S}$: state space), receives a reward $r : \mathcal{A} \times \mathcal{S} \rightarrow \mathbb{R}$ for its choice and transits to the next state $s^{n+1}$. Each transition is stored in the so-called transition tuple $\mathcal{T} = (s^n, a, r, s^{n+1})$. Here, the policy $\pi : \mathcal{S} \rightarrow \mathcal{A}$ is chosen to be deterministic. In this sense, the agents aims to maximize the accumulated discounted reward over time, designated as the return $R = \sum_{i=n}^{\infty} \gamma^{i-n} r(s^i, a^i)$, and in turn find the optimal policy $\pi_{\boldsymbol{\phi}}(s^n) = \pi(s^n; \boldsymbol{\phi})$, with parameters $\boldsymbol{\phi} \in \mathbb{R}^{n_\pi},\ n_\pi \in \mathbb{N}_{>0}$. The discount factor $0 < \gamma < 1$ assures that immediate returns are weighted more than future ones [22]. To find the optimal policy, we employ a policy gradient algorithm, where the parameters of the policy are adjusted in the direction of the gradient of the performance objective, $J(\pi_{\boldsymbol{\phi}}) = \mathbb{E}[R|\pi_{\boldsymbol{\phi}}]$, with the expectation operator $\mathbb{E}[\cdot]$. The gradient of the performance objective is given by the deterministic policy gradient theorem [28] as

$$\nabla_{\boldsymbol{\phi}} J(\pi_{\boldsymbol{\phi}}) = \mathbb{E}_{s^n \sim \rho^\pi} [\nabla_a Q^\pi(s^n, a)|_{a=\pi_{\boldsymbol{\phi}}(s)} \nabla_{\boldsymbol{\phi}} \pi_{\boldsymbol{\phi}}(s^n)], \tag{1}$$

where $\mathbb{E}_{s^n \sim \rho^\pi}[\cdot]$ is the expected value as defined in [28] and $Q^\pi(s^n, a)$ is the action-value function, i.e., the expected return for an action $a = \pi_{\boldsymbol{\phi}}(s^n)$ originating from $s^n$. The unknown value $Q^\pi(s^n, a)$ can be obtained from a policy evaluation algorithm, e.g., temporal difference (TD) learning [28], where the true value is approximated by $Q^\pi(s^n, a) \approx Q_{\boldsymbol{\theta}}(s^n, a) = Q(s^n, a; \boldsymbol{\theta})$ with the parameters $\boldsymbol{\theta} \in \mathbb{R}^{n_Q},\ n_Q \in \mathbb{N}_{>0}$. In TD learning, $Q_{\boldsymbol{\theta}}(s^n, a)$ is estimated by the so-called Bellman equation, see e.g. [29], given as $Q_{\boldsymbol{\theta}}(s^n, a) = r + \gamma\ Q_{\boldsymbol{\theta}}(s^{n+1}, \pi_{\boldsymbol{\phi}}(s^{n+1}))$.

A popular choice of policy gradient methods based on Eq. (1) and TD learning are actor-critic algorithms [22, 28]. An actor-critic algorithm is equipped with an actor to update the parameters of $\pi_{\boldsymbol{\phi}}(s^n)$ and a critic to assess $Q_{\boldsymbol{\theta}}(s^n, a)$, e.g., via TD learning. Thereby, the policy can be divided into a target and a behavior policy. The (stochastic) behavior policy $\tilde{\pi}_{\boldsymbol{\phi}} = \pi(s^n, a; \boldsymbol{\phi})$ with $a$ from $\pi_{\boldsymbol{\phi}}$, is used to draw the action during training, while the target of both the actor and the critic are determined via the (deterministic) target policy, $\pi_{\boldsymbol{\phi}} = \pi(s^n; \boldsymbol{\phi})$. Such methods are called off-policy algorithms. The advantage of these methods is that transitions can be written to an experience replay buffer, see e.g. [29]. To handle continuous action and state spaces, both the actor and the critic are approximated by neural networks.

In this work, we utilize a state of the art off-policy actor-critic algorithm, called twin delayed deep deterministic policy gradient (TD3) [29]. TD3 is based on a clipped variant of the original double Q-learning [30], where additional target networks with parameters $\boldsymbol{\theta}'$ and $\boldsymbol{\phi}'$ for both the critic and the actor are introduced to enable a less biased estimate of $\pi_{\boldsymbol{\phi}}$ and $Q_{\boldsymbol{\theta}}$. Moreover, two critic networks, $Q_{\boldsymbol{\theta}_p}$, $p = 1, 2$, are used and fitted via the mean square ($L_2$) error between their prediction and the objective, where the latter is given by

$$y = r + \gamma \min_{p=1,2} Q_{\boldsymbol{\theta}'_p}(s^{n+1}, \pi_{\boldsymbol{\phi}'}(s^{n+1}) + \epsilon), \ \epsilon \sim \text{clip}(\mathcal{N}(0, \tilde{\sigma}), -c, c), \quad (2)$$

where $\mathcal{N}(\mu, \sigma)$ is the Gaussian distribution with variance $\sigma$ and mean $\mu$, $\gamma = 0.99$, $c = 0.5$ and $\tilde{\sigma} = 0.2$, as suggested in [29]. Following [29], the update of the actor is delayed to the critic, e.g., every $\tilde{d}$ iterations, to reduce the error per update. Further details on the algorithm and its performance are given in [29].

## 3 The *RL*Indi2D Framework

In the following, we briefly present the methodology developed to train an RL-based slope limiter. The structure of the algorithm is illustrated in Algorithm 1.

At the beginning of the algorithm, the networks and a replay buffer $\mathcal{B} \in \mathbb{R}^{n_{RB}}$ of size $n_{RB} = 5 \cdot 10^4$ are initialized. The number of training epochs is chosen as $n_{\text{epoch}} = 1000$. The actor and its target are convolutional neural networks with two layers, each composed of 32 filters with a filter size of $2 \times 2$ (no padding). Both critics, as well as their targets, are multilayer perceptrons with two hidden layers, each composed of 300 nodes. The *SiLU* function [31] is applied between the layers. After the output of the actor, a *tanh* function is utilized.

In each training epoch, arbitrarily initialized two-dimensional Riemann problems for the Euler equations (ideal gas) are solved on the computational domain $\Omega = [0, 1]^2$ until a predefined end time $t_{\text{end}} = 0.2$ is reached. Following [32], the initial conditions are given as vector of unknowns consisting of density, velocities and pressure, $(\rho_q, u_{x,q}, u_{y,q}, p_q)^T$ for $q = 1, \ldots, 4$, where $q = 1$ corresponds to the first quadrant ($x \geq 0.5$, $y \geq 0.5$) and the indexing is counterclockwise. The density is chosen to be $\rho_q \sim \mathcal{U}[0.01, 4]$, the velocities $u_{x,q}, u_{y,q} \sim \mathcal{U}[-2, 2]$ and the pressure $p_q \sim \mathcal{U}[0.01, 10]$. The operator $\mathcal{U}[a, b]$ denotes a uniform distribution in the interval $I = [a, b]$. The training is performed with a representative discretization of $N_x \times N_y = 50 \times 50$ elements ($N = 50^2$) and CFL $= 0.99$.

In each epoch, the procedure begins with the initialization of the test case and the calculation of the time step $dt$. Then, the slope $\delta\mathbf{q}_k$, $k = 1, 2$ $(1 : x, 2 : y)$, with $\mathbf{q} = (\rho, u_1, u_2, p)^T$, in cell $i$ for each direction is calculated as, see e.g. [14],

$$\delta\mathbf{q}_k = 0.5\left[(1 - a_{k,i})(\mathbf{q}_{i+1} - \mathbf{q}_i) + (1 + a_{k,i})(\mathbf{q}_i - \mathbf{q}_{i-1})\right],\ a_{k,i} \in [-1, 1] \tag{3}$$

by the use of the actions predicted by the actor network $a_{k,i} = \text{clip}(\pi_\phi(\tilde{\mathbf{s}}_i^n) + \epsilon,\ a_{\text{low}} = -1,\ a_{\text{up}} = 1)$, to which an exploration noise $\epsilon \sim \mathcal{N}(0, \sigma), \sigma \in \mathbb{R}_{>0}$ is added, here $\sigma = 0.1$, as suggested in [29]. The definition of the operator clip$(\cdot)$ is given in [29]. The input $\tilde{\mathbf{s}}_i^n$ to the actor is a nine-point stencil composed of the normalized solution at the current time level $t^n$ in cell $i$ and its direct Voronoi neighbors, i.e., the adjacent and the diagonal cells. For this, the physical state

$$\mathbf{s}_i^n = \left(\left(V^n_{l^-,m^+}, V^n_{l,m^+}, V^n_{l^+,m^+}\right), \left(V^n_{l^-,m}, V^n_{l,m}, V^n_{l^+,m}\right), \left(V^n_{l^-,m^+}, V^n_{l,m^-}, V^n_{l^+,m^-}\right)\right)^T$$

with $l = 1, \ldots, N_x, m = 1, \ldots, N_y, i = 1, \ldots, N$ and $V = \{\rho, p\}$ is normalized by a min-max normalization NORMALIZE$(\mathbf{s}_i) : \mathbf{s}_i \in \mathbb{R}^{3\times 3} \mapsto \tilde{\mathbf{s}}_i \in [0, 1]^2$, see [27]. The superscripts $(\cdot)^-$ and $(\cdot)^+$ indicate $(\cdot) - 1$ or $(\cdot) + 1$, respectively. For states which are nearly constant, $\mathbf{s}_{c,i} = (\max(\mathbf{s}_i) - \min(\mathbf{s}_i) \leq \epsilon_1 \max(\mathbf{s}_i))$, the action is set to $a_{k,i} = 1$ for cell $i$ and each direction $k$. This can be interpreted as a restrictive oscillation indicator [17], here $\epsilon_1 = 10^{-5}$. With the computed slopes, the solution $\mathbf{q}^n$ can be advanced in time by the MUSCL-Hancock scheme, yielding $\mathbf{q}^{n+1}$ at $t^{n+1} = t^n + dt$; as a Riemann solver, the HLL approximate flux [33] is applied. Since RL algorithms are based on MDP, transition tuples $\mathcal{T} = (\tilde{\mathbf{s}}^n, \mathbf{a}, \mathbf{r}, \tilde{\mathbf{s}}^{n+1}, \mathbb{S})$ are stored in the replay buffer $\mathcal{B}$ for each discrete $dt$. These tuples are composed of the normalized states at $t^n$, $\tilde{\mathbf{s}}^n = \{\tilde{\mathbf{s}}_i^n\}_{i=1}^N$, the normalized states at the next, now current, time level $t^{n+1}$, $\tilde{\mathbf{s}}^{n+1}$, the actions $\mathbf{a} = (a_1, a_2)^T$, composed of the actions in each direction, the rewards $\mathbf{r} =$ REWARD$(\mathbf{s}^{n+1}, \mathbf{s}^n, \mathbf{a})$ and the positivity condition of the new state $\mathbb{S} =$SANITY$(\mathbf{s}^{n+1})$, see [27]. For $\mathbf{s}_c^n$, the slope is known and the corresponding transition tuple can be neglected during training.

An adequate definition of the reward is generally critical for the success of the training. In this work, the reward is derived under the following considerations. First, samples are penalized by a negative reward if the positivity of their solution cannot be guaranteed. Second, it has to be ensured that the action predicted by the actor yields the maximum permissible slope for which oscillations are bounded over time. With this in mind, the reward in each direction is defined as

$$\text{REW}(\mathbf{s}_i^{n+1}, \mathbf{s}_i^n, \mathbf{a}_i, k) = \begin{cases} c_1 & : \mathbb{S}_i = 1, \\ c_2|a_{k,i}| & : \mathbb{M}_i = 1 \wedge \neg\mathbb{J}_k, \\ |a_{k,i}| & : \mathbb{M}_i = 1 \wedge \mathbb{J}_k, \\ a_{k,i} & : \mathbb{M}_i = 0 \wedge |\hat{r}_k| > 1, \\ -a_{k,i} & : \mathbb{M}_i = 0 \wedge |\hat{r}_k| < 1,\ i = 1, \ldots, N, \end{cases} \tag{4}$$

**Algorithm 1** *RL*Indi2D-Framework

Initialize critic $Q_{\theta_1}$, $Q_{\theta_2}$ and actor $\pi_{\phi}$ networks with random parameters $\theta_1, \theta_2, \phi \sim \mathcal{N}(0, 1)$, target networks as in [29], and the replay buffer $\mathcal{B}$
**for** $i_{\text{epoch}} = 1$ **to** $n_{\text{epoch}}$ **do**
  Initialize test case, $t^n = t_0$
  **while** $t < t_{\text{end}}$ **and** $\forall j \in \mathbb{S} : j = 0$ **do**
    Compute $dt$ and state vector $\mathbf{s}^n = \{\mathbf{s}_i^n\}_{i=1}^N$ and $\tilde{\mathbf{s}}^n = \text{NORMALIZE}(\mathbf{s}^n)$
    Select actions $\mathbf{a} = \text{clip}(\pi_{\phi}(\tilde{\mathbf{s}}^n) + \epsilon, -1, 1)$, $\epsilon \sim \mathcal{N}(0, \sigma)$
    Compute slopes $\delta \mathbf{q} = f(\mathbf{s}^n, \mathbf{a})$ according to Eq. (3)
    Update solution to $t^{n+1} = t^n + dt$
    Compute state vector $\mathbf{s}_i^{n+1} = \{\mathbf{s}_i^{n+1}\}_{i=1}^N$ and $\tilde{\mathbf{s}}^{n+1} = \text{NORMALIZE}(\mathbf{s}^{n+1})$
    Check positivity $\mathbb{S} = \text{SANITY}(\mathbf{s}^{n+1})$ and compute $\mathbf{r} = \text{REWARD}(\mathbf{s}^{n+1}, \mathbf{s}^n, \mathbf{a})$
    Store transition tuples $(\tilde{\mathbf{s}}^n, \mathbf{a}, \mathbf{r}, \tilde{\mathbf{s}}^{n+1}, \mathbb{S})$ in $\mathcal{B}$
  **end while**
  Sample mini-batches of $m$ transitions from $\mathcal{B}$
  Update critics $\theta_{1,2} \leftarrow \eta m^{-1} \sum_m (\mathbf{y} - Q_{\theta_{1,2}}(\tilde{\mathbf{s}}^n, \mathbf{a}))^2$, $\mathbf{y}$ from Eq. (2)
  **if** $i_{\text{epoch}}$ mod $\tilde{d}$ **then**
    Update the actor, Eq. (1), and the target networks, see [29]
  **end if**
**end for**

where $\mathbb{J}_k = \big((a_{k,i} < -1 + \epsilon_3 : |\hat{r}_k| > 1) \vee (a_{k,i} > 1 - \epsilon_3 : |\hat{r}_k| < 1)\big)$. In Eq. (4), the ratio $\hat{r}_k$ is defined as, e.g., $\hat{r}_{k=1} = f(\mathbf{s}_i^n) = \frac{\rho_I - \rho_{I-}}{\rho_{I+} - \rho_I}$ and $c_1 = -50$, $c_2 = -10$, $\epsilon_3 = 10^{-1}$ are tuning parameters, which may depend on the considered $\mathcal{S}$. Moreover, an oscillation indicator, denoted by $\mathbb{M}_i$, is utilized to aid the training process, see [27] for details. The final reward is the sum of the rewards in each direction, i.e., $\mathbf{r} = \text{REWARD}(\mathbf{s}_i^{n+1}, \mathbf{s}_i^n, \mathbf{a}_i) = \sum_k \text{REW}(\mathbf{s}_i^{n+1}, \mathbf{s}_i^n, \mathbf{a}_i, k)$.

The first condition in Eq. (4) accounts for invalid states. The other cases are the following: First, state-action pairs which exhibit an oscillating solution, i.e., $\mathbb{M}_i = 1$, receive a negative reward, scaled by the used action (second case). However, it has to be guaranteed that these pairs are not penalized if their actions yield the lowest admissible slope, i.e., the slope predicted by the *minmod* limiter (third case). Second, for valid states, i.e., a less oscillatory or non-oscillating solution, the best action in terms of the maximum permissible slope depends on the magnitude of $\hat{r}_k$. According to [13], the optimal action for $\hat{r}_k > 1$ is $a^* \to 1$, otherwise $a^* \to -1$. Thus, for $\hat{r}_k > 1$ or $\hat{r}_k < 1$, an action with a negative (fourth case) or positive (fifth case) sign, respectively, causes a negative reward and enforces the actor to converge to $a^*$. Since constant samples are not stored in $\mathcal{B}$, the condition $\hat{r}_k = 1$ is never true.

Then, mini-batches of size $m = 100$ are randomly drawn from $\mathcal{B}$. The Adam algorithm [34] with a learning rate of $\eta = 10^{-3}$ is employed to optimize the parameters of both critic networks and the ones of the actor. The actor and the target networks are updated every $\tilde{d} = 2$ epochs, see [29] for details. The action is randomly drawn from $\mathbf{a} \sim \mathcal{U}[-1, 1]$ in the first 100 epochs to ensure that the training procedure features enough exploration such that the networks are able to find the best action according to Eq. (4). The networks are trained in $\leq 2h$ on a NVIDIA Tesla K40c.

## 4 Numerical Results

For the validation of the methodology, the trained actor is applied to selected well-known 2D Riemann problems, configuration 4, F and K in [32], and the results are compared to the ones obtained with the *OSPRE* limiter [35]. Configuration 4 features four shock waves, configuration F two shock waves and two contact discontinuities and configuration K consists of one shock wave, two contact discontinuities and one rarefaction wave. The computational domain $\Omega = [0, 1]^2$ was discretized with a grid spacing of $h = 1/500$ in each direction, i.e., a ten times finer grid spacing than during training. The results of the *RL*Indi2D-limiter are in agreement to those obtained with the MUSCL-Hancock scheme and the robust low-dissipative *OSPRE* limiter [35], as illustrated in Fig. 1. However, the *RL*Indi2D-limiter is able to resolve finer structures and tends to allow higher reconstructed slopes at discontinuities while maintaining the stability of the scheme, especially for configuration K.

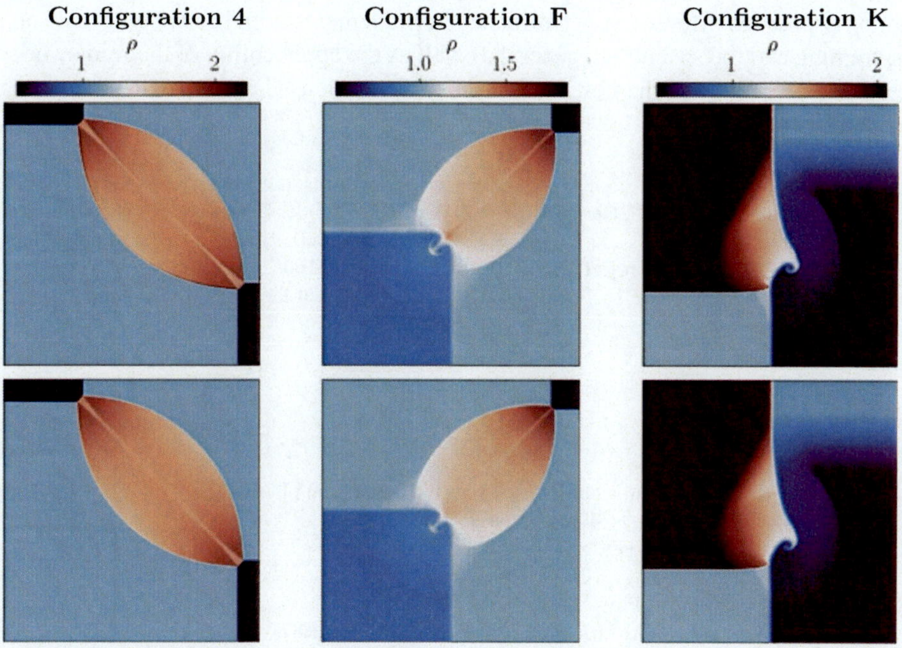


**Fig. 1** Density of the 2D Riemann problems for configuration 4, F and K (left to right) at $t_{\text{end}} = 0.25$ with $500 \times 500$ elements. Top: *RL*Indi2D, bottom: *OSPRE*

## 5 Conclusion and Outlook

In this work, we proposed an RL-based slope limiter embedded in a second-order FV solver. The reward function was defined in an a posteriori manner through the use of admissibility constraints based on the positivity of the solution and a relaxed discrete maximum principle. Hence, the long-term stability properties of the proposed a priori limiter converge towards those of the a posteriori limiter. The framework was trained on arbitrarily initialized two-dimensional Riemann problems to enable a wide applicability, but generalization to arbitrary 2D problems is limited, which is a common problem in machine learning. Furthermore, the off-policy RL algorithm in combination with the replay buffer is fundamental to guarantee an unbiased training set in terms of the reward function. The successful application of the proposed framework to selected well-known Riemann problems demonstrates the potential of RL in numerics.

Future works may cover the extension of the framework to, e.g., a larger subset of initial conditions, arbitrary resolutions and unstructured grids to improve its generalization properties. Moreover, performance improvements with special emphasis on the training and the speedup in general as well as the applicability of the framework to other discretization schemes, such as the discontinuous Galerkin method, will be tackled in the future.

**Acknowledgements** The research presented in this paper was funded in parts by Deutsche Forschungsgemeinschaft (DFG, German Research Foundation) under Germany's Excellence Strategy - EXC 2075 - 390740016 and by the DFG Rebound - 420603919. S. C. acknowledges the support of the DFG via the project DROPIT, grant no. GRK 2160/2, and funding from Project HPC-EUROPA3 (INFRAIA-2016-1-730897). S. C. is a member of the INdAM GNCS group.

## References

1. Dafermos, C.M.: Hyperbolic conservation laws in continuum physics (2010)
2. Harten, A.: J. Comput. Phys. **49**(3), 357–393 (1983)
3. Reed, W.H., Hill, T.R.: Los Alamos Sci. Lab. Report LA-UR-73-479 (1973)
4. Lele, S.K.: J. Comput. Phys. **103**(1), 16–42 (1992)
5. van Leer, B.: J. Comput. Phys. **32**(1), 101–136 (1979)
6. Harten, A., Osher, S., SIAM,: J. Num. Anal. **24**, 279–309 (1987)
7. Clain, S., Diot, S., Loubère, R.: J. Comput. Phys. **230**(10), 4028–4050 (2011)
8. Hesthaven, J.S., Kirby, R.M.: Math. Comput. **77**(263), 1425–1452 (2008)
9. Jameson, A., Schmidt, W., Turkel, E.: AIAA Paper, 81-1259 (1981)
10. Kuzmin, D., Möller, M.: Sci. Comput. 207–250 (2005)
11. Sonntag, M., Munz, C.D.: SIAM: J. Sci. Comput. **70**(3), 1262–1289 (2017)
12. Godunov. S.K., Bohachevsky, I.: Matematičeskij sbornik **47**(89)(3) (1959)
13. LeVeque, R.J.: Finite Volume Methods for Hyperbolic Problems. CUP (2002)
14. Toro, E.F.: Riemann Solvers and Numerical Methods for Fluid Dynamics (2009)
15. Shu, C.: NASA/CR-97-206253 ICASE Report No.97-65 (1997)
16. Van Leer, B.: SIAM J. Sci. Comput. (1997)
17. Dumbser, M., Zanotti, O., Loubère, R., Diot, S.: J. Comput. Phys. **278** (2014)

18. Ray, D., Hesthaven, J.S.: J. Comput. Phys. **397**, 108845 (2019)
19. Beck, A., Zeifang, J., Schwarz, A., Flad, D.: J. Comput. Phys. **423**, 109824 (2020)
20. Han Veiga, M., Abgrall, R.: ECCM ECFD VII, Glasgow, 2525–2550 (2018)
21. Yu, J., Hesthaven, J.S.: Comput. Fluids **245**, 105592 (2022)
22. Sutton, R.S., McAllester, D., Singh, S., Mansour, Y.: NeurIPS **12** (1999)
23. Silver, D., Hubert, T., Schrittwieser, J., et al.: Science **362**(6419), 1140–1144 (2018)
24. Fan, D., Yang, L., Wang, Z., et al.: PNAS **117**(42), 26091–26098 (2020)
25. Wang, Y., Shen, Z., Long, Z., Dong B.: (2020). arXiv:1905.11079
26. Novati, G., de Laroussilhe, H.L., et al.: Nat. Mach. Intell. **3**, 87–96 (2021)
27. Schwarz, A., Keim, J., Chiocchetti, S., Beck, A.: GAMM22. PAMM22 (2023)
28. Silver, D., Lever, G., Heess, N., Degris, T. et al.: In: ICML'14. PMLR (2014)
29. Fujimoto, S., van Hoof, H., Meger, D.: ICML **2018**(4), 2587–2601 (2018)
30. Hasselt, H.V.: Double Q-learning, NeurIPS, 2613–2621 (2010)
31. Hendrycks, D., Gimpel, K.: Gaussian Error Linear Units (GELUs) (2016)
32. Schulz-Rinne, C., Collins, J., Glaz, H.: SIAM J. Sci. Comput. **14**(6) (1993)
33. Harten, A., Lax, P., van Leer, B.: SIAM Rev. **25**, 35–61 (1983)
34. Kingma, D.P., Bapp, J.: Adam: A Method for Stochastic Optimization (2014)
35. Waterson, N., Deconinck, H.: Num. Meth. Laminar Turb. Flow **9** (1995)

# Essentially Non-oscillatory Schemes Using the Entropy Rate Criterion

**Simon-Christian Klein**

**Abstract** A new definition of numerical fluxes for Finite-Volume schemes is given. These numerical flux functions directly use the information from a recovery procedure, i.e. a point value and a recovery error estimate, to calculate a numerical flux. This flux implements the maximal entropy rate criterion pioneered by Dafermos in a numerical setting. The schemes show essentially non-oscillatory behavior during shock calculations and converge with a high order of accuracy for smooth solutions. Tests are carried out for Burgers' equation and the Buckley-Leverett equation.

**Keywords** Finite-volume · Optimal recovery · Reconstruction · Entropy stability · Entropy rate criterion

## 1 Introduction

We are interested in solving hyperbolic conservation laws [1]

$$\partial_t u + \partial_x f(u(x,t)) = 0 \tag{1}$$

using Finite-Volume methods. Here $f : \mathbb{R}^K \to \mathbb{R}^K$ shall be the flux function. If a convex function $U : \mathbb{R}^K \to \mathbb{R}$ together with a second function $F : \mathbb{R}^K \to \mathbb{R}$ satisfies $\partial_u U(u)\partial_u f(u) = \partial_u F(u)$ we call the pair $(U, F)$ an entropy flux pair [2]. One can show that in this case the additional conservation law $\partial_t U \circ u(x,t) + \partial_x F \circ u(x,t) = 0$ follows for smooth solutions [2], while vanishing viscosity solutions satisfy $\partial_t U(u(x,t)) + \partial_x F(u(x,t)) \leqslant 0$. In our case using a Finite-Volume (FV) method boils down to trying to predict the average values of the conserved variables $u_k^n = \frac{1}{x_{k+\frac{1}{2}} - x_{k-\frac{1}{2}}} \int_{x_{k-\frac{1}{2}}}^{x_{k+\frac{1}{2}}} u(x, t_n)\, \mathrm{d}x$ in a set of cells $[x_{k-\frac{1}{2}}, x_{k+\frac{1}{2}}]$. Integrating over the conservation law shows that the evolution of these mean values is deter-

S.-C. Klein (✉)
TU Braunschweig, Universitätsplatz 2, 38106 Braunschweig, Germany
e-mail: simon-christian.klein@tu-bs.de

 
E. Franck et al. (eds.), *Finite Volumes for Complex Applications X—Volume 2, Hyperbolic and Related Problems*, Springer Proceedings in Mathematics & Statistics 433,
https://doi.org/10.1007/978-3-031-40860-1_23

mined by the integral of the flux function over the cell edges [3]. Dividing by the time-step and going over to the limit leaves us with the semi-discrete version $\partial_t u_k = \frac{f\left(x_{k-\frac{1}{2}},t\right)-f\left(x_{k+\frac{1}{2}},t\right)}{x_{k+\frac{1}{2}}-x_{k-\frac{1}{2}}}$, and our job is to approximate the unknown values of the flux at $x_{k+\frac{1}{2}}$ while only having the mean values at hand. Popular solutions to this problem are the (W)ENO schemes [4, 5]. Apart from accuracy and a non-oscillatory solution one is also interested in building entropy dissipative schemes to enforce entropy inequalities in the limit [3, 6]. One hopes to single out a physically meaningful weak solution out of the many possible weak solutions using this technique. We will show a different method to approximate the unknown fluxes and will make use of a different entropy criterion. The entropy rate criterion was introduced by Dafermos [7]. It states that the total entropy $E_u(t) = \int U(u(x,t))\,\mathrm{d}x$ of the selected weak solution $u$ should diminish faster than the entropy of all other weak solutions $\tilde{u}$, i.e.: $\forall t \geqslant 0 : \partial_t E_u \leqslant \partial_t E_{\tilde{u}}$.

## 2 Variational Fluxes

One can define a total entropy for FV methods based on the cell averages as $E^{\mathrm{FV}} = E^{\mathrm{FV}}_{(u_k)_k}(t) = \sum_k U\left(u_k^n(t)\right)\Delta x_k$. Therefore, the derivative of this total entropy depends on the selected flux approximations

$$\partial_t E^{\mathrm{FV}} = \sum_k \langle \partial_u U(u_k) | \partial_t u_k \rangle\, \Delta x_k = \sum_k \left\langle \partial_u U(u_k) \middle| f_{k-\frac{1}{2}} - f_{k+\frac{1}{2}} \right\rangle.$$

Demanding that this entropy rate is minimized (maximally negative) results in the following variational condition on $f_{l+\frac{1}{2}}$

$$\begin{aligned}
f_{l+\frac{1}{2}} &= \underset{f_{l+\frac{1}{2}}}{\arg\min}\, \partial_t E^{\mathrm{FV}} = \underset{f_{l+\frac{1}{2}}}{\arg\min} \underbrace{\sum_{k \neq l, l+1} \left\langle \partial_u U(u_k) \middle| f_{k-\frac{1}{2}} - f_{k+\frac{1}{2}} \right\rangle}_{\text{const. w.r.t. } f_{l+\frac{1}{2}}} \\
&\quad + \left\langle \partial_u U(u_l) \middle| f_{l-\frac{1}{2}} - f_{l+\frac{1}{2}} \right\rangle + \left\langle \partial_u U(u_{l+1}) \middle| f_{l+\frac{1}{2}} - f_{l+\frac{3}{2}} \right\rangle \\
&= \underset{f_{l+\frac{1}{2}}}{\arg\min} \left\langle \partial_u U(u_{l+1}) - \partial_u U(u_l) \middle| f_{l+\frac{1}{2}} \right\rangle + \underbrace{\left\langle \partial_u U(u_l) \middle| f_{l-\frac{1}{2}} \right\rangle + \left\langle \partial_u U(u_{l+1}) \middle| f_{l+\frac{3}{2}} \right\rangle}_{\text{const. w.r.t. } f_{l+\frac{1}{2}}} \\
&= \underset{f_{l+\frac{1}{2}}}{\arg\min} \left\langle \partial_u U(u_{l+1}) - \partial_u U(u_l) \middle| f_{l+\frac{1}{2}} \right\rangle.
\end{aligned}$$

In [8] it was shown that Godunov's flux is the exact solution to this problem $f^{\mathrm{G}}(u_l, u_r) = \arg\min_{u \in \mathrm{conv}(u_l, u_r)} \langle \partial_u U(u_r) - \partial_u U(u_l) | f(u) \rangle$ for scalar conservation laws under a suitable restriction on the set $\mathcal{U} = \operatorname{conv}(u_l, u_r)$ of values $u \in \mathcal{U}$ that can be entered into the analytical flux function. We will from now on denote a solution to the problem $f_{\mathcal{U}}^{\mathrm{D}}(u_l, u_r) = \arg\min_{u\in\mathcal{U}} \langle \partial_u U(u_r) - \partial_u U(u_l) | f(u)\rangle$ as Dafermos flux, as the identity above is in general untrue for systems of conservation laws, and consider different sets than $\operatorname{conv}(u_l, u_r)$ for the restriction. In particular, we will try to recover a set $\mathcal{U} = V(\mathcal{I}\,u)$ of admissible values, and it will be convenient to approximate this set as a ball around a value $u_c$ with radius $r$.

## 3 Recovery and Recovery Error Indicators

Given a set of average values $(u_k)_k$ one can try to recover point values of the function $u$. This recovery, even when optimal [9], will in general not be exact. Examples of optimal recoveries, i.e. recoveries minimizing the error, in this context are the polynomial recoveries of (W)ENO methods [10]. We denote with $\mathcal{I} : \mathrm{L}^1 \to \mathbb{R}^{N\times K}$ the information operator that maps a given function $u$ onto its cell-average values, i.e. the information available at our disposal. When $B(\mathcal{I}\,u)$ is the set of all functions $v$ with the same information as $\mathcal{I}\,u$, under suitable restrictions, one can define the diameter and radius of the information $\mathcal{I}\,u$.

**Definition 1** (*Local information radius*) The local information diameter shall be defined as

$$\operatorname{diam}(x, \mathcal{I}, W, u) = \sup_{v\in B(\mathcal{I}\,u)} |v(x) - u(x)|\,,$$

together with the information radius defined by

$$\operatorname{rad}(x, \mathcal{I}, W, \mathcal{I}\,u) = \inf_{y\in\mathbb{R}} \sup_{v\in B(\mathcal{I}\,u)} |v(x) - y| = \operatorname{rad}(V(\mathcal{I}\,u)).$$

These give bounds on the errors that have to be expected when one tries to approximate point values of $u$ with only the information $\mathcal{I}\,u$ available [11]. Note that the name radius and diameter mirrors their connection to the radius and diameter of sets. In particular can the radius be interpreted as the radius of the set $V(\mathcal{I}\,u) = \{v(x) | v \in B(\mathcal{I}\,u)\}$. The operator $\mathcal{R}$ shall be a recovery operator that approximates those point values, and we are interested in the corresponding radius, as this allows us to estimate $|u(x) - \mathcal{R}\mathcal{I}\,u| \leqslant 2\operatorname{rad}(x, \mathcal{I}, W, \mathcal{I}\,u)$, when $\mathcal{R}$ evaluates a function out of $B(\mathcal{I}\,u)$ at $x$. A simple but effective algorithm to recover a point value and a radius estimate is by using several base recovery operators $\mathcal{R}_k$, defined as the point evaluation of polynomials having the same averages as the given information $\mathcal{I}\,u$ on different stencils $S_k$, i.e.

$$R_k u\left(x_{n+\frac{1}{2}}\right) = p\left(x_{n+\frac{1}{2}}\right), \quad \forall n \in S_k : \frac{1}{x_{n+\frac{1}{2}} - x_{n-\frac{1}{2}}} \int_{x_{n-\frac{1}{2}}}^{x_{n+\frac{1}{2}}} p(x)\,\mathrm{d}x = u_n(t).$$

After these base recoveries were evaluated we can take their center

$$c = \arg\min_{c \in \mathbb{R}^n} \max_{k=1,\dots,K} \|\mathcal{R}_k \mathcal{I} u - c\|, \quad r = \max_{k=1,\dots,K} \|\mathcal{R}_k \mathcal{I} u - c\|$$

as the recovery and the radius of their bounding sphere as information radius estimate.

While this recovery and radius estimate is usable, its performance for shock calculations can be improved via the removal of bad base recoveries $\mathcal{R}_k$. In fact, removing some recoveries $\mathcal{R}_k \mathcal{I} u$ that lie on the surface of the bounding sphere is sufficient. This is done in the following sense. Given a set of recoveries $\{\mathcal{R}_1 u, \mathcal{R}_2 u, \dots, \mathcal{R}_K u\}$ we can calculate the distances of these recoveries from the center $c$ as $d_k = \|\mathcal{R}_1 u - c\|$ and order the recoveries by their distance from the center. We can then discard the recoveries with the highest distances $d_k$. The center $c$ and radius $r$ can be recalculated from this restricted set of recoveries.

## 4 Approximate Solution of the Variational Problem

Approximate Riemann solvers, like the (local-)Lax-Friedrichs flux [12] are used to save computational effort by swapping them in place of the exact Godunov flux in FV solvers. We are also interested in numerical fluxes that can be evaluated efficiently and will therefore design approximate solutions to the variational problem defining Dafermos' flux. A numerical flux that can be interpreted as approximate solution is the modified Lax-Friedrichs flux below.

**Definition 2** (*Modified Lax-Friedrichs flux*) Let $f$ be a smooth flux function. We define the *modified Lax-Friedrichs* flux with admissible set $\mathcal{U}$ as

$$f_{\mathcal{U}}^{\mathrm{MLF}}(u_l, u_r) = f(u_c) + \frac{\partial_u U(u_l) - \partial_u U(u_r)}{\|\partial_u U(u_l) - \partial_u U(u_r)\|} \max_{u \in \mathcal{U}} \|\partial_u f\| \operatorname{rad}(\mathcal{U}),$$

where $u_c$ is a center of the set $\mathcal{U}$.

**Lemma 1** *The modified Lax-Friedrichs flux is more dissipative than Dafermos' flux calculated using the same admissible set $\mathcal{U}$.*

$$\left\langle \partial_u U(u_r) - \partial_u U(u_l) \middle| f_{\mathcal{U}}^{\mathrm{MLF}} \right\rangle \leqslant \left\langle \partial_u U(u_r) - \partial_u U(u_l) \middle| f_{\mathcal{U}}^{\mathrm{D}} \right\rangle$$

***Proof*** Let $u^* \in \mathcal{U}$ denote an argument minimizing the functional in the variational problem of Dafermos' flux. As $u_c$ is a center of the admissible set $\mathcal{U}$ the estimate $\|u_c - u^*\| \leqslant \sup_{u \in \mathcal{U}} \|u_c - u\| = \operatorname{rad}(\mathcal{U})$ is satisfied. The flux function is locally Lipschitz-continuous on the admissible set $\mathcal{U}$ with constant $L_f = \max_{u \in \mathcal{U}} \|\partial_u f(u)\|$. We can therefore conclude that $\|f(u_c) - f(u^*)\| \leqslant L_f \|u_c - u^*\| \leqslant L_f \operatorname{rad}(\mathcal{U})$ allows us to bound the error in the flux stemming from using the center of $\mathcal{U}$ as a replacement for the most dissipative value. The corresponding error in the entropy dissipation is therefore bounded by

$$
\begin{aligned}
\left|\left\langle \partial_u U(u_r) - \partial_u U(u_l) \middle| f(u_c) - f(u^*) \right\rangle\right| &\leqslant \|\partial_u U(u_r) - \partial_u U(u_l)\| \, \|f(u_c) - f(u^*)\| \\
&\leqslant \|\partial_u U(u_r) - \partial_u U(u_l)\| \, L_f \operatorname{rad}(\mathcal{U}).
\end{aligned}
$$

This contribution can be counteracted by adding a correction to the flux for which we will use the most dissipative direction possible. This is exactly the difference in the entropy variables and yields

$$
\begin{aligned}
&\left\langle \partial_u U(u_r) - \partial_u U(u_l) \middle| f(u^c) - L_f \operatorname{rad}(\mathcal{U}) \frac{\partial_u U(u_r) - \partial_u U(u_l)}{\|\partial_u U(u_r) - \partial_u U(u_l)\|} \right\rangle \\
&\leqslant \left\langle \partial_u U(u_r) - \partial_u U(u_l) \middle| f(u^*) \right\rangle \\
&\quad + \underbrace{\|\partial_u U(u_l) - \partial_u U(u_r)\| \, L_f \operatorname{rad}(\mathcal{U}) - \|\partial_u U(u_l) - \partial_u U(u_r)\| \, L_f \operatorname{rad}(\mathcal{U})}_{=0}.
\end{aligned}
$$

The claim follows, as the first expression exactly describes the entropy dissipation of the modified Lax-Friedrichs flux while the second expression in the inequality equals the entropy dissipation of Dafermos' flux. □

To achieve high order accurate schemes not only the local truncation error of the schemes is important, but also the long time behavior of the solutions of the schemes. If high-order modes are excited these can even at small amplitudes destroy the high accuracy of polynomial reconstructions, and this was experienced in numerical experiments with our schemes. Using the MLF flux one can define a viscosity coefficient $\mu$

$$
f_{\mathcal{U}}^{\mathrm{MLF}}(u_l, u_r) = f(u_c) + (\partial_u U(u_l) - \partial_u U(u_r)) \underbrace{\frac{\max_{u \in \mathcal{U}} \|\partial_u f\| \operatorname{rad}(\mathcal{U})}{\|\partial_u U(u_l) - \partial_u U(u_r)\|}}_{\mu}
$$

and this viscosity is in general unbounded and of low regularity [13], as $u_l \to u_r$ does in general not imply $\operatorname{rad} U \to 0$ when a recovery operator with a wide stencil is used. Therefore, forward Euler time steps of finite size can have an anti-dissipative effect and a regularization is needed to achieve a bounded viscosity distribution. This regularization will be provided by a special mollification operator applied to the viscosity distribution. While standard mollifications are conservative with respect to integrals, i.e. $\int \mu \, \mathrm{d}x = \int \sigma * \mu \, \mathrm{d}x$, we are interested in mollifications that are conservative or even augmenting with respect to the entropy dissipation induced by the viscosity distribution. The *additional* dissipation induced by a viscosity $\sigma_l$ with support around $[-p, p]$ entered into the MLF flux in viscosity form

$$
f_{\mu}^{\mathrm{MLF}}(u_l, u_r; u_c) = f(u_c) + \mu(\partial_u U(u_l) - \partial_u U(u_r))
$$

is given by

$$\partial_t E_{\sigma\, k+\frac{1}{2}} = \sum_{l=-p}^{p} \langle \partial_u U(u_{k+l+1}) - \partial_u U(u_{k+l}) | \sigma_l \left(\partial_u U(u_{k+l}) - \partial_u U(u_{k+l+1})\right)\rangle$$
$$= - \sum_{l=-p}^{p} \sigma_l \left\| \partial_u U(u_{k+l+1}) - \partial_u U(u_{k+l}) \right\|^2 .$$

And we can therefore enforce that the smoothed viscosity distribution induces the same entropy dissipation as $\mu_{k+\frac{1}{2}}$ by defining

$$\left(\sigma \underset{\partial_u U}{\widehat{*}} \mu\right)_{k+\frac{1}{2}} = A \sum_{l=-p}^{p} \sigma_l \frac{- \left\| \partial_u U(u_{k+l+1}) - \partial_u U(u_{k+l}) \right\|^2 \mu_{k+l+\frac{1}{2}}}{\partial_t E_{\sigma\, k+l+\frac{1}{2}}},$$

with $A = 1$. One can show that this leads to a bounded viscosity distribution [13]. This viscosity distribution will be more dissipative than $\mu_{k+\frac{1}{2}}$ for all $A \geqslant 1$, and we choose $A = 1/\sigma_0$. When $u_k$ is constant with exception to one jump, i.e. a shock, this enforces that the viscosity at the jump will have the same amplitude as before, while additional viscosity is added around the jump.

## 5 Numerical Tests

The scheme was tested on Burgers' equation and the Buckley-Leverett equation. The source code is available under https://github.com/simonius/EAR.

### 5.1 *Example I: Burgers Equation*

Tests were carried out for Burgers' equation with the square entropy

$$\partial_t u + \partial_x f = 0, \text{ with } f(u) = \frac{u^2}{2}, \quad U(u) = u^2/2.$$

The modified Lax-Friedrichs flux, applied to an admissible set $\mathcal{U} = \{u \in \mathbb{R} | \, |u - u_c| \leqslant r\}$ takes the form

$$f_{\mathcal{U}}^{\text{MLF,Burgers}}(u_l, u_r) = f(u_c) + \frac{u_l - u_r}{|u_l - u_r|}(|u_c| + r)r = f(u_c) + \mu(u_l - u_r)$$

when the estimate $\max_{u \in \mathcal{U}} \|\partial_u f\| = |u_c| + r$ is entered into the definition together with the derivative of the entropy $U(u)$. For $\mu$ the regularized viscosity

$\left(\sigma \overset{*}{\underset{\partial_u U}{\frown}} \mu\right)_{k+\frac{1}{2}}$ shall be used. Three initial conditions were selected to showcase the abilities of the scheme,

$$u_1(x,0) = \sin(\pi x), \quad u_2(x,0) = \begin{cases} -1 & x < 1 \\ 1 & x > 1 \end{cases}, \quad u_3(x,0) = 1 + \frac{\sin(\pi x)}{50},$$

i.e. $u_1$ developing into a shock, $u_2$ showcasing a rarefaction and $u_3$ staying smooth for a prolonged time to allow us to carry out a convergence analysis. The scheme was run with a CFL number of 0.1 and integrated in time using the DP8 method, as presented by Hairer, Norsett and Wanner. Base recoveries were reconstruction polynomials of degree 4, from which two recoveries were discarded as described at the end of Sect. 3. The Hann window function [14] was used as mollifier $\sigma_l$ to regularize the viscosity distribution. In Fig. 1d several solutions were calculated with 50 cells to demonstrate the shock-capturing ability of the scheme. The total entropy of the solution to $u_1$ over time can be seen in Fig. 1e together with the entropy of an extremely fine calculation using a Godunov scheme. The total entropy is constant until a shock forms and the entropy dissipation happens fast enough to keep the total entropy of the new scheme below the entropy of the Godunov scheme. Albeit the scheme is able to dissipate entropy with high speed it is also able to calculate solutions with a high order of accuracy, as can be seen in the convergence analysis

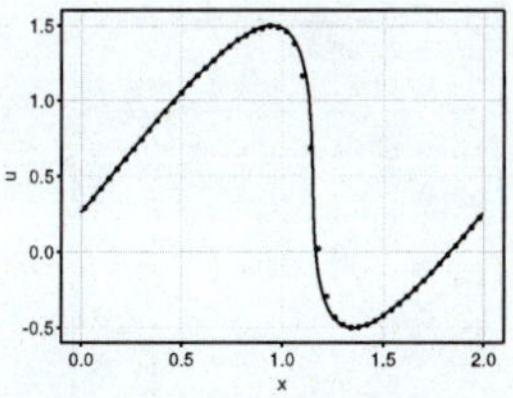

(a) Solution to $u_1(x,0)$ at $t = 0.3$

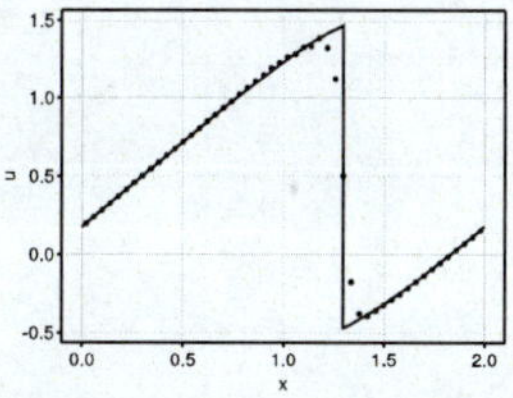

(b) Solution to $u_1(x,0)$ at $t = 0.6$

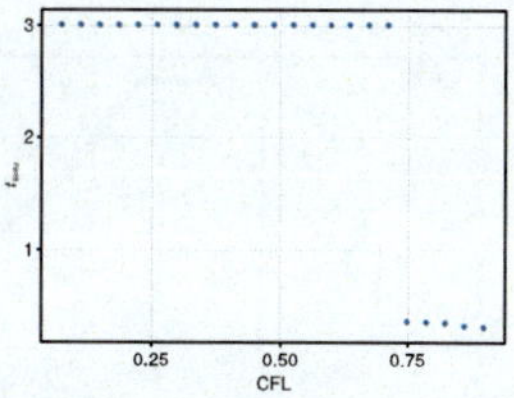

(c) Maximal end time using SSPRK(3,3).

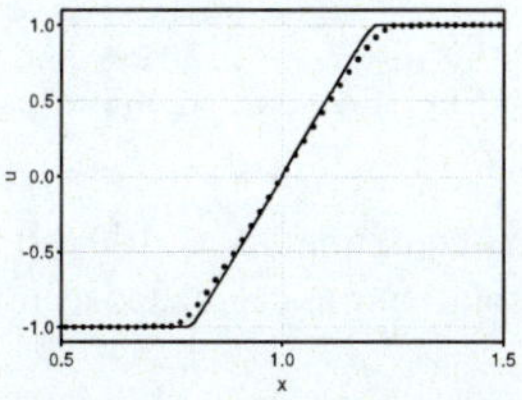

(d) Solution to $u_2(x,0)$ at $t = 0.2$

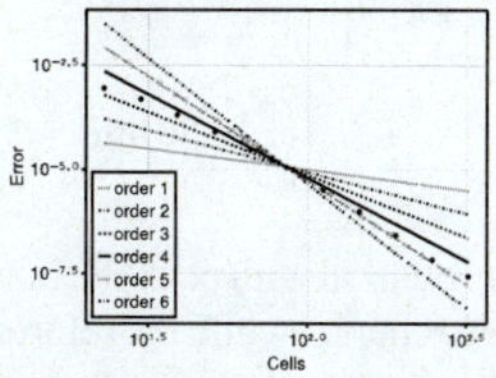


(e) Convergence analysis for polynomial order p=4 and the bounding box surface discard information radius predictor.

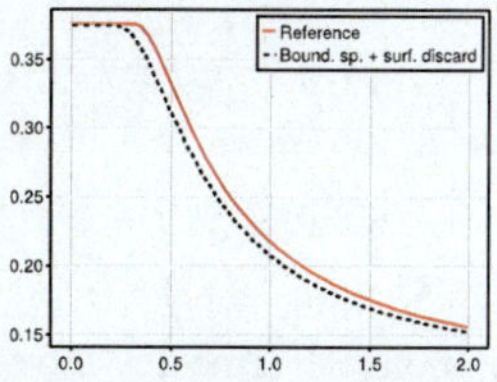


(f) Total entropies of all tested solvers for the initial condition $u_1$ and $N = 50$ cells. Obviously, all entropies lie below the reference entropy calculated from the solution of a Godunov scheme with $N = 5000$ cells.

**Fig. 1** Solution, convergence analysis and total entropy analysis for the scheme applied to Burgers' equation in Fig. 1e. The maximal possible time-step size was tested in Fig. 1c. There, the time until a breakdown of the calculation is shown in relation to the used CFL number for the SSPRK(3, 3) time integrator. Please note that at $t = 3.0$ the simulation was stopped when it did not crash up to that point. One can see that CFL numbers of 0.7 can be reached without problems.

## *5.2 Example II: Buckley-Leverett Equation*

The Buckley-Leverett equation given by the flux in conjunction with the initial condition below was tested as a second example

$$f(u) = \frac{u^2}{u^2 + a(1-u)^2}, \quad a = \frac{1}{2}, \quad u(x,0) = \begin{cases} 1 & x \leqslant 0 \\ 0 & x > 0 \end{cases}, \quad U(u) = \frac{u^2}{2},$$

with periodic boundary conditions. This initial condition leads to the well known shock immediately followed by a rarefaction, that can be seen also in Fig. 2. A plot of the total entropy shows that at least compared to an extremely fine Lax-Friedrichs scheme, that was used to calculate the reference solution, satisfies the new scheme Dafermos' entropy rate criterion also for the Buckley-Leverett equation.

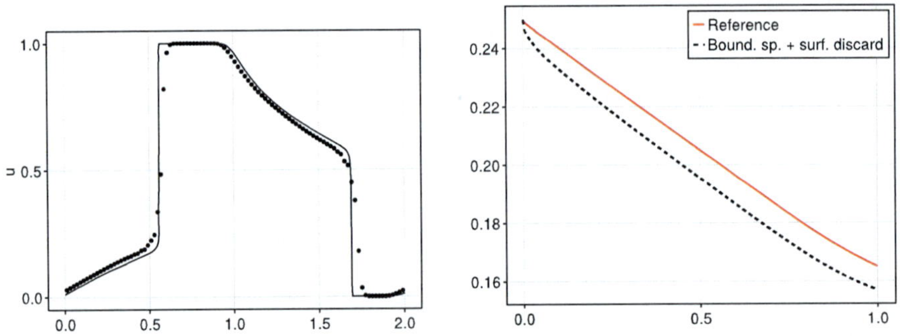


(a) Solution at $t = 0.5$.

(b) Total entropy of the reference solution and the solution calculated by the new scheme over time.

**Fig. 2** Solution to the Buckley-Leverett equation with periodic boundary conditions. $N = 100$ cells were used for the calculation with the new scheme, while the reference solution was calculated using a Lax-Friedrichs scheme with $N = 10^4$ cells

## 6 Conclusion

We designed a scheme that tries to implement the entropy rate criterion for reconstruction based Finite-Volume methods. Key elements were the introduction of an error indicator for the recovery, that denotes a ball around the recovered value. By selecting the most dissipative flux possible in this ball the identity discovered by Dafermos in [8] is generalized to the case of recovery based high-order schemes. Numerical tests indicate that the entropy rate criterion is satisfied and the scheme is high-order accurate. Future work will be focused around tests for systems of conservation laws. The definition of the modified Lax-Friedrichs flux and the radius of information also applies for systems when a suitable norm is used, but it is unclear if it should be used without modification, as in this case a big recovery error in one conserved quantity can induce a high amount of viscosity that applies for all components.

Afterwards the idea will be tested in several space dimensions. There one of the hoped for advantages of the new construction could come up. While classical (W)ENO schemes need to consider prohibitively large numbers of different stencils to achieve high order in several space dimensions is it plausible that the new construction only needs to consider a stencil into every direction away from the point where one would like to recover to find a suitable information radius estimate. If this conjecture turns out true this could pave the way towards less costly high order Finite-Volume schemes on unstructured grids.

## References


1. Dafermos, C.M.: Hyperbolic conservation laws in continuum physics
2. Lax, P.D.: Shock waves and entropy. English (1971)
3. Tadmor, E.: English. Math. Comput. **43**, 369–381 (1984)
4. Harten, A., Engquist, B., Osher, S., Chakravarthy, S.: English. J. Comput. Phys. **71**, 231–303 (1987)
5. Liu, X.-D., Osher, S., Chan, T.: J. Comput. Phys. **115**, 200–212 (1994)
6. Tadmor, E.: English. Math. Comput. **49**, 91–103 (1987)
7. Dafermos, C.M.: J. Differ. Equ. 202–212 (1972)
8. Dafermos, C.M.: English. Discret. Contin. Dyn. Syst. **23**, 185–195 (2009)
9. Sonar, T.: English. IMA J. Numer. Anal. **16**, 549–581 (1996)
10. Abgrall, R., Sonar, T.: English. Numer. Math. **76**, 1–25 (1997)
11. Micchelli, C.A., Rivlin, T.J.: Lectures on optimal recovery (1984)
12. Harten, A., Lax, P.D., van Leer, B.: English. SIAM Rev. **25**, 35–61 (1983)
13. Klein, S.-C., Sonar, T.: Entropy-aware non-oscillatory high-order finite volume methods using the Dafermos entropy rate criterion (2023)
14. Blackman, R.B., Tukey, J.W.: Bell Syst. Tech. J. **37**, 185–282 (1958)

# Hybrid Kinetic/Fluid Numerical Method for the Vlasov-Poisson-BGK Equation in the Diffusive Scaling

**Tino Laidin and Thomas Rey**

**Abstract** This short note presents an extension of the hybrid, model-adaptation method introduced in T. Laidin, *arXiv 2202.03696*, 2022 for linear collisional kinetic equations in a diffusive scaling to the nonlinear mean-field Vlasov-Pois-son-BGK model. The aim of the approach is to reduce the computational cost by taking advantage of the lower dimensionality of the asymptotic model while reducing the overall error. It relies on two criteria motivated by a perturbative approach to obtain a dynamic domain adaptation. The performance of the method and the conservation of mass are illustrated through numerical examples.

**Keywords** Vlasov-Poisson-BGK equations · Diffusion scaling · Asymptotic preserving scheme · Micro-macro decomposition · Hybrid solver

## 1 The Model

In this work we are interested in the kinetic description of a system of particles interacting via both mean-field electromagnetic interaction and collisions. Such system can be modeled using the Vlasov-Poisson-BGK equation. The unknown is the probability distribution function $f = f(t, x, v) \in \mathbb{R}^+$ solution to:

$$\begin{cases} \dfrac{\partial}{\partial t} f^\varepsilon + \dfrac{v_x}{\varepsilon}\, \partial_x f^\varepsilon + \dfrac{E^\varepsilon}{\varepsilon}\, \partial_{v_x} f^\varepsilon = \dfrac{1}{\varepsilon^2}\mathcal{Q}(f^\varepsilon), \\ f(0, x, v) = f_0(x, v), \end{cases} \tag{1}$$

where $(t, x, v) \in \mathbb{R}^+ \times [0, x_*] \times \mathbb{R}^{d_v}$ with periodic boundary conditions in the physical space and $v = (v_x, v_y, v_z)$. The long range interactions are modeled via the self-consistent electrical field $E = E(t, x)$ solution to the Poisson equation

The authors were partially funded by Labex CEMPI (ANR-11-LABX-0007-01) and the MSCA DN-2022 program DATAHYKING.

T. Laidin (✉) · T. Rey
University of Lille, CNRS, Inria, UMR 8524 - Laboratoire Paul Painlevé, 59000 Lille, France
e-mail: tino.laidin@univ-lille.f



E. Franck et al. (eds.), *Finite Volumes for Complex Applications X—Volume 2, Hyperbolic and Related Problems*, Springer Proceedings in Mathematics & Statistics 433,
https://doi.org/10.1007/978-3-031-40860-1_24

$$\partial_x E^\varepsilon = \rho^\varepsilon - \bar{\rho} \quad \text{with} \quad \bar{\rho} = \iint f_0 \, \mathrm{d}x \, \mathrm{d}v. \tag{2}$$

The short-range interactions between particles are taken into account through the linear BGK-like relaxation collision operator

$$\mathcal{Q}(f^\varepsilon)(t, x, v) = \rho^\varepsilon(t, x)\mathcal{M}(v) - f^\varepsilon(t, x, v), \quad \forall (t, x, v) \in \mathbb{R}^+ \times [0, x_*] \times \mathbb{R}^{d_v},$$

where the global Maxwellian and local density are respectively defined as

$$\mathcal{M}(v) = \frac{e^{-|v|^2/2}}{(2\pi)^{d_v/2}}, \quad \rho^\varepsilon(t, x) = \int f^\varepsilon(t, x, v) \, \mathrm{d}v =: \langle f^\varepsilon(t, x, \cdot) \rangle.$$

In (1), the scaling parameter $\varepsilon$ is related to the Knudsen number: the ratio between the mean free path of the particles and the length scale of observation. It is now well known [4] that the limit case $\varepsilon = 0$ is described by a drift-diffusion equation on the density $\rho$: when $\varepsilon \to 0$, the distribution function $f^\varepsilon$ converges towards a Maxwellian distribution $\rho\mathcal{M}$ whose density $\rho$ is solution to

$$\begin{cases} \partial_t \rho - \partial_x J = 0, & J = \partial_x \rho - E\rho, \\ \rho(0, x) = \rho_0(x), & \forall x \in [0, x_*]. \end{cases} \tag{P}$$

The approximation of solutions to (1)–(2) can be computationally expensive because of the high dimensionality of the kinetic phase space. Nevertheless, using the more accurate kinetic description of the system may not be necessary in the whole computing domain because of the local validity of the fluid description (the system being in a state of thermodynamic equilibrium). This latter is less accurate in describing the kinetic zones, but computationally less costly. The aim of this work is therefore to design a hybrid kinetic/fluid scheme with an automatic domain adaptation method. It relies on a robust numerical scheme for the kinetic equation, on relevant criteria to carefully determine fluid and kinetic regions and on a smart implementation.

## 2 Macroscopic Models

The aim of this section is to derive a higher-order macroscopic model from which we deduce a macroscopic coupling criterion. It generalizes the approach presented in [5]. Let us now introduce the truncated Chapman-Enskog expansion of the distribution function $f^\varepsilon$ at order $K \in \mathbb{N}^*$:

$$f^\varepsilon(t, x, v) = \rho^\varepsilon(t, x)\mathcal{M}(v) + \sum_{k=1}^{K} \varepsilon^k h^{(k)}(t, x, v). \tag{3}$$

By inserting (3) into the original Eq. (1), one can identify powers of epsilon to obtain

$$k = 0: \quad h^{(1)} = -\mathcal{T}(\rho^\varepsilon \mathcal{M}), \tag{4a}$$

$$k = 1: \quad h^{(2)} = -\,\partial_t\,(\rho^\varepsilon \mathcal{M}) - \mathcal{T}(h^{(1)}), \tag{4b}$$

$$2 \le k \le K-1: \; h^{(k+1)} = -\,\partial_t\, h^{(k-1)} - \mathcal{T}(h^{(k)}), \tag{4c}$$

where $\mathcal{T} f = v_x\,\partial_x f + E\,\partial_{v_x} f$ is the transport operator. To obtain a hierarchy of macroscopic models, one considers different truncation orders $K$, then plugs (3) into (1) and integrates in velocity. The order $K = 1$ allows us to (formally) recover the asymptotic model $(P)$.

Let us now present the idea behind the computations in a $1D/1D$ setting for the case $K = 3$. Note that the same method can be applied up to the full $3D/3D$ setting. The set of Eqs. (4) allows us to compute the functions $h^{(k)}$, $k = 1, 2, 3$. Using the identity $\partial_v\,\mathcal{M} = -v\mathcal{M}$, one has

$$\begin{aligned} h^{(1)} &= -v\mathcal{M}J^\varepsilon, \quad J^\varepsilon = \partial_x\,\rho^\varepsilon - E^\varepsilon\rho^\varepsilon,\\ h^{(2)} &= -\mathcal{M}\,\partial_t\,\rho^\varepsilon + v^2\mathcal{M}\,\partial_x\,J^\varepsilon + (1-v^2)\mathcal{M}E^\varepsilon J^\varepsilon,\\ h^{(3)} &= v\mathcal{M}J^\varepsilon + v\mathcal{M}\,\partial_x\,(\partial_t\,\rho^\varepsilon) - v^3\mathcal{M}\,\partial_{xx}\,J^\varepsilon - (v-v^3)\mathcal{M}\,\partial_x\,(E^\varepsilon J)\\ &\quad -v\mathcal{M}E^\varepsilon\,\partial_t\,\rho^\varepsilon - (2v-v^3)\mathcal{M}E^\varepsilon\,\partial_x\,J^\varepsilon - (v^3-3v)(E^\varepsilon)^2 J^\varepsilon. \end{aligned}$$

Integrating in velocity yields

$$\partial_t\,\rho^\varepsilon + \partial_x\,\langle vh^{(1)}\rangle + \varepsilon^2\,\partial_x\,\langle vh^{(3)}\rangle = \mathcal{O}(\varepsilon^4). \tag{5}$$

It remains to compute the quantities $\partial_x\,\langle vh^{(1)}\rangle$ and $\partial_x\,\langle vh^{(3)}\rangle$. By the definition of $\mathcal{M}$, one has $m_2 = 1$, $m_4 = 3$ and $\partial_x\,\langle vh^{(1)}\rangle = \partial_x\,J^\varepsilon$. Moreover, to avoid some approximation of mixed derivatives, we observe from (5) that

$$\partial_t\,\rho^\varepsilon = \partial_x\,J^\varepsilon + \mathcal{O}(\varepsilon^2), \tag{6}$$

$$\partial_t\,J^\varepsilon = \partial_{xx}\,J^\varepsilon - E^\varepsilon\,\partial_x\,J^\varepsilon - \rho^\varepsilon\,\partial_t\,E^\varepsilon + \mathcal{O}(\varepsilon^2). \tag{7}$$

Replacing the time derivatives by their approximations (6) and (7) yields

$$\begin{aligned} \partial_x\,\langle vh^{(3)}\rangle = &-\,\partial_{xxx}\,J^\varepsilon + \partial_x\,\rho^\varepsilon\,\partial_t\,E^\varepsilon - \rho^\varepsilon\,\partial_t\,(\partial_x\,E^\varepsilon) + 3\,\partial_x\,E^\varepsilon\,\partial_x\,J^\varepsilon\\ &+ E^\varepsilon\,\partial_{xx}\,J^\varepsilon + 2J^\varepsilon\,\partial_x\,(\partial_x\,E) + \mathcal{O}(\varepsilon^2). \end{aligned}$$

Finally, using the Poisson equation (2) and rearranging the terms, we obtain a higher order macroscopic model:

$$\partial_t\,\rho^\varepsilon - \partial_x\,J^\varepsilon = -\varepsilon^2\mathcal{R} + \mathcal{O}(\varepsilon^4), \tag{8}$$

where the remainder $\mathcal{R}^\uparrow$ is given by

$$\mathcal{R}^\varepsilon = -\,\partial_{xxx} J^\varepsilon + E\,\partial_{xx} J^\varepsilon + (2\rho^\varepsilon - 3\bar{\rho})\,\partial_x J^\varepsilon + 2J^\varepsilon\,\partial_x \rho^\varepsilon - \partial_x \rho^\varepsilon\,\partial_t E^\varepsilon. \quad (9)$$

Let us emphasize that this term does not depend on the velocity variable but on $\rho$ and $E$. It quantifies the deviation from the thermodynamical equilibrium.

## 3 Numerical Scheme

In this part, we briefly recall the derivation of the micro-macro model for (1) introduced in [2]. Then, we present a micro-macro finite volume scheme that enjoys the property of being Asymptotic Preserving (AP), which is a crucial point of the hybrid method we present.

Let us decompose the distribution $f^\varepsilon$ as follows:

$$f^\varepsilon = \rho^\varepsilon \mathcal{M} + g^\varepsilon. \quad (10)$$

We introduce the orthogonal projector $\Pi$ in $L^2(\,\mathrm{d}x\,\mathrm{d}v\mathcal{M}^{-1})$ onto $Ker(\mathcal{Q})$ defined for all $f \in L^2(\,\mathrm{d}x\,\mathrm{d}v\mathcal{M}^{-1})$ by $\Pi f = \langle f \rangle \mathcal{M}$. The so-called micro equation is obtained by plugging (10) into (1) and applying $(I - \Pi)$. Moreover, plugging (10) into (1) and applying the projection $\Pi$ yields the macro equation. The micro-macro model is then given by:

$$\partial_t g^\varepsilon + \frac{1}{\varepsilon}(\mathcal{T} g^\varepsilon - \partial_x \langle v_x g^\varepsilon \rangle \mathcal{M} + v_x \mathcal{M} J^\varepsilon) = \frac{-1}{\varepsilon^2} g^\varepsilon, \quad (Micro)$$

$$\partial_t \rho^\varepsilon + \frac{1}{\varepsilon}\,\partial_x \langle v_x g^\varepsilon \rangle = 0. \quad (Macro)$$

Let us now present the discretization of the $(Micro)$-$(Macro)$ model. We shall adopt a finite volume approach to discretize the phase space and present only the $1D/3D$ case that we shall consider in our numerical simulations.

*The mesh.* The velocity domain is restricted to the bounded symmetric cube $[-v_\star, v_\star]^3$. We consider a Cartesian mesh of the phase space composed of $N_v = 2L$ velocity cells in each direction arranged symmetrically around $v = 0$. Let $\mathcal{J} = \{-L+1, \ldots, L\}$ and $j = (j_x, j_y, j_z) \in \mathcal{J}^3$ be a multi-index. The cells of the velocity mesh are denoted by $\mathcal{V}_j$ for $j \in \mathcal{J}^3$. Each cell $\mathcal{V}_j$ has a constant volume $\Delta v^3$ and midpoint $v_j$.

The physical domain, a torus $\mathbb{T}$ of length $x_*$, is discretized into $N_x$ primal cells $\mathcal{X}_i = (x_{i-\frac{1}{2}}, x_{i+\frac{1}{2}})$ for $i \in \mathcal{I} = \mathbb{Z}/N_x\mathbb{Z}$, of constant length $\Delta x$ and centers $x_i$. We also define dual cells $\mathcal{X}_{i+\frac{1}{2}} = (x_i, x_{i+1})$ for $i \in \mathcal{I}$, of constant length $\Delta x$ and centers $x_{i+\frac{1}{2}}$. This defines the so-called primal and staggered meshes as

$$K_{ij} = \mathcal{X}_i \times \mathcal{V}_j \text{ and } K_{i+\frac{1}{2},j} = \mathcal{X}_{i+\frac{1}{2}} \times \mathcal{V}_j, \quad \forall (i,j) \in \mathcal{I} \times \mathcal{J}^3.$$

Finally, we set a time step $\Delta t > 0$ and define $t^n = n\Delta t$ for $n \in \mathbb{N}$. The Maxwellian is discretized as a product of 1-dimensional Gaussians in such a way that it satisfies discrete counterparts of its continuous properties, namely parity, positivity, and unit mass.

The transport terms in $(Micro)$ are approximated using a first order upwind scheme and the time derivative is dealt with using a first order exponential time integrator. Regarding the macro equation, we take advantage of approximating the perturbation on the dual cells in position for the space derivative. The usual choice [1] to obtain an AP scheme is to implicit the stiff term of $(Macro)$. The equation on the electrical field is solved using a centered finite difference scheme. The numerical scheme reads as follows:

**Proposition 1** *Let $n \in \mathbb{N}$. Let $\left(g^{\varepsilon,n}_{i+\frac{1}{2},j}\right)_{ij}$ and $\left(\rho^{\varepsilon,n}_i\right)_i$ be given by the following micro-macro finite volume scheme:*

$$g^{\varepsilon,n+1}_{i+\frac{1}{2},j} = g^{\varepsilon,n}_{i+\frac{1}{2},j} e^{-\Delta t/\varepsilon^2} - \varepsilon(1 - e^{-\Delta t/\varepsilon^2}) \left( \frac{T^{\varepsilon,n}_{i+\frac{1}{2},j}}{\Delta x \Delta v^3} + \xi_j \mathcal{M}_j J^{\varepsilon,n}_{i+\frac{1}{2}} \right),$$

$$\rho^{\varepsilon,n+1}_i = \rho^{\varepsilon,n}_i - \frac{\Delta t}{\varepsilon \Delta x} \left( \langle \xi g^{\varepsilon,n+1}_{i+\frac{1}{2}} \rangle_\Delta - \langle \xi g^{\varepsilon,n+1}_{i-\frac{1}{2}} \rangle_\Delta \right),$$

$$E^{\varepsilon,n}_{i+\frac{1}{2}} - E^{\varepsilon,n}_{i-\frac{1}{2}} = (\rho^{\varepsilon,n}_i - \bar{\rho})\Delta x,$$

*where $\xi_j = \xi_{(j_x,j_y,j_z)} = v_{j_x}$ $\forall j \in \mathcal{J}^3$ and $T^{\varepsilon,n}_{i+\frac{1}{2},j}$ is the discretization of the transport terms. Assuming some uniform bounds in $\varepsilon$ on $\rho^\varepsilon$ and for a fixed mesh size $\Delta x$, $\Delta v > 0$, the scheme enjoys the AP property in the diffusion limit. This property does not depend on the initial data, and the associated limit scheme reads for $m_2^{\Delta v} = \sum_{l\in\mathcal{J}} v_l^2 M_l \Delta v_l$ as*

$$\rho^{n+1}_i = \rho^n_i + m_2^{\Delta v} \frac{\Delta t}{\Delta x} \left( J^n_{i+\frac{1}{2}} - J^n_{i-\frac{1}{2}} \right), \quad J^n_{i+\frac{1}{2}} = \frac{\rho^n_{i+1} - \rho^n_i}{\Delta x} - E_{i+\frac{1}{2}} \rho^n_{i+\frac{1}{2}}$$

$$E^n_{i+\frac{1}{2}} - E^n_{i-\frac{1}{2}} = (\rho^n_i - \bar{\rho})\Delta x.$$

## 4 Hybrid Method

Let us now introduce the main contribution of this work, namely the hybrid method between kinetic and fluid schemes. It consists in a coupled solver that is faster than a full kinetic one to solve (1) while still being accurate. Following [3, 5] we first construct a hybrid kinetic/fluid solver with a dynamic domain adaptation method and present its implementation.

*Coupling criteria* The idea of the dynamic domain adaptation method is twofold. On the one hand, the subdomains must accurately describe the state of the solution. In particular, the fluid model has to be used only where the solution is near the local velocity equilibrium. On the other hand, the method has to be dynamic, in the sense that the subdomains are adapted at each time step. To determine in which domain each cell lies, we use two criteria based on the higher order macroscopic model (8) introduced in Sect. 2 and the norm of the perturbation $g^\varepsilon = f^\varepsilon - \rho^\varepsilon \mathcal{M}$. Let $\mathcal{R}_i^{\varepsilon,n}$ be a discretization of the remainder (9) at time $t^n$ in cell $\mathcal{X}_i$.

The coupling procedure will unfold as follows: $\mathcal{R}_i^{\varepsilon,n}$ is computed using both the kinetic density $\rho^\varepsilon$ in kinetic cells and the fluid density $\rho$ in fluid cells. Cell changes happen when one of the following situations occurs, depending on a coupling threshold investigated in the previous work [5]:

- If $\mathcal{R}_i^{\varepsilon,n}$ is small (w.r.t. a fixed threshold), the solution (8) is close to the limit model $(P)$ and one must use it;
- If $g^\varepsilon$ is small (w.r.t. another fixed threshold), the solution is close to the local equilibrium in velocity, and one must consider the fluid description.

If any of those two conditions are not met then, depending on its current state, the cell either stays or becomes kinetic.

**Remark 1** Note that in a kinetic cell, the criterion on the norm of $g^\varepsilon$ is mandatory. Indeed, the remainder $\mathcal{R}_i^{\varepsilon,n}$ could be small because of small gradients, but the perturbation large. In this situation, one does not want to change from kinetic to fluid. As an example, one could take a distribution function at equilibrium in position and far from the Maxwellian in velocity (see [3] for details).

*Implementation* The way the method is implemented is crucial. Indeed, the main goal of the method is to avoid the update of the perturbation which is the most computationally expensive part of the code. Therefore, from an implementation point of view, $g^\varepsilon$ is not updated in fluid regions and is set to 0 only when needed. In particular, it occurs only when a fluid cell becomes kinetic and when saving data. The interface conditions between kinetic and fluid cells are treated in the same way as in [5].

Note that we want to start the resolution with the approach containing the full information on the system. Hence, the domain is initialized as fully kinetic. Moreover, let us emphasize that in practice the kinetic fluxes are explicitly computed. Therefore, the hybrid setting is an explicit method.

## 5 Numerical Results

Let us now present some numerical simulations with our approach for the 1D/3D case. We shall start with the validation of the AP property of the micro-macro Vlasov-Poisson-BGK solver with exponential integrator presented in Sect. 3. It is a combination of the methods presented in [2, 6] that have never been implemented in any work, to the best of our knowledge.

Figure 1 presents the time evolution of the $L^2$ norm of the energy of solutions to Eqs. (1)–(2) with different values of the relaxation parameter $\varepsilon$, for the seminal weak

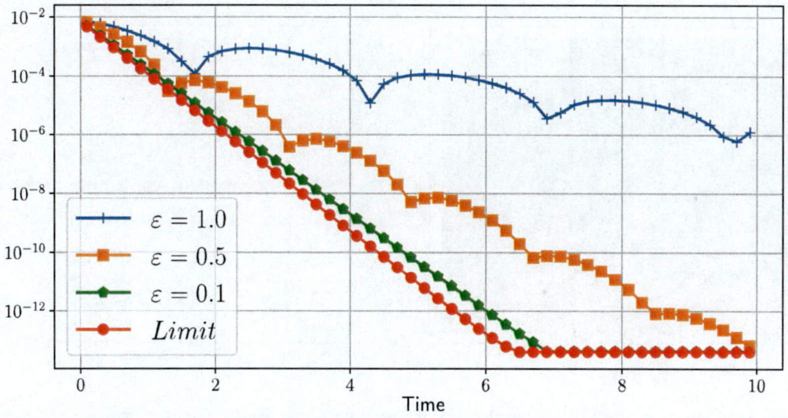


**Fig. 1 Fully kinetic scheme: Landau damping.** Time evolution of $||E(t)||_2$ for different values of $\varepsilon$

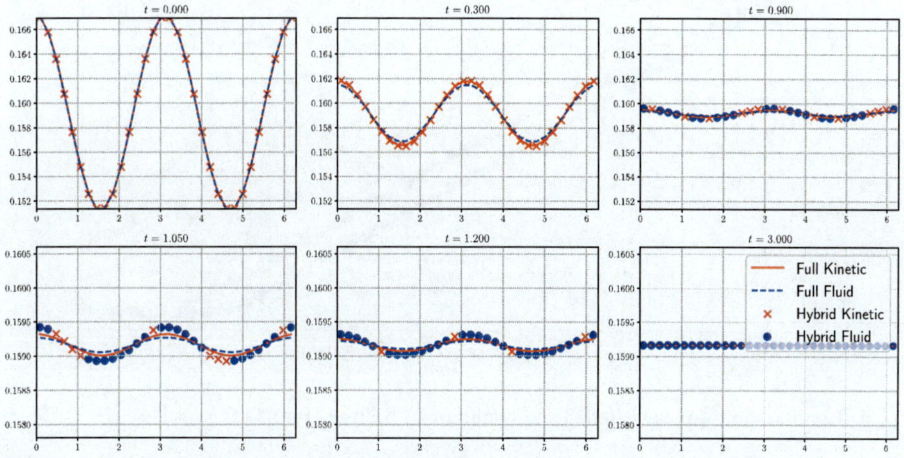


**Fig. 2 Comparison of the solvers.** Time evolution of the space densities for $\varepsilon = 0.1$ with a smooth initial data

Landau damping initial data from [2]. We observe the convergence with respect to $\varepsilon$ of the energy. The oscillations due to the Vlasov-Poisson transport term $\mathcal{T}$ occur only in the kinetic regime, when $\varepsilon$ is large. They are then damped by the linear BGK term for smaller values of $\varepsilon$, where exponential decay of the electric field occurs.

We now turn our attention to the hybrid method. Figure 2 assesses the validity of this new method by computing the time evolution of the density of a smooth solution. The initial condition is a Maxwellian distribution $\rho\mathcal{M}$, whose initial space dependent density is given by $\rho(x) = 1 + 0.05\cos(2x)$ for $x_* = 2\pi$ with a fixed $\varepsilon = 0.1$. Although the solution is far from the fluid description because of this mild value of $\varepsilon$, we observe an almost perfect agreement between the fully kinetic and the hybrid kinetic-fluid solvers. We also observe the back-and-forth phenomenon

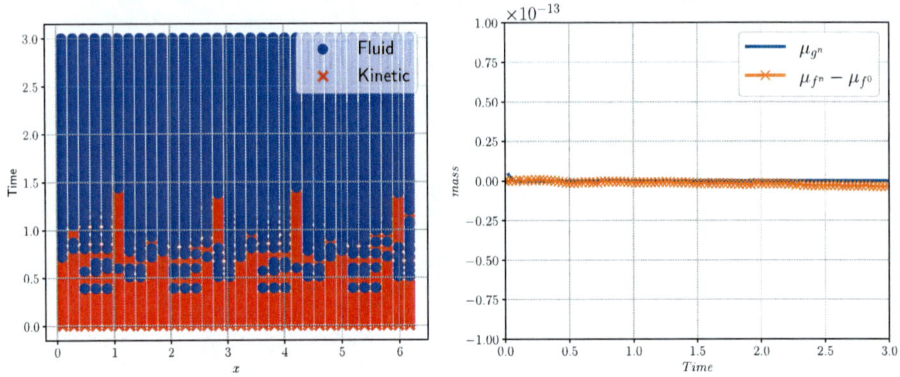


**Fig. 3 Hybrid scheme.** *Left.* Time evolution of the state of the cells. *Right.* Evolution of the mass variation (orange crosses) and mass of $g^\varepsilon$ (solid blue line) for $\varepsilon = 0.1$

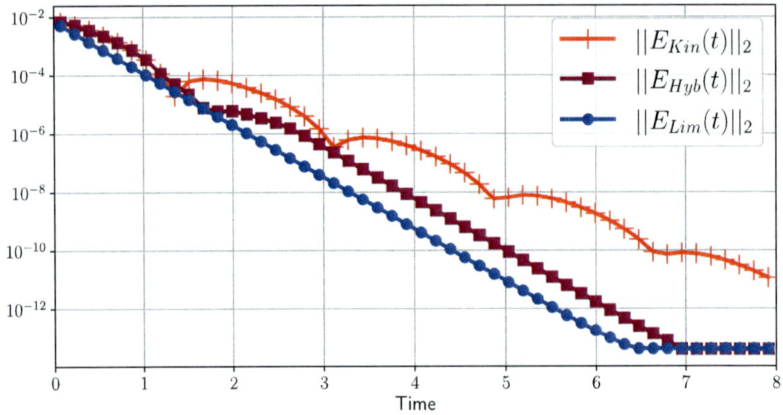


**Fig. 4 Landau damping revisited.** Time Evolution of $||E(t)||_2$ for $\varepsilon = 0.5$

between kinetic and fluid cells, resulting in large time in a full fluid (albeit accurate with the kinetic equation) solver for the global equilibrium. The speed-up tables are similar to those presented in [5], where factors up to 400 have been observed between the hybrid and the fully kinetic solver.

Then this domain adaptation is investigated in Fig. 3. One can observe the precise domain adaptation during time. We notice in particular the quick vanishing of the kinetic cells in favor of the fluid ones (and hence a computational speedup). This adaptation phenomenon can bring mass variation as noticed in [5], but we observe that it remains very close to the machine precision.

Finally, we investigate again the case of the weak Landau damping using the new hybrid solver. We observe in Fig. 4 that this new method is able to accurately capture the oscillations induced by the transport operator $\mathcal{T}$ in short time. Nevertheless, these oscillations are destroyed by the switch to a full fluid solver, which relaxes exponentially because of its relaxation structure.

## References

1. Bennoune, M., Lemou, M., Mieussens, L.: Uniformly stable numerical schemes for the Boltzmann equation preserving the compressible Navier-Stokes asymptotics. J. Comput. Phys. **227** (2008)
2. Crouseilles, N., Lemou, M.: An asymptotic preserving scheme based on a micro-macro decomposition for collisional Vlasov equations: diffusion and high-field scaling limits. Kinet. Relat. Mod. **4**(2), 441–477 (2011)
3. Filbet, F., Rey, T.: A hierarchy of hybrid numerical methods for multi-scale kinetic equations. SIAM J. Sci. Comput. **37**(3), A1218–A1247 (2015)
4. Goudon, T., Poupaud, F.: Approximation by homogenization and diffusion of kinetic equations. Comm. Part. Diff. Euq. (2001)
5. Laidin, T.: Hybrid Kinetic/Fluid numerical method for the Vlasov-BGK equation in the diffusive scaling (2022). arXiv: 2202.03696
6. Lemou, M.: Relaxed micro–macro schemes for kinetic equations. C. R. Acad. Sci. Paris, Ser. I **348**(7–8), 455–460 (2010)

# Composition Schemes for the Guiding-Center Model

**Michel Mehrenberger, Laurent Navoret, and Anh-Tuan Vu**

**Abstract** We present composition schemes for solving non-linear transport equations, like the guiding-center model. This is based on direct and adjoint transport steps. The adjoint steps being implicit, they are replaced by explicit approximated adjoint steps using a given number of fixed point iterations. This does not modify the order of accuracy. Several numerical tests assess the performance of the method.

**Keywords** Composition scheme · Splitting · Semi-lagrangian · Guiding-center model

## 1 Introduction

The guiding-center model is a non-linear two-dimensional transport equation that describes a specific drift undergone by charged particles confined in a Tokamak by a strong magnetic field. More precisely, denoting $f = f(t, x, y)$ the density of the charged particles and $\Phi = \Phi(t, x, y)$ the electrostatic potential, the guiding-center model writes:

$$\begin{cases} \partial_t f + \partial_y \Phi \, \partial_x f - \partial_x \Phi \, \partial_y f = 0, \\ -\partial_{xx} \Phi - \partial_{yy} \Phi = f, \end{cases} \tag{1}$$

with $t \in [0, T]$, $x \in [0, L_1]$, $y \in [0, L_2]$ and $T > 0$, $L_1 > 0$, $L_2 > 0$. This is simplified model for turbulence in a Tokamak. In order to make accurate predictions, high-order methods are essential.

M. Mehrenberger (✉) · A.-T. Vu
Aix Marseille Univ, CNRS, I2M, Marseille, France
e-mail: michel.mehrenberger@univ-amu.fr

L. Navoret
IRMA, Université de Strasbourg et CNRS, UMR 7501, Strasbourg, France

INRIA Nancy-Grand Est, TONUS Project, Strasbourg, France



E. Franck et al. (eds.), *Finite Volumes for Complex Applications X—Volume 2, Hyperbolic and Related Problems*, Springer Proceedings in Mathematics & Statistics 433,
https://doi.org/10.1007/978-3-031-40860-1_25

In this work, we focus on the time discretization. Different schemes have been used in the literature, especially in the context of semi-Lagrangian methods [2, 4, 6, 7, 10–12]. We propose here to use a composition scheme in the spirit of [7]. Composition schemes consist in chaining direct and adjoint steps. One interest is that such methods can benefit from the development of optimized algorithms as pointed out in [9]. Direct steps are explicit and can be either a 2d transport scheme or a splitting scheme, which is the composition of two 1d transport along the two directions. For these two kind of direct steps, we can identify their associated adjoint steps.

Since the adjoint steps are actually implicit, we propose to replace them by a fixed point algorithm as in [3]. Note that it is sufficient to take the number of iterations equal to the desired order of accuracy. Contrary to [7], we consider composition schemes with non-symmetric coefficients. Indeed, taking benefit from the equivalence from composition and splitting schemes [5], we use the coefficients obtained from the second order O2 and the fourth order O4 splitting schemes. This enables to optimize the number of steps at a given accuracy.

The rest of the article is as follows. In Sect. 2, we first detail the direct and adjoint steps, before describing the composition scheme in Sect. 3. Numerical results are given in Sect. 4, before concluding in Sect. 5.

## 2 Direct and Adjoint Steps

In this section, we first define the elementary steps involved in the time discretization of the guiding center equation (1).

First, we define the transport operator over a time interval $\Delta t$ starting from the initial condition $g_2$ and with advection potential obtained from $g_1$:

$$\mathcal{T}_{\Delta t}^{\alpha\beta}[g_1]g_2 = g(\Delta t, \cdot, \cdot),$$

where $g$ is defined as the exact solution to:

$$\begin{cases} \partial_t g + \alpha\, \partial_y \Phi\, \partial_x g - \beta\, \partial_x \Phi\, \partial_y g = 0, \\ -\partial_{xx}\Phi - \partial_{yy}\Phi = g_1, \\ g_{|t=0} = g_2. \end{cases}$$

Let now consider $t_0 = 0 < t_1 < \cdots < t_N = T$ and $f^n \simeq f(t_n, ., .)$. Then, at iteration $n \in \mathbb{N}$, denoting $\Delta t = t_{n+1} - t_n$, the first order direct $2d$ step writes

$$f^{n+1} = \mathcal{S}_{\Delta t}^{2d} f^n = \mathcal{T}_{\Delta t}^{11}[f^n]f^n, \tag{2}$$

while the direct splitting step is given by

$$f^{n+1} = \mathcal{S}^{xy}_{\Delta t} f^n = \mathcal{T}^{01}_{\Delta t}[f^n]\mathcal{T}^{10}_{\Delta t}[f^n]f^n, \tag{3}$$

which corresponds to first make the advection in $x$ and then in $y$. This splitting version is more adapted to parallelization. The two schemes can be written:

$$f^{n+1} = \mathcal{R}_{\Delta t}[f^n]f^n,$$

with $\mathcal{R}_{\Delta t}[g] = \mathcal{T}^{11}_{\Delta t}[g]$ for the 2d scheme and $\mathcal{R}_{\Delta t}[g] = \mathcal{T}^{01}_{\Delta t}[g]\mathcal{T}^{10}_{\Delta t}[g]$ for the splitting scheme. The transport elementary blocks $\mathcal{T}^{\alpha\beta}_{\Delta t}$ will be solved using a high order accurate semi-Lagrangian scheme.

Given a scheme $\mathcal{S}_{\Delta t}$, its adjoint is defined as follows: $\mathcal{S}^*_{\Delta t} = (\mathcal{S}_{-\Delta t})^{-1}$.

**Proposition 1** *The adjoints of the first order 2d and splitting steps are:*

$$f^{n+1} = \mathcal{S}^{2d,*}_{\Delta t} f^n = \mathcal{T}^{11}_{\Delta t}[f^{n+1}]f^n,$$
$$f^{n+1} = \mathcal{S}^{xy,*}_{\Delta t} f^n = \mathcal{T}^{10}_{\Delta t}[f^{n+1}]\mathcal{T}^{01}_{\Delta t}[f^{n+1}]f^n.$$

*They can be written:*

$$f^{n+1} = \mathcal{R}^*_{\Delta t}[f^{n+1}]f^n,$$

*with $\mathcal{R}^*_{\Delta t}[g] = \mathcal{T}^{11}_{\Delta t}[g]$ for the 2d adjoint step and $\mathcal{R}^*_{\Delta t}[g] = \mathcal{T}^{10}_{\Delta t}[g]\mathcal{T}^{01}_{\Delta t}[g]$ for the splitting adjoint step.*

***Proof*** We give the proof for the splitting scheme. The proof for the direct scheme is similar. Defining $\mathcal{I}^{xy}_{\Delta t} g = \mathcal{T}^{10}_{-\Delta t}[(\mathcal{S}^{xy}_{\Delta t})^{-1} g]\mathcal{T}^{01}_{-\Delta t}[(\mathcal{S}^{xy}_{\Delta t})^{-1} g]g$, we have

$$\begin{aligned}\mathcal{I}^{xy}_{\Delta t}\mathcal{S}^{xy}_{\Delta t} g &= \mathcal{T}^{10}_{-\Delta t}[(\mathcal{S}^{xy}_{\Delta t})^{-1}\mathcal{S}^{xy}_{\Delta t} g]\mathcal{T}^{01}_{-\Delta t}[(\mathcal{S}^{xy}_{\Delta t})^{-1}\mathcal{S}^{xy}_{\Delta t} g]\mathcal{S}^{xy}_{\Delta t} g \\ &= \mathcal{T}^{10}_{-\Delta t}[g]\mathcal{T}^{01}_{-\Delta t}[g]\mathcal{T}^{xy}_{\Delta t} g = \mathcal{T}^{10}_{-\Delta t}[g]\mathcal{T}^{01}_{-\Delta t}[g]\mathcal{T}^{01}_{\Delta t}[g]\mathcal{T}^{10}_{\Delta t}[g]g = g.\end{aligned}$$

This leads to

$$\mathcal{S}^{xy,*}_{\Delta t} f^n = \mathcal{I}^{xy}_{-\Delta t} f^n = \mathcal{T}^{10}_{\Delta t}[(\mathcal{S}^{xy}_{-\Delta t})^{-1} f^n]\mathcal{T}^{01}_{\Delta t}[(\mathcal{S}^{xy}_{-\Delta t})^{-1} f^n]f^n = \mathcal{T}^{10}_{\Delta t}[f^{n+1}]\mathcal{T}^{01}_{\Delta t}[f^{n+1}]f^n.$$

As proposed in [3], the implicit adjoint steps can be approximated by $K$ iterations of fixed point algorithm:

$$f^{n+1} = f^{n+1,K}, \quad f^{n+1,k} = \mathcal{R}^*_{\Delta t}[f^{n+1,k-1}]f^n, \; k = 1, \ldots, K, \quad f^{n+1,0} = f^n.$$

This provides a $K$th order approximation of the solution of the adjoints steps. We denote the two corresponding approximated scheme:

$$f^{n+1} = \tilde{\mathcal{S}}^{2d,*}_{\Delta t} f^n, \quad f^{n+1} = \tilde{\mathcal{S}}^{xy,*}_{\Delta t} f^n.$$

Finally, we recall that composing a scheme with its adjoint, $\mathcal{S}^*_{\Delta t}\mathcal{S}_{\Delta t}$, results in a second order accurate scheme [5]. We will consider in the next Section more general composition methods.

## 3 Composition Scheme

Composition schemes aim at constructing high-order time discretizations by composing different elementary steps. Let $s \in \mathbb{N}^*$ and composition coefficients $\alpha_1, \ldots, \alpha_s$, $\beta_1, \ldots, \beta_s \in \mathbb{R}$. The $n$th iteration of the composition scheme reads:

$$f^{n+1} = \mathcal{S}_{\alpha_s \Delta t} \tilde{\mathcal{S}}^*_{\beta_s \Delta t} \cdots \mathcal{S}_{\alpha_2 \Delta t} \tilde{\mathcal{S}}^*_{\beta_2 \Delta t} \mathcal{S}_{\alpha_1 \Delta t} \tilde{\mathcal{S}}^*_{\beta_1 \Delta t} f^n,$$

where $\mathcal{S}$ and $\tilde{\mathcal{S}}^*$ refer either to the 2d or the splitting direct and approximated adjoint steps defined in the previous section. This can be equivalently written $f_0^{n+1} = f^n$:

$$\text{for } i = 1, \ldots, s, \quad f_{2i-1}^{n+1} = \tilde{\mathcal{S}}^*_{\beta_i \Delta t} f_{2(i-1)}^{n+1}, \quad f_{2i}^{n+1} = \mathcal{S}_{\alpha_i \Delta t} f_{2i-1}^{n+1},$$

and $f^{n+1} = f_{2s}^{n+1}$. Note that we use the approximated adjoint step with $K$ substeps, as detailed in Sect. 2, instead of the implicit exact adjoint step.

A possible choice of composition coefficients is to impose $\alpha_i = \beta_i$ for any $i = 1, \ldots, s$, which makes the scheme the composition of second-order steps. However, according to [8], the composition coefficients can also be more generally computed from splitting coefficients $a_0, a_1, \ldots, a_s$ and $b_1, \ldots, b_s$ through the relations $\beta_1 = a_0,\ \alpha_1 = b_1 - \beta_1,\ \beta_i = a_i - \alpha_{i-1},\ \alpha_i = b_i - \beta_i,\ i = 2, \ldots, s$. We will focus on a second order scheme $O2$, taking $s = 1$ $a_0 = a_1 = 1/2,\ b_1 = 1$ Namely, the O2 scheme corresponds to $\alpha_1 = 1/2, \beta_1 = 1/2$ and thus writes:

$$f^{n,1} = \mathcal{R}^*_{\Delta t/2}[f^n]f^n; \quad f^{n,2} = \mathcal{R}^*_{\Delta t/2}[f^{n,1}]f^n; \quad f^{n+1} = \mathcal{R}_{\Delta t/2}[f^{n,2}]f^{n,2}.$$

We will also consider a fourth order scheme $O4$ [1], taking $s = 6$ and $a_0 = 0.0792036964311957$, $a_1 = 0.353172906049774$, $a_2 = -0.0420650803577195$, $b_1 = 0.209515106613362$, $b_2 = -0.143851773179818$, completing by symmetry: $a_3 = 1 - 2(a_0 + a_1 + a_2)$, $a_{6-i} = a_i$, $i = 0, 1, 2, b_3 = 1/2 - b_1 - b_2$, $b_{7-i} = b_i$, $i = 1, 2, 3$. By default, we will take for $K$ the order of the corresponding composition scheme, that is $K = 2$, for $O2$ and $K = 4$ for $O4$, thanks to the following property, due to [3] in the symmetric case.

**Proposition 2** *Let* $(E, \|\cdot\|)$ *a normed space, and* $f^0 \in E$. *Let* $\mathcal{R}^*_{\Delta t}[g], \mathcal{R}_{\Delta t}[g] : E \to E$, *Lipschitz, Lipschitz in* $\Delta t \in \mathbb{R}$ *and* $O(\Delta t)$*-Lipschitz*[1] *in* $g \in E$, *first order approximations of* $\mathcal{T}^{exact}_{\Delta t}$. *Let* $s \in \mathbb{N}^*$ *and* $\alpha_1, \ldots, \alpha_s, \beta_1, \ldots, \beta_s \in \mathbb{R}$. *We suppose that the composition method is given by* $f_0^1 = f^0 \in E$

$$f_{2i-1}^1 = \mathcal{R}^*_{\beta_i \Delta t}[f_{2i-1}^1] f_{2(i-1)}^1 \tag{4}$$

$$f_{2i}^1 = \mathcal{R}_{\alpha_i \Delta t}[f_{2i-1}^1] f_{2i-1}^1 \tag{5}$$

[1] $\|\mathcal{R}_{\gamma\Delta t}[g]f_1 - \mathcal{R}_{\gamma\Delta t}[g]f_2\| \le C_1\|f_1 - f_2\|,\ f_1, f_2 \in E,$ $\|\mathcal{R}_{\gamma_1\Delta t}[g] - \mathcal{R}_{\gamma_2\Delta t}[g]\| \le C_2\|\gamma_1\Delta t - \gamma_2\Delta t\|,\ \gamma_1, \gamma_2 \ge 0,\ \|\mathcal{R}_{\Delta t}[g_1] - \mathcal{R}_{\Delta t}[g_2]\| \le C_4\Delta t\|g_1 - g_2\|,\ g_1, g_2 \in E.$

*for* $i = 1, \ldots, s$ *and* $f^1 = f^1_{2s}$*. We suppose that the method is of order* $\geq p$*:* $\|f^1 - T^{exact}_{\Delta t} f^0\| = O(\Delta t^{p+1})$*. Then, if the adjoint step (4) is replaced by its* $K$*th fixed point iteration, with* $K \geq p$*, the modified method is still of order* $\geq p$*.*

***Proof*** Let us denote $\tilde{f}^1_j$ the solution to the approximated scheme. At the $2i-1$th iteration, we have $\tilde{f}^{1,k}_{2i-1} = \mathcal{R}^*_{\beta_i \Delta t}[\tilde{f}^{1,k-1}_{2i-1}]\tilde{f}^1_{2(i-1)}$, $k = 1, \ldots, K$, and $f^1_{2i-1} = \mathcal{R}^*_{\beta_i \Delta t}[f^1_{2i-1}]f^1_{2(i-1)}$. Using the $O(\Delta t)$-Lipschitz in $g_1$ and Lipschitz property in $g_2$ of $\mathcal{R}^*_{\beta_i \Delta t}[g_1]g_2$, we get

$$\begin{aligned}
\|\tilde{f}^{1,K}_{2i-1} - f^1_{2i-1}\| &\leqslant \|\mathcal{R}^*_{\beta_i \Delta t}[\tilde{f}^{1,K-1}_{2i-1}]\tilde{f}^1_{2(i-1)} - \mathcal{R}^*_{\beta_i \Delta t}[f^1_{2i-1}]\tilde{f}^1_{2(i-1)}\| \\
&\quad + \|\mathcal{R}^*_{\beta_i \Delta t}[f^1_{2i-1}]\tilde{f}^1_{2(i-1)} - \mathcal{R}^*_{\beta_i \Delta t}[f^1_{2i-1}]f^1_{2(i-1)}\| \\
&\leqslant C_1 \Delta t \|\tilde{f}^{1,K-1}_{2i-1} - f^1_{2i-1}\| + C_2 \|\tilde{f}^1_{2(i-1)} - f^1_{2(i-1)}\|.
\end{aligned}$$

Consequently, iteratively, we have:

$$\begin{aligned}
\|\tilde{f}^1_{2i-1} - f^1_{2i-1}\| &\leqslant (C_1 \Delta t)^K \|\tilde{f}^1_{2(i-1)} - f^1_{2i-1}\| + \frac{C_2}{1 - C_1 \Delta t}\|\tilde{f}^1_{2(i-1)} - f^1_{2(i-1)}\| \\
&\leq (C_1 \Delta t)^K \|f^1_{2(i-1)} - f^1_{2i-1}\| + (\frac{C_2}{1 - C_1 \Delta t} + (C_1 \Delta t)^K)\|\tilde{f}^1_{2(i-1)} - f^1_{2(i-1)}\|.
\end{aligned}$$

Since, using the Lipschitz property in $\Delta t$, $\|f^1_{2(i-1)} - f^1_{2i-1}\| = \|\mathcal{R}^*_0[f^1_{2i-1}]f^1_{2(i-1)} - \mathcal{R}^*_{\beta_i \Delta t}[f^1_{2i-1}]f^1_{2(i-1)}\| = O(\Delta t)$, we obtain that $||\tilde{f}^1_{2i-1} - f^1_{2i-1}|| = O(\Delta t^{p+1})$ as soon as $||\tilde{f}^1_{2(i-1)} - f^1_{2(i-1)}|| = O(\Delta t^{p+1})$. The property is also true at the $2i$th iteration, and this leads to the result, by using the triangular inequality.

## 4 Numerical Results

**The May model.** Before considering the guiding center model, we test the composition scheme on the May model [3], which is the following non-linear differential system:

$$x'(t) = a\,x(t)(1 - \frac{x(t)}{b}) - c\,\frac{x(t)y(t)}{(x(t) + d)}, \quad y'(t) = e\,y(t) - \frac{y^2(t)}{(f x(t))}.$$

We consider the following initial condition $x(0) = 100$, $y(0) = 20$ and the following parameters: $a = 0.6$, $b = 10$, $c = 0.5$, $d = e = 0.1$, $f = 2$. We compute the error in $L^\infty$ norm on $\mathbb{R}^2$ and consider as reference solution at time $T = 5$ $(x(T), y(T)) = (4.463072920803955, 3.133825837839541)$, obtained with a RK4 scheme using $\Delta t = 1.25 \cdot 10^{-6}$.

Here the equivalent "transport operator" is defined by $T^{\alpha\beta}_{\Delta t}[(u^*, v^*)](u^0, v^0) = (u(\Delta t), v(\Delta t))$, where $u$ is solution to:

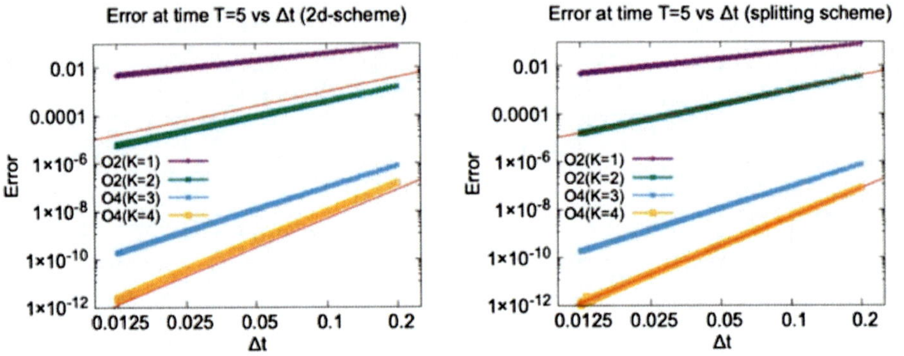


**Fig. 1** Error at time $T = 5$ versus $\Delta t$ for $2d$-scheme (left) and splitting scheme (right) for the May model

$$u'(t) = \alpha a\, u(t)(1 - \frac{u(t)}{b}) - c\,\frac{u(t)v^*}{(u^* + d)}, \quad v'(t) = \alpha e\, v(t) - \frac{v(t)v^*}{(fu^*)}.$$

with $(u(0), v(0)) = (u^0, v^0)$. The exact formula is used:

$$v(t) = v^0 e^{((\alpha e - \beta v^*/(fu^*))t)}, \; u(t) = b\, e^{(\alpha a - \frac{\beta c v^*}{u^*+d})t)} / \left(\gamma\, (e^{((\alpha a - \frac{\beta c v^*}{u^*+d})t)} - 1) + \frac{b}{u^0}\right),$$

with $\gamma = \alpha a/(\alpha a - \frac{\beta c v^*}{u^*+d})$. Numerical results are shown on Fig. 1. The functions $0.1\Delta t^2$ and $5 \cdot 10^{-5}\Delta t^4$ are also plotted for comparison. The O4 scheme behaves as $5 \cdot 10^{-5}\Delta t^4$ and we can check that the RK4 scheme behaves asymptotically as $3.5\Delta t^4$, meaning that the $O4$ scheme RK4 scheme with a time step $\Delta t/16$ has around the same accuracy as the $O4$ scheme using $K = 4$ and time step $\Delta t$. The $O2$ $2d$-scheme (which is in fact the classical predictor-corrector scheme) is more accurate than the $O2$ splitting scheme for a given $\Delta t$, and for the $O4$ scheme, the splitting scheme is little more accurate. Concerning the efficiency, the $O4$ $2d$-scheme with $K = 4$ is around 50 (resp. 30) slower than the RK4 scheme, in our python (resp. $C$) implementation, meaning that it is around $50/16 \approx 3$ (resp. $30/16 \approx 2$) times slower than the classical RK4 scheme for a given accuracy.

**The guiding center model.** We now consider the guiding center model. A semi-Lagrangian scheme is used to solve the transport elementary blocks with centered Lagrange interpolation of degree 17 (ref. [2]) for the spatial interpolation and the characteristics (assumed to be exact in the previous analysis) are approximated by the RK4 method, for the $2d$-scheme and by a RK2 method for the splitting scheme (similar results are obtained with a RK4 scheme). Initial condition is $f(t = 0, x, y) = \sin(y) + 0.015\cos(x/2), \;\; x \in [0, 4\pi], \;\; y \in [0, 2\pi]$, periodic boundary conditions are considered for $f$, Dirichlet boundary conditions for $\Phi$. Numerical results are shown on Figs. 2, 3 and 4. We consider only the $O4$ scheme with $K = 4$: the $2d$-scheme and the splitting scheme. The reference solution has been computed with the $O4$ $2d$ scheme with $256 \times 256$ spatial discretization points and a time step

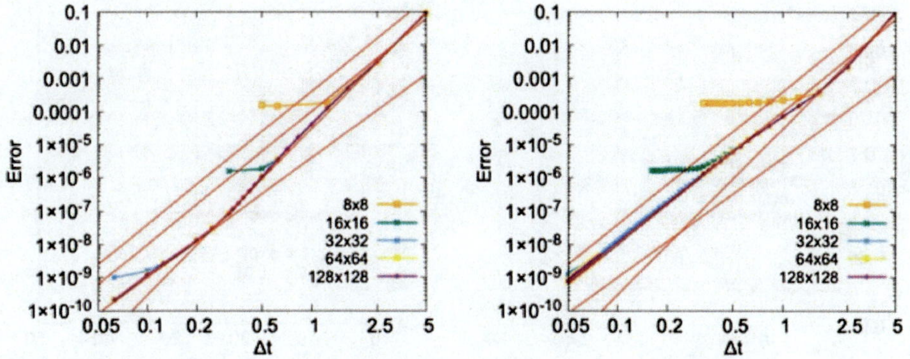


Fig. 2 $L^\infty$ error at time $T = 5$ versus $\Delta t$ for $2d$-scheme (left) and splitting scheme (right) for the guiding center model for different spatial discretizations ($8 \times 8, 16 \times 16, 32 \times 32, 64 \times 64, 128 \times 128$). Each point corresponds to one simulation. The functions $3 \cdot 10^{-5} \Delta t^5$, $4 \cdot 10^{-4} \Delta t^4$, $10^{-4} \Delta t^4$ and $4 \cdot 10^{-6} \Delta t^{3.6}$ are also plotted for comparison.

$\Delta t = 0.0078125$. Both schemes exhibit a convergence order around 4, with enhanced convergence for the $2d$-scheme for large time steps and convergence order around 5. For comparison, we consider also the fourth order time-scheme (RKEI4) developed in [12], whose error is approximated by the function $4 \cdot 10^{-4} \Delta t^4$. The cost of the $O4$ $2d$-scheme is around 6 times slower, while the error is divided by 40. So again, we see that the scheme is around $2-3$ times slower than the RKEI4 scheme for a given accuracy. On the other hand, it seems that the O4 schemes seem to behave well regarding the mass and energy conservation (see Fig. 3), and they have both similar accuracy. This is rather usual for splitting schemes. The $L^1$ et $L^2$ norms are on the contrary better conserved by the RKEI4 scheme. Note also that the splitting version, despite the known inherent disadvantage of breaking the conservative/advective form of the equations, is faster (around 6 times faster than the 2d version, in our implementation, in the O4 case), can be easily parallelized, and conservative versions are easier to handle. The fact that it gives quite good results for a high order time discretization (O4 here) may thus enlarge its range of applicability.

## 5 Conclusion and Perspectives

In this paper, we present second and fourth order composition schemes for the guiding-center model. They are based on explicit steps and approximated adjoint steps, which are all composition of 2d or 1d transports. Note that the total number of steps is quite large: for instance, the fourth order method requires 30 steps per iteration (counting all the steps of the fixed point iterations). However, the numerical simulations show that it is only 2–3 slower than the RKEI4 scheme [12] and the mass and energy are shown to be better preserved, in accordance to the optimized composition coefficients which reduce the errors for a fixed cost. Moreover,

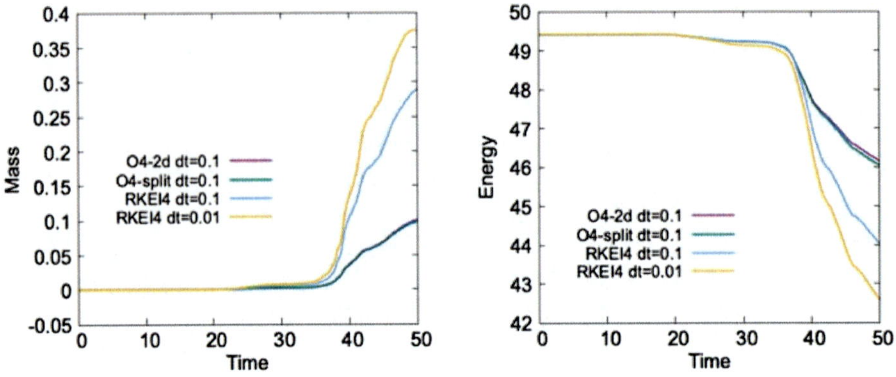


**Fig. 3** Time evolution of mass (left) and energy (right). Spatial discretization: $128 \times 128$.

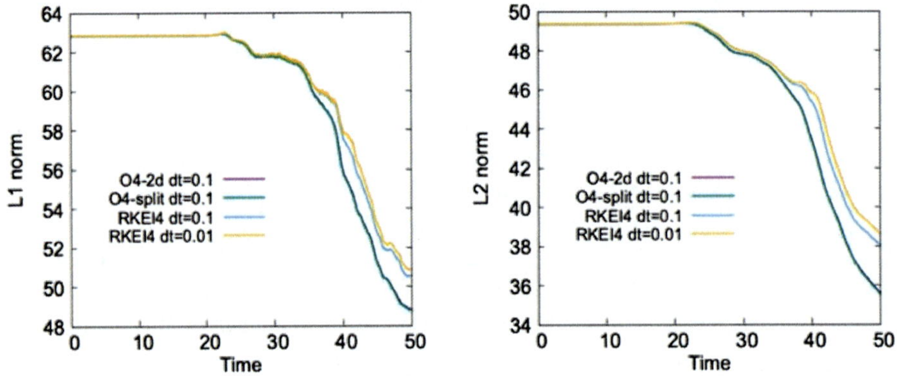


**Fig. 4** Time evolution of $L^1$ norm (left) and $L^2$ norm (right). Spatial discretization: $128 \times 128$.

the method is quite generic. Here a semi-Lagrangian scheme has been used for the spatial discretization but conservative schemes, like finite-volume or discontinuous Galerkin schemes, could be considered instead. The method could then be tested to more complex models.

**Acknowledgements** This work has been carried out within the framework of the EUROfusion Consortium, funded by the European Union via the Euratom Research and Training Programme (Grant Agreement No 101052200 EUROfusion). Views and opinions expressed are however those of the authors only and do not necessarily reflect those of the European Union or the European Commission. Neither the European Union nor the European Commission can be held responsible for them.

## References

1. Blanes, S., Moan, P.C.: Practical symplectic partitioned Runge-Kutta and Runge-Kutta-Nyström methods. J. Comput. Appl. Math. **142**(2), 313–330 (2002)

2. Crouseilles, N., Mehrenberger, M., Sonnendrücker, E.: Conservative semi-Lagrangian schemes for Vlasov equations. J. Comput. Phys. **229**(6), 1927–1953 (2010)
3. Einkemmer, L., Ostermann, A.: An almost symmetric Strang splitting scheme for the construction of high order composition methods. J. Comput. Appl. Math. **271**, 307–318 (2014)
4. Filbet, F., Prouveur, C.: High order time discretization for backward semi-Lagrangian methods. J. Comput. Appl. Math. **303**, 171–188 (2016)
5. Hairer, E., Wanner, G., Lubich, C.: Geometric numerical integration. Structure-Preserving Algorithms for Ordinary Differential Equations. Springer Series in Computational Mathematics (SSCM, volume 31) (2006)
6. Hamiaz, A., Mehrenberger, M., Back, A., Navaro, P.: Guiding center simulations on curvilinear grids. ESAIM: Proc. Surv. **53**, 99–119 (2016)
7. Hélie, R., Helluy, P., Franck, E., Navoret, L.: Kinetic over-relaxation method for the convection equation with Fourier solver. In: Finite Volumes for Complex Applications IX-Methods, Theoretical Aspects, Examples: FVCA 9, Bergen, Norway, June 2020 IX, pp. 745–753. Springer International Publishing (2020)
8. McLachlan, R.I.: On the numerical integration of ordinary differential equations by symmetric composition methods. SIAM J. Sci. Comput. **16**(1), 151–168 (1995)
9. McLachlan, R.I.: Tuning symplectic integrators is easy and worthwhile. Commun. Comput. Phys. **31**(3), 987–996 (2022)
10. Piao, X., Kim, P., Kim, D.: One-step L($\alpha$)-stable temporal integration for the backward semi-Lagrangian scheme and its application in guiding center problems. J. Comput. Phys. **366**, 327–340 (2018)
11. Sonnendrücker, E., Roche, J., Bertrand, P., Ghizzo, A.: The semi-Lagrangian method for the numerical resolution of the Vlasov equation. J. Comput. Phys. **149**(2), 201–220 (1999)
12. Zheng, N., Cai, X., Qiu, J. M., Qiu, J.: Fourth-order conservative non-splitting semi-Lagrangian Hermite WENO schemes for kinetic and fluid simulations (2022). arXiv:2208.03890

# TVD Analysis of a (Pseudo-)Staggered Scheme for the Isentropic Euler Equations

**Katia Ait-Ameur and Michaël Ndjinga**

**Abstract** In this paper, we build and analyze the stability of a collocated scheme, involving a specific numerical diffusion operator, for the isentropic Euler equations. This scheme is based on the numerical diffusion operator of a family of staggered finite volume schemes introduced in [1]. The properties of this operator allowed to understand the $L^2$-stability of staggered finite volume methods. Staggered schemes are popular in the thermal hydraulics community for their reported robustness and lack of spurious oscillations. The contributions of this paper are twofold: we firstly build a colocated scheme with a staggered-based numerical diffusion operator, hence the proposed terminology of "pseudo-staggered" scheme, and we secondly present a rigorous TVD analysis, in order to explain the non-oscillatory behaviour of staggered discretisations.

**Keywords** Euler equations · Compressible flows · Finite volumes · Staggered grids · TVD analysis

## 1 Introduction

Staggered finite volume discretisations were introduced in the 1960s for incompressible flows [5], then nearly incompressible flows [2]. Although non conservative, they are nowadays very popular in the thermal hydraulics community of two phase flows [4], and used in many nuclear safety simulation platforms. However, their stability analysis remained based on heuristics [3], and practical simulations. Starting from 2008, entropy-stable staggered discretisations have been proposed for the Euler sys-

K. Ait-Ameur
CMAP, CNRS, École polytechnique, Institut Polytechnique de Paris, Palaiseau, France
e-mail: katia.ait-ameur@polytechnique.edu

M. Ndjinga (✉)
Université Paris-Saclay, CEA Saclay, DES, ISAS, DM2S, STMF, Gif-sur-Yvette, France
e-mail: michael.ndjinga@cea.fr

 
E. Franck et al. (eds.), *Finite Volumes for Complex Applications X—Volume 2, Hyperbolic and Related Problems*, Springer Proceedings in Mathematics & Statistics 433,
https://doi.org/10.1007/978-3-031-40860-1_26

tem [6]. In [1], a framework is proposed for the rigourous analysis of the linearised $L^2$-stability of these staggered schemes [4] in view of designing robust conservative $L^2$-stable for multiphase flows.

Beyond stability, one of the purposed advantages of the staggered discretisation is their apparent lack of spurious oscillations. This paper proposes a TVD analysis to explain the good behaviour of staggered discretisation for low Mach flows. We perform a rigourous TVD analysis in the case of the 1D pseudo-staggered scheme introduced in [1] for the isentropic Euler equations as a first step before studying other discretisations and more advanced models. In Sect. 2 we recall the numercal method, then we compute in Sect. 3 the equations giving the evolution of the discrete Riemann invariants. Finaly in Sect. 4 we give a partial result obtained in the case of low Mach number flows. Some first numerical results are given in Sect. 5.

## 2 A Pseudo-Staggered Scheme

We consider the following 1D isentropic Euler equations in conservative form:

$$\begin{cases} \partial_t \rho + \partial_x q = 0 \\ \partial_t q + \partial_x \frac{q^2}{\rho} + \partial_x p = 0. \end{cases} \tag{1}$$

where $t$ stands for the time, $\rho$, $q$ and $p$ are the density, momentum and pressure respectively. The pressure $p$ is a function of the density $p = f(\rho)$ with $f$ a strictly increasing function. System (1) takes the conservative form

$$\partial_t U + \partial_x F(U) = 0, \quad U = \begin{pmatrix} \rho \\ q \end{pmatrix}, \quad F(U) = \begin{pmatrix} q \\ \frac{q^2}{\rho} + p \end{pmatrix}. \tag{2}$$

We propose a conservative finite volume discretisation of (1)

$$\partial_t U_i + \frac{F_{i,i+1} - F_{i-1,i}}{\Delta x} = 0, \quad U_i = \begin{pmatrix} \rho_i \\ q_i \end{pmatrix}, \tag{3}$$

where the numerical flux takes the form

$$F_{i,i+1}(U_i, U_{i+1}) = \frac{F(U_i) + F(U_{i+1})}{2} + D_{Stag}(U_i, U_{i+1}) \frac{U_i - U_{i+1}}{2}, \tag{4}$$

where the matrix $D_{Stag}$ is the diffusion operator of the scheme and is the same as the one of the staggered scheme [1, Corollary 1]

$$D_{Stag}(U_i, U_j) = \begin{pmatrix} |u_{ij}| - u_{ij} & 1 \\ -c^2 - u_{ij}^2 & 2u_{ij} \end{pmatrix}. \tag{5}$$

where $c$ is the sound speed, and $u_{ij}$ is the Roe average of the velocity between the two states $U_i$ and $U_j$. This matrix has complex eigenvalues $|u_{ij}| \pm ic$ but still satisfies the stability condition introduced in [1]: $\tilde{P}(D_{Stag} + {}^t D_{Stag})\tilde{P}^{-1} = 2|u_{ij}|\mathbb{I}_2 \geq 0$, where $\tilde{P} = \begin{pmatrix} c & 0 \\ -u_{ij} & 1 \end{pmatrix}$ is a transfer matrix that symmetrises $\nabla F$.

Hence the final form of the semi-discrete finite volume scheme

$$\partial_t U_i + \frac{A^-(U_i, U_{i+1})(U_{i+1} - U_i) + A^+(U_{i-1}, U_i)(U_i - U_{i-1})}{\Delta x} = 0. \tag{6}$$

The matrices $A^+$ and $A^-$ are defined by:

$$A^+(U_i, U_j) = \frac{A(U_i, U_j) + D_{Stag}(U_i, U_j)}{2} = \begin{pmatrix} \frac{|u_{ij}| - u_{ij}}{2} & 1 \\ -u_{ij}^2 & u_{ij} + \frac{|u_{ij}| + u_{ij}}{2} \end{pmatrix}, \tag{7}$$

$$A^-(U_i, U_j) = \frac{A(U_i, U_j) - D_{Stag}(U_i, U_j)}{2} = \begin{pmatrix} -\frac{|u_{ij}| - u_{ij}}{2} & 0 \\ c^2 & u_{ij} - \frac{|u_{ij}| + u_{ij}}{2} \end{pmatrix}, \tag{8}$$

where $A(U_i, U_j)$ is the Roe matrix between the states $U_i$ and $U_j$.

In order to analyse the total variation of schemes (6), we study the behaviour of Riemann invariants that are connected to the left eigenvectors of $\nabla F$ (see [7, Chap. I.5]). Indeed for 1D smooth flows, the Riemann invariants $l_1, l_2$ follow a transport equation $\partial_t l_k + \lambda_k \partial_x l_k$, and therefore should follow a TVD property: $\partial_t TV(l_k)(t) = 0, \quad k = 1, 2$ (see Kruzkov's theorem [7, Sect. 3.3]).

## 3 Evolution of the Discrete Riemann Invariants

We linearise the Euler system (2) around a constant state $U_0 = (\rho_0, q_0)$:

$$\partial_t U + \nabla F(U_0)\partial_x U = 0, \tag{9}$$

as well as the (pseudo-)staggered scheme (6) which becomes

$$\partial_t U_i + \frac{A^-(U_0, U_0)(U_{i+1} - U_i) + A^+(U_0, U_0)(U_i - U_{i-1})}{\Delta x} = 0, \tag{10}$$

$\nabla F(U_0)$ is diagonalisable with left eigenvectors $L_1(U_0) = (u_0 + c, -1)$. and $L_2(U_0) = (-u_0 + c, 1)$. The transition matrix $P(U_0)$ for diagonalising $A(U_0, U_0)$ and its inverse $P^{-1}(U_0)$ are given by

$$P(U_0) = \begin{pmatrix} u_0 + c & -1 \\ -u_0 + c & 1 \end{pmatrix}, \quad P^{-1}(U_0) = \frac{1}{2c}\begin{pmatrix} 1 & 1 \\ u_0 - c & u_0 + c \end{pmatrix}. \tag{11}$$

In the linear case, the Riemann invariants are:

$$l_1(U_0, U) = \rho u_0 + \rho c - q, \quad l_2(U_0, U) = -\rho u_0 + \rho c + q. \tag{12}$$

We compute the evolution of the discrete invariants by multiplying (10) by the transition matrix $P(U_0)$,

- $U_i$ becomes $P(U_0)U_i = (l_{1,i}, l_{2,i})$,
- $A(U_0, U_0)$ becomes $P(U_0)A(U_0, U_0)P^{-1}(U_0) = \begin{pmatrix} u_0 - c & 0 \\ 0 & u_0 + c \end{pmatrix}$,
- $D_{Stag}(U_0, U_0)$ becomes

$$\bar{D}_{Stag}(U_0, U_0) = P(U_0)D_{Stag}(U_0, U_0)P^{-1}(U_0) = \begin{pmatrix} |u_0| & c \\ -c & |u_0| \end{pmatrix}, \tag{13}$$

- $A^+(U_0, U_0)$ becomes $\frac{1}{2}\begin{pmatrix} u_0 + |u_0| - c & c \\ -c & u_0 + |u_0| + c \end{pmatrix}$,
- $A^-(U_0, U_0)$ becomes $\frac{1}{2}\begin{pmatrix} u_0 - |u_0| - c & -c \\ c & u_0 - |u_0| + c \end{pmatrix}$.

Hence the evolution of the discrete Riemann invariants is given by

$$\partial_t l_{1,i} + (u_0^- - \frac{c}{2})\frac{[l_1]_i^{i+1}}{\Delta x} - \frac{c}{2}\frac{[l_2]_i^{i+1}}{\Delta x} + (u_0^+ - \frac{c}{2})\frac{[l_1]_{i-1}^i}{\Delta x} + \frac{c}{2}\frac{[l_2]_{i-1}^i}{\Delta x} = 0 \tag{14}$$

$$\partial_t l_{2,i} + \frac{c}{2}\frac{[l_1]_i^{i+1}}{\Delta x} + (u_0^- + \frac{c}{2})\frac{[l_2]_i^{i+1}}{\Delta x} - \frac{c}{2}\frac{[l_1]_{i-1}^i}{\Delta x} + (u_0^+ + \frac{c}{2})\frac{[l_2]_{i-1}^i}{\Delta x} = 0. \tag{15}$$

where $[X]_j^i$ is the jump between cells $j$ and $i$ of the $X$ variable: $[X]_j^i = X_i - X_j$.

## 4 Low Mach TVD Analysis

The following theorem on the low Mach TVD analysis follows from the fact that for large sound speeds ($c \gg |u|$) we have $|l_1 - l_2| \ll \min\{l_1, l_2\}$.

**Theorem 1** (Low Mach TVD analysis) *The numerical scheme (3)–(4) based on the diffusion matrix (5) satisfies*

$$\partial_t \sum_i |\, [l_1]_i^{i+1} \,| \leq 0$$

*provided*

- *Assumption 1: $c$ (resp. $M$) is large enough (resp. small enough)*
- *Assumption 2: $[\rho]_i^{i+1} = 0$ implies $2|[q]_i^{i+1}| + u^+|[\rho]_{i-1}^i| - [q - \rho u_0]_{i-1}^i \geq 0$.*

***Proof*** We focus on $TV(l_1)$, but the same approach can be used to obtain $TV(l_2)$. We first compute the evolutions of $[l_1]_i^{i+1}$ from (14) for two cells $i$ and $i+1$.

$$\partial_t [l_1]_i^{i+1} + (u^- - \frac{c}{2})\frac{[l_1]_{i+1}^{i+2} - [l_1]_i^{i+1}}{\Delta x} + (u^+ - \frac{c}{2})\frac{[l_1]_i^{i+1} - [l_1]_{i-1}^{i}}{\Delta x} - \frac{c}{2}\frac{[l_2]_{i+1}^{i+2} - [l_2]_i^{i+1}}{\Delta x} + \frac{c}{2}\frac{[l_2]_i^{i+1} - [l_1]_{i-1}^{i}}{\Delta x} = 0. \tag{16}$$

We multiply (16) by $sign([l_1]_i^{i+1})$ to obtain the evolution of $|\,[l_1]_i^{i+1}\,|$ and sum over $i$ to obtain the evolution of the variations:

$$\partial_t \sum_i |\,[l_1]_i^{i+1}\,| + (u^- - \frac{c}{2})\frac{\sum_i sign([l_1]_i^{i+1})\,[l_1]_{i+1}^{i+2} - \sum_i |\,[l_1]_i^{i+1}\,|}{\Delta x}$$
$$+(u^+ - \frac{c}{2})\frac{\sum_i |\,[l_1]_i^{i+1}\,| - \sum_i sign([l_1]_i^{i+1})\,[l_1]_{i-1}^{i}}{\Delta x}$$
$$-\frac{c}{2}\sum_i sign([l_1]_i^{i+1})\frac{[l_2]_{i+1}^{i+2} - [l_2]_i^{i+1}}{\Delta x} + \frac{c}{2}\sum_i sign([l_1]_i^{i+1})\frac{[l_2]_i^{i+1} - [l_2]_{i-1}^{i}}{\Delta x} = 0.$$

We perform the shift of indices $i+1 \to i$ and $i-1 \to i$, and after some factorisation we get:

$$\partial_t \sum_i |\,[l_1]_i^{i+1}\,| + (u^- - \frac{c}{2})\sum_i |\,[l_1]_i^{i+1}\,|\frac{sign([l_1]_{i-1}^{i})sign([l_1]_i^{i+1}) - 1}{\Delta x}$$
$$+(u^+ - \frac{c}{2})\sum_i |\,[l_1]_i^{i+1}\,|\frac{1 - sign([l_1]_{i+1}^{i+2})sign([l_1]_i^{i+1})}{\Delta x}$$
$$-\frac{c}{2}\sum_i [l_2]_i^{i+1}\frac{sign([l_1]_{i-1}^{i}) - sign([l_1]_i^{i+1})}{\Delta x} + \frac{c}{2}\sum_i [l_2]_i^{i+1}\frac{sign([l_1]_i^{i+1}) - sign([l_1]_{i+1}^{i+2})}{\Delta x} = 0.$$

We now expand $u^{\pm} - \frac{c}{2}$ and introduce the difference $[l_1]_i^{i+1}$ in the last two lines:

$$\partial_t \sum_i |\,[l_1]_i^{i+1}\,| - \frac{c}{\Delta x}\sum_i |\,[l_1]_i^{i+1}\,|(sign([l_1]_i^{i+1})sign([l_1]_{i-1}^{i}) - 1)$$
$$+\sum_i \left(u^-|\,[l_1]_i^{i+1}\,| - sign([l_1]_i^{i+1})\frac{c}{2}\,[l_1 - l_2]_i^{i+1}\right)\frac{sign([l_1]_{i-1}^{i})sign([l_1]_i^{i+1}) - 1}{\Delta x}$$
$$+\sum_i \left(u^+|\,[l_1]_i^{i+1}\,| - sign([l_1]_i^{i+1})\frac{c}{2}\,[l_1 - l_2]_i^{i+1}\right)\frac{1 - sign([l_1]_{i+1}^{i+2})sign([l_1]_i^{i+1})}{\Delta x} = 0.$$

Performing a shift of indices $i \to i-1$ in the last line we find

$$\partial_t \sum_i |[l_1]_i^{i+1}| - \frac{c}{\Delta x}\sum_i |[l_1]_i^{i+1}|(sign([l_1]_i^{i+1})sign([l_1]_{i-1}^i) - 1)$$
$$+\sum_i \left(u^-|[l_1]_i^{i+1}| - sign([l_1]_i^{i+1})\frac{c}{2}[l_1 - l_2]_i^{i+1}\right)\frac{sign([l_1]_{i-1}^i)sign([l_1]_i^{i+1}) - 1}{\Delta x}$$
$$+\sum_i \left(u^+|[l_1]_{i-1}^i| - sign([l_1]_{i-1}^i)\frac{c}{2}[l_1 - l_2]_{i-1}^i\right)\frac{1 - sign([l_1]_i^{i+1})sign([l_1]_{i-1}^i)}{\Delta x} = 0,$$

and therefore

$$\partial_t \sum_i |[l_1]_i^{i+1}| - \frac{1}{\Delta x}\sum_i (sign([l_1]_i^{i+1})sign([l_1]_{i-1}^i) - 1)$$
$$\Big(c|[l_1]_i^{i+1}| - u^-|[l_1]_i^{i+1}| + u^+|[l_1]_{i-1}^i| + sign([l_1]_i^{i+1})\frac{c}{2}[l_1 - l_2]_i^{i+1}$$
$$-sign([l_1]_{i-1}^i)\frac{c}{2}[l_1 - l_2]_{i-1}^i\Big) = 0. \tag{17}$$

In order to complete the proof we need to prove that the second line in (17) is positive. Note that the first three terms in (17) are positive. Thus there is competition between the respective jumps of $l_1$ and $l_1 - l_2$ in absolute value.

Now consider the limit in (17) when $c \to \infty$. From the definition of the Riemann invariants (12)

$$[l_1]_i^{i+1} = c[\rho]_i^{i+1} + [q - \rho u_0]_i^{i+1} \tag{18}$$
$$[l_1 - l_2]_i^{i+1} = 2[q - \rho u_0]_i^{i+1}, \tag{19}$$

there are two cases:

- if $[\rho]_i^{i+1} \neq 0$ then $[l_1 - l_2]_i^{i+1}$ and $[l_1 - l_2]_{i-1}^i$ are negligible (since order one in $c$) compared to $[l_1]_i^{i+1}$ (order two in $c$) as $c \to \infty$, hence the Assumption 1
- if $[\rho]_i^{i+1} = 0$ then $[l_1]_i^{i+1} = [q]_i^{i+1}$, $[l_1 - l_2]_i^{i+1} = 2[q]_i^{i+1}$ and the second line in (17) reduces to

$$c|[q]_i^{i+1}| - u^-|[q]_i^{i+1}| + u^+|[l_1]_{i-1}^i| + sign([q]_i^{i+1})c\,[q]_i^{i+1}$$
$$-sign([l_1]_{i-1}^i)c\,[q - \rho u_0]_{i-1}^i$$
$$= 2c|[q]_i^{i+1}| - u^-|[q]_i^{i+1}| + u^+|[l_1]_{i-1}^i| - sign([l_1]_{i-1}^i)c\,[q - \rho u_0]_{i-1}^i$$

The leading term for large $c$ is $2|[q]_i^{i+1}| + u^+|[\rho]_{i-1}^i| - [q - \rho u_0]_{i-1}^i$, hence the Assumption 2.

□

**Remark 1** The Assumption 1 in Theorem 1 is met for flows at low Mach number although we do not have a specific bound on the Mach number to be met in practice. The Assumption 2 is more tricky as it has to be checked for every numerical method.

## 5 Numerical Results

As an illustration of our main result, we present the numerical simulation of a Riemann problem for a perfect gas with isothermal gas law $p = \rho c^2$ with $c = 300\,\mathrm{m/s}$. The initial data is made of two constant states separated by a discontinuity: a left state: $(\rho_L = 1, q_L = 10\rho_L)$, and a right state: $(\rho_R = 2, q_R = -10\rho_L)$.

We compare the numerical result obtained with three fully implicit finite volume methods taking the form (3) where the numerical flux takes the form

$$F_{i,i+1}(U_i, U_{i+1}) = \frac{F(U_i) + F(U_{i+1})}{2} + D(U_i, U_{i+1})\frac{U_i - U_{i+1}}{2}, \qquad (20)$$

where the matrix $D$ can be $D_{upwind} = |\nabla F|$ for the upwind scheme, $D_{centered} = 0$ for the centered scheme, or $D = D_{stag}$ (Eq. 5) for the staggered scheme.

Figure 1 shows the numerical results obtained with CFL number 1, and Fig. 2a shows the evolution of the Riemann invariant $l_1 = \frac{u}{c} + \ln(\rho)$. Figure 1 shows as expected that the three conservative schemes capture the three constant states separated by two waves. The result of the upwind and staggered schemes are very similar. However the centered scheme oscillates since it is not TVD.

On Fig. 2a we observe as expected that the total variation of the centered scheme grows to a large final value. The total variation of the upwind scheme remains almost constant. The total variation of the staggered scheme is very close to that of the upwind scheme, however it increases in the first time step.

With the initial data chosen, the second assumption of Theorem 1 is satisfied. We therefore expected that the scheme would become TVD as one increases the sound speed. However looking at Fig. 2b where we increased the sound speed from 30 to 30000, the total variation of the pseudo-staggered scheme never appears to

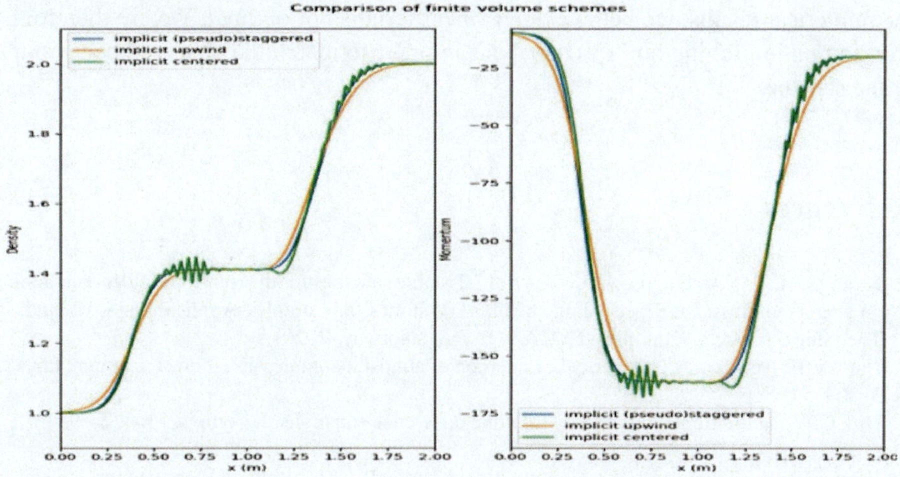


**Fig. 1** Solution of the Riemann problem at $t = \frac{nx}{4}$

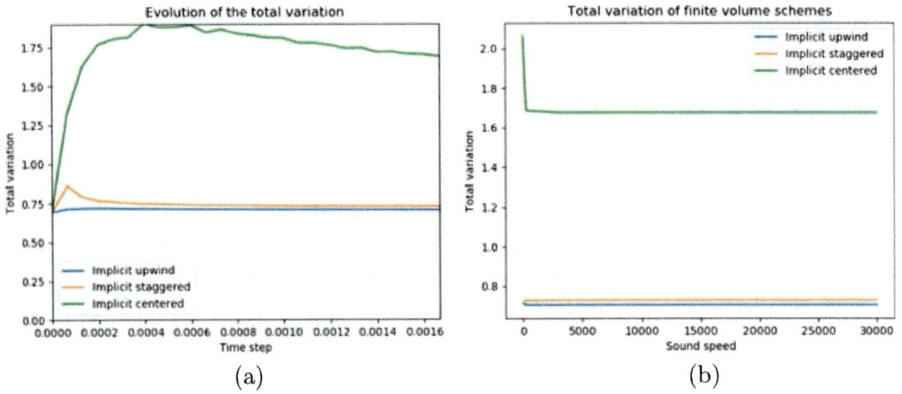


**Fig. 2** **a** Evolution of the total variation of the Riemannin variant $l_1$. **b** Evolution of the total variation of the Riemannin variant $l_1$

be diminishing. One possible explanation is the fact that we analysed the linearised Euler system but simulated and plotted the nonlinear invariant. Another possible reason is that our theoretical analysis is based on a continuous in time semi-discrete equation whereas the numerical results are obtained with a fully discrete scheme.

## 6 Conclusion

We performed a TVD analysis for a stable conservative scheme inspired from staggered discretisations used in the multiphase flow community. The Riemann invariants were analysed and their total variation decreases provided the two conditions of Theorem 1 are met. These assumptions are not necessarily met in practice, and given the numerical results, we believe our TVD criteria is not optimal. We are therefore working on improving our TVD analysis in order to match the observed behaviour of the scheme.

## References

1. Ndjinga, M., Ait-Ameur, K.: A new class of L2-stable schemes for the isentropic euler equations on staggered grids. In: Proceeedings of Finite Volumes for Complex Applications—Methods, Theoretical Aspects, Examples: FVCA 9. Bergen, Norway (2020)
2. Harlow, F., Amsden, A.: Numerical calculation of almost incompressible flow. J. Comput. Phys. **3**, 8093 (1968)
3. Hirt, C.W.: Heuristic stability theory for finite difference equations. J. Comput. Phys. **2**, 339–355 (1968)
4. Prosperetti, A., Tryggvason, G.: Computational Methods for Multiphase Flow. Cambridge University Press (2009)

5. Harlow, F.H., Welch, J.E.: Numerical calculation of time-dependent viscous in-compressible flow of fluid with free surface. Phys Fluids **8**(12), 2182–2189 (1965)
6. Gallouët, T., Gastaldo, L., Herbin, R., Latché, J.-C.: An unconditionally stable pressure correction scheme for the compressible barotropic Navier-Stokes equations. Math. Model. Numer. Anal. **42**(2), 303–331 (2008)
7. Godlewski, E., Raviart, P.A.: Numerical approximation of hyperbolic systems of conservation laws. Applied Mathematical Sciences, vol. 118. Springer, New York (1996)

# Backward Reconstruction for Non Resonant Triangular Systems of Conservation Laws

**Florian Peru**

**Abstract** We consider a special class of triangular systems of conservation laws which essentially consist of a scalar nonlinear strictly convex conservation law coupled with one (or more) linear continuity equation. We propose a finite volumes numerical scheme extending the numerical results in [3], on the attainable profiles for such systems, and in [10], on the initial data identification problem for scalar convex conservation laws. For any given attainable profile of the system at time $T > 0$, and $k \in \mathbb{N}$, our numerical approach produces an initial condition containing $k$ jumps discontinuity which lead to the given target profile in time $T$. In this contribution we describe our procedure, give an example and verify the numerical convergence of the finite volumes scheme.

**Keywords** Systems of conservation laws · Triangular systems · Finite volumes scheme · Backward reconstruction · Initial data identification

## 1 Preliminaries

Our focus is the backward reconstruction and problem of initial data identification for triangular systems of the form

$$\begin{cases} \partial_t u + \partial_x f(u) = 0, & t > 0,\ x \in \mathbb{R}, \\ \partial_t v + \partial_x (g(u)v) = 0, & t > 0,\ x \in \mathbb{R}, \\ u_T(x) = u(T, x), & x \in \mathbb{R}, \\ v_T = v(T, x), & x \in \mathbb{R}, \end{cases} \tag{1}$$

where the $f : \mathbb{R} \to \mathbb{R}$ and $g : \mathbb{R} \to \mathbb{R}$ satisfy the following assumptions:

F. Peru (✉)
Laboratoire de mathématiques de Besançon, UMR CNRS 6623, Université de Franche-Comté, 16 route de Gray, 25000 Besançon, France
e-mail: florian.peru@univ-fcomte.fr



E. Franck et al. (eds.), *Finite Volumes for Complex Applications X—Volume 2, Hyperbolic and Related Problems*, Springer Proceedings in Mathematics & Statistics 433,
https://doi.org/10.1007/978-3-031-40860-1_27

(F1) *Regularity of the flux*: $f \in C^2(\mathbb{R})$;
(F2) *Local uniform convexity of the flux*: for all compact sets $K \in \mathbb{R}$, there exists $\alpha_K > 0$ such that $f''_{|K} \geq \alpha_K$;
(G1) *Regularity of the transport coefficient*: $g \in C^1(\mathbb{R})$;
(NR) *Non-resonance*: $f'(\xi) \neq g(\xi)$ for all $\xi \in \mathbb{R}$.

**Remark 1** To keep the presentation as light as possible, we take one transport equation, but all the presentation is valid for a system with more transport equations which satisfies (G1) and (NR).

Entropy solutions for a strictly hyperbolic system of conservation laws form a Lipschitz continuous semigroup

$$\begin{array}{rcl} S^+ : \mathbb{R}^+ \times (L^\infty(\mathbb{R}))^2 & \to & (L^\infty(\mathbb{R}))^2 \\ (t, U_0) & \mapsto & S_t^+(U_0)\,, \end{array} \tag{2}$$

i.e. $S_t^+(U_0) = U(t, .) : \mathbb{R} \to \mathbb{R}$ where $U$ is the entropy solution of

$$\begin{cases} \partial_t U + \partial_x F(U) = 0, & t > 0,\ x \in \mathbb{R}, \\ U(0, x) = U_0(x), & t > 0,\ x \in \mathbb{R}, \end{cases}$$

$F : \mathbb{R}^2 \to \mathbb{R}^2$, $U : \mathbb{R}^+ \times \mathbb{R} \to \mathbb{R}^2$, see [4, 6].

The initial data identification problem consists of three steps. First, for a given $U_T = (u_T, v_T) \in L^\infty(\mathbb{R}^2)$ we verify that $U_T$ is an attainable profile for the system, i.e. that there exists an initial condition $U_0$ such that $S_T^+(U_0) = U_T$. Second, if this is the case, we describe the set of initial conditions which leads to $U_T$, $\mathcal{I}(U_T)$, and investigate its topological properties. Finally, we define optimal control problems and we look for minimizers within the set $\mathcal{I}(U_T)$.

In this paper we focus on a backward-forward numerical procedure which can be applied in relation with the above problem. For the scalar conservation law $\partial_t u + \partial_x f(u) = 0$, the set of states in $L^\infty(\mathbb{R})$ attainable by entropy solutions at time $T > 0$ was characterized in [1, 2] as follows:

$$\mathcal{A}_T(\mathbb{R}, f) = \Big\{u \in L^\infty(\mathbb{R}) : \exists \rho : \mathbb{R} \to \mathbb{R}, \text{ right continuous, nondecreasing such that } f'(u) = \frac{x - \rho(x)}{T}\Big\}.$$

In [3] , we have the following characterization of the set of attainable states for the system (1):

**Theorem 1** (Attainable states for the triangular system) *Assume that $f$ satisfies (F1)–(F2), $g$ satisfies (G1), and that the system satisfies the non-resonance condition (NR) in the range $[a, b]$ of values of $u$, i.e., for all $u \in [a, b]$, $f'(u) \neq g(u)$.*

*Given $T > 0$, we consider a target state $U_T = (u_T, v_T)$ where $u_T \in \mathcal{A}_T(\mathbb{R}, f)$ takes values in $[a, b]$, and $v_T \in L^\infty(\mathbb{R})$.*

*Then there exists an initial datum* $U_0 = (u_0, v_0)$ *such that* (1) *admits an entropic solution* $U = (u, v)$ *on* $(0, T) \times \mathbb{R}$ *with* $U(0, \cdot) = U_0$ *and* $U(T, \cdot) = U_T$.

*On the contrary, whenever* $u_T \notin \mathcal{A}_T(\mathbb{R}, f)$, *for any* $v_T \in L^\infty(\mathbb{R})$ *the terminal state* $U_T = (u_T, v_T)$ *is not attainable at time T by any entropy solution of* (1).

In other words, from [3], it follows that—under assumptions (F1), (F2), (G1), and (NR)—the set of states in $L^\infty(\mathbb{R})$ which are attainable by entropy solutions of system (1) is given by

$$\mathfrak{A}_T(\mathbb{R}) = \mathcal{A}_T(\mathbb{R}, f) \times L^\infty(\mathbb{R}).$$

In [5, 11] the authors describe the set $\mathcal{I}(u_T) = \{u_0 \in L^\infty(\mathbb{R}) : K_T^+(u_0) = u_T\}$ of initial conditions which leads to the target profile $u_T$ where $K^+$ is the operator defining the semigroup of entropy solutions of the scalar conservation law.

We define the backward semigroup $K^-$ :

$$\begin{array}{rcl} K^- : \mathbb{R}^+ \times L^\infty(\mathbb{R}) & \to & L^\infty(\mathbb{R}) \\ (t, u_T) & \mapsto & K_t^+ (u_T \circ r) (r(\cdot)), \end{array} \tag{3}$$

where $r(x) = -x$. If $u_T \in \mathcal{A}_T$, this semigroup gives the unique isentropic solution of the Cauchy problem

$$\begin{cases} \partial_t u + \partial_x f(u) = 0, & t > 0,\ x \in \mathbb{R}, \\ u_T(x) = u(T, x), & x \in \mathbb{R}. \end{cases} \tag{4}$$

The papers [5, 11] contain the following result:

**Theorem 2** (Inverse design for scalar conservation laws, [11]) *Let* $T > 0$ *and* $u_T \in \mathcal{A}_T$ : *Then, the initial data* $\tilde{u}_0 \in L^\infty(\mathbb{R})$ *verifies* $K_T^+(\tilde{u}_0) = u_T$ *if and only if the following statement holds:*

*(1) For any* $(x, y) \in X(u_T) \times \mathbb{R}$

$$\int_{x - T f'(u_T(x))}^{y} K_T^-(u_T)(s)\,\mathrm{d}s \le \int_{x - T f'(u_T(x))}^{y} \tilde{u}_0(s)\,\mathrm{d}s, \tag{5}$$

*(2) For any* $(x, y) \in X(u_T)^2$

$$\int_{x - T f'(u_T(x))}^{y - T f'(u_T(y))} K_T^-(u_T)(s)\,\mathrm{d}s = \int_{x - T f'(u_T(x))}^{y - T f'(u_T(y))} \tilde{u}_0(s)\,\mathrm{d}s, \tag{6}$$

*where* $X(u_T) = \{x \in \mathbb{R}, u_T(x_-) = u_T(x_+)\}$, $u_T(x_-) = \lim\limits_{y \to x, y < x} u_T(y)$ *and* $u_T(x_+) = \lim\limits_{y \to x, y > x} u_T(y)$.

**Remark 2** (Geometric interpretation of Theorem 2, [10]) There is a geometrical way to describe the set of initial states which leads to the state $u_T$.

Let $f \in \mathcal{C}^1$ a strictly convex function, $u_T$ an admissible state, with a discontinuity at $\overline{x} \in \mathbb{R}$, $u_L > u_R$ such that $u_T(\overline{x}_-) = u_L$ and $u_T(\overline{x}_+) = u_R$, $T > 0$. We call $D_T$ the interval $[\overline{x} - Tf'(u_L), \overline{x} - Tf'(u_R)]$. We introduce the set $\Gamma(u_L, u_R, \overline{x}, T)$ such that $\gamma \in \Gamma(u_L, u_R, \overline{x}, T)$ if

1. $\gamma : D_T \to \mathbb{R} \in W^{1,1}$,
2. $\dot{\gamma} \in \mathrm{BV}(\mathcal{D})$ the space of functions with bounded variations,
3. $\gamma(\overline{x} - Tf'(u_L)) = 0$,
4. $\gamma(\overline{x} - Tf'(u_R)) = T(u_L f'(u_L) - f(u_L) - u_R f'(u_R) + f(u_R))$,
5. for every $x \in [\overline{x} - Tf'(u_L), \overline{x} - Tf'(u_R)]$,

$$\gamma(x) \geq \gamma_*(x) = -T \int_{u_L}^{(f')^{-1}\left(\frac{\overline{x}-x}{T}\right)} s f''(s)\, \mathrm{d}s.$$

Then, to find an initial state $\tilde{u}_0$, we pick a path $\gamma \in \Gamma(u_L, u_R, \overline{x}, T)$, and the initial data $\tilde{u}_0$ corresponding to $\gamma$ is defined by

$$\tilde{u}_0(x) = \begin{cases} \dot{\gamma}(x) & \text{for a.e } x \in D_T, \\ u_0(x) & \text{for every } x \text{ in } \mathbb{R} \setminus D_T, \end{cases}$$

where $u_0$ is the isentropic solution.

## 2 Numerical Analysis

In this section, we present some results on the numerical backward reconstruction of system (1) on $[0, T] \times J$, where $J$ is a given close interval $[A_1, A_2] \subset \mathbb{R}$, $A_1 < A_2$, $T > 0$ and where the final state $U_T = (u_T, v_T)$ belongs to $\mathcal{A}_T \times L^\infty(J)$. Firstly, we describe the numerical method.

For the first equation of (1), we use the backward semigroup for the first equation $K^-$ (3) Then, the numerical backward resolution of the second equation of (1) is based on the numerical resolution of the forward problem

$$\partial_t v + \partial_x(-g(u)v) = 0, \tag{1}$$

where $u$ is the isentropic solution of the first equation. To perform the computations, we need to determine the domains of dependence of the system, interval in which the solution changes due to a finite propagation speed, which is given by $I = [C_1, C_2]$, where

$$B_{1,1} = A_1 - Tf'(u_T(A_1)), \quad B_{1,2} = A_2 + Tf'(u_T(A_2)),$$
$$B_{2,1} = A_1 - Tg(u_T(A_1)), \quad B_{2,2} = A_2 + Tg(u_T(A_2)),$$

and

$$C_1 = \min\{A_1, B_{1,1}, B_{2,1}\},\ C_2 = \max\{A_2, B_{1,2}, B_{2,2}\}.$$

To compute the solution, we use the finite volume method. We introduce a regular grid,

$$C_1 = x_{1/2} < x_{3/2} < \cdots < x_{M-1/2} < x_{M+1/2} = C_2,$$

and define the cells $K_j = [x_{j-1/2}, x_{j+1/2}]$ with centers at $x_j = \frac{1}{2}(x_{j-1/2} + x_{j+1/2})$, for $1 \leq j \leq M$, and a constant space step $\Delta x = x_{3/2} - x_{1/2}$. We define a time discretization $(t^n)_{n=0}^{N+1}$, where $t^0 = 0$, $t^{N+1} = T$, and $t^n = n\Delta t$ for a constant time-step $\Delta t$.

With this notation in place, we define $u_j^n$ and $v_j^n$ as the approximations of the averages $u(t^n, \cdot)$ and $v(t^n, \cdot)$ on $K_j$. Namely,

$$u_j^n \simeq \frac{1}{\Delta x}\int_{K_j} u(t^n, x)\,\mathrm{d}x, \qquad v_j^n \simeq \frac{1}{\Delta x}\int_{K_j} v(t^n, x)\,\mathrm{d}x$$
$$u_j^{N+1} \simeq \frac{1}{\Delta x}\int_{K_j} u(T, x)\,\mathrm{d}x, \qquad v_j^{N+1} \simeq \frac{1}{\Delta x}\int_{K_j} v(T, x)\,\mathrm{d}x.$$

For the first equation, we consider the numerical backward solver $K_n^-\left(u_T, (x_j)_{j=1}^M\right) = (u_j^{N+1-n})_{j=1}^M$, which gives us an approximation of the isentropic solution of the first equation, such that

$$u_j^n = u_j^{n+1} - \frac{\Delta t}{\Delta x}\left(\mathcal{F}(u_j^{n+1}, u_{j-1}^{n+1}) - \mathcal{F}(u_{j+1}^{n+1}, u_j^{n+1})\right),$$

$\mathcal{F}$ being the *Godunov numerical flux*, see [7, 9]

$$\mathcal{F}(u_l, u_r) = \begin{cases} \min\limits_{u\in[u_l,u_r]} f(u) & \text{if } u_l \leq u_r, \\ \max\limits_{u\in[u_l,u_r]} f(u) & \text{otherwise.} \end{cases}$$

Having obtained $u_j^n$, we then update $v_j^n$ by using an upwind-type scheme, see [8]:

$$\begin{aligned} v_j^n = v_j^{n+1} - \frac{\Delta t}{\Delta x}[\{(-g_{j+1/2}^{n+1})^+ v_j^{n+1} - (-g_{j-1/2}^{n+1})^+ v_{j-1}^{n+1}\} \\ + \{(-g_{j+1/2}^{n+1})^- v_{j+1}^{n+1} - (-g_{j-1/2}^{n+1})^- v_j^{n+1}\}], \end{aligned}$$

with

$$g_{j+1/2}^{n+1} = (1-\theta)g(u_j^{n+1}) + \theta g(u_{j+1}^{n+1}),$$

where $\theta \in (0, 1)$ and $\xi^+ = \max\{0, \xi\}$, and $\xi^- = \min\{0, \xi\}$. For the numerical simulations in Sect. 2.1, we take $\theta = 1/2$.

For the boundary conditions, we set, for $1 \leq n \leq N+1$

$$(u_0^n, v_0^n) = (u_T(A_1), v_T(A_1)), \quad (u_{M+1}^n, v_{M+1}^n) = (u_T(A_2), v_T(A_2)).$$

This procedure yields the isentropic solution $U = (u, v)$ of the system (1) corresponding to the attainable profile $U_T = (u_T, v_T)$. In order to find other initial data leading to $U_T$, we rely of the Theorem 2 and the geometric interpretation Remark 2 to change the solution of the scalar conservation law. Thanks to this procedure, we have a new profile $\tilde{u}$, which we can use to solve the second equation backward and obtain, in conclusion, another set of initial data $\tilde{U}_0 = (\tilde{u}_0, \tilde{v}_0)$ that leads to $U_T$.

## 2.1 Numerical Experiments

We now present a numerical experiment on the approximation of solutions to the backward resolution. We take the following system

$$\begin{cases} \partial_t u + \partial_x \frac{u^2}{2} = 0, \\ \partial_t v + \partial_x \left(\cos(\frac{\pi u}{8}) v\right) = 0. \end{cases}$$

We set it on the domain $[0, 1] \times [-2.2, 1.5]$. We consider as final state for the first equation

$$u_T(x) = \begin{cases} 0 \text{ if } x < -1, \\ 0.5x + 0.5 \text{ if } -1 \leq x < 0, \\ 0.5x - 0.5 \text{ if } 0 \leq x < 1, \\ 0 \text{ if } 1 \leq x, \end{cases}$$

and we choose for the transport equation

$$v_T(x) = \begin{cases} 0.25 \text{ if } x \leq -1, \\ 0.5\cos(\frac{\pi}{4}(x-1)) + 0.25 \text{ if } -1 < x \leq 1, \\ 0.75 \text{ if } 1 < x. \end{cases}$$

First of all, we construct a random path $\gamma \in \Gamma(u_L, u_R, \overline{x}, T)$ as in Fig. 1 with $u_L = 0.5, u_R = -0.5, \overline{x} = 0, T = 1$ and $k = 7$ the number of jumps that we chose (due to the geometric interpretation Remark 2, we can choose the number of jumps that we want in $D_T$).

The iterative construction is initiated with $(X_0, Y_0) = (\overline{x} - Tf'(u_L), 0)$. Then, we construct randomly $(X_i, Y_i)_{i=1,\ldots,k-2}$ such that for all $i$, $X_i < X_{i+1}$ and $X_i \in \mathring{D}_T$. For $(Y_i)_{i=1,\ldots,k-2}$, we pick randomly $Y_{i+1}$ such that for $x \in [X_i, X_{i+1}]$, the line with slope $\frac{Y_{i+1}-Y_i}{X_{i+1}-X_i}$ have to be above the path $\gamma_*$, see Fig. 2. After that, we take $(X_{k-1}, Y_{k-1})$ such that $X_{k-1} \in (X_{k-2}, \overline{x} - Tf'(u_R))$ and add the condition $Y_{k-1}$ have to be above the

**Fig. 1** $\gamma_*$ and a random path $\gamma_1$

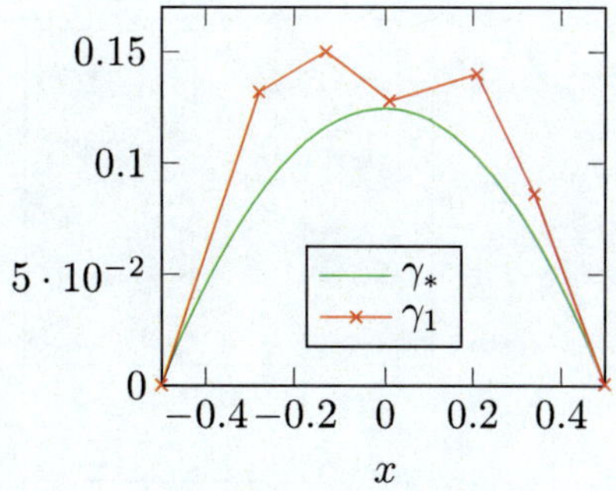


**Fig. 2** Admissible $(X_{i+1}, Y_{i+1})$

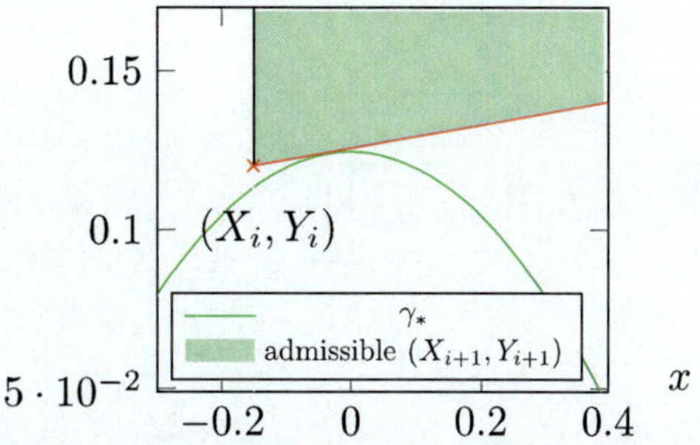


**Fig. 3** Admissible $(X_{k-1}, Y_{k-1})$

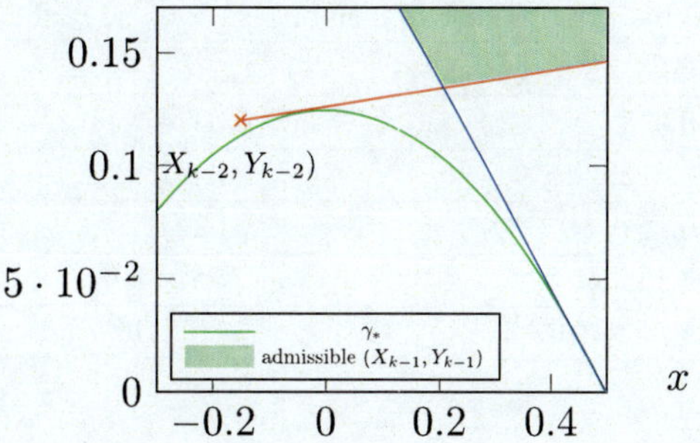


tangent to the path $\gamma_*$ at the point $\overline{x} - Tf'(u_R)$, see Fig. 3. We finished the iterative construction with $(X_k, Y_k) = (\overline{x} - Tf'(u_R), \gamma_*(\overline{x} - Tf'(u_R)))$.

Then, we do the backward algorithm for $u_T$ described in the previous part, which leads us to the following initial state $\tilde{u}_0$.

With this new initial state $\tilde{u}_0$, we can compute a classical forward Godunov scheme to have $\tilde{u}$ in the entire domain, and we do the backward algorithm for the transport equation with $\tilde{u}$ and $v_T$. It goes to the following initial state $\tilde{v}_0$ and we check that the datum $\tilde{u}_0$ and $\tilde{v}_0$ are good approximations.

To do the computations, we choose $\Delta x = 1.56 \times 10^{-3}$ and $\Delta t = \frac{\Delta x}{2}$ with the path in Fig. 1. This path leads us to the situation illustrated in Fig. 4.

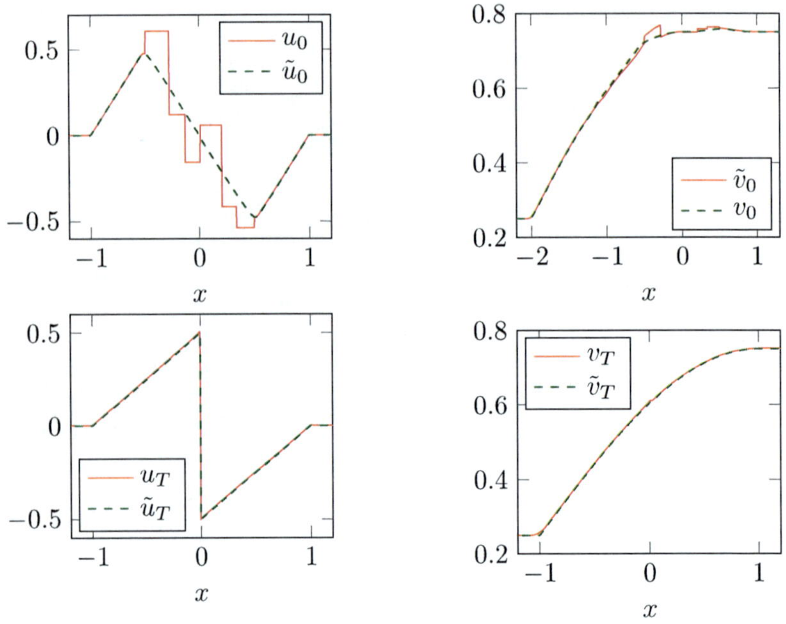


**Fig. 4** Initial states and comparison between target and approximate profiles

**Table 1** Error table of $u$ and $v$

| $\Delta x$ | $e_u$ | $e_v$ |
|---|---|---|
| 0.025 | 0.0374 | $7.58 \times 10^{-3}$ |
| 0.0125 | 0.0352 | $4.02 \times 10^{-3}$ |
| $6.25 \times 10^{-3}$ | 0.0224 | $2.20 \times 10^{-3}$ |
| $3.125 \times 10^{-3}$ | $7.75 \times 10^{-3}$ | $1.22 \times 10^{-3}$ |
| $1.56 \times 10^{-3}$ | $2.85 \times 10^{-3}$ | $6.65 \times 10^{-4}$ |

To investigate the numerical convergence of our scheme, we consider the relative $L^1$-discrete error norms

$$e_u = \frac{\sum_{j=1}^{m} |u_T(x_j) - u_j^N|}{\sum_{j=1}^{m} |u_T(x_j)|}, \qquad e_v = \frac{\sum_{j=1}^{m} |v_T(x_j) - v_j^N|}{\sum_{j=1}^{m} |v_T(x_j)|}.$$

We observe that the error norms $e_u$ and $e_v$ converge to zero when $\Delta x$ goes to zero, as reported in Table 1 and illustrated in Fig. 5.

**Fig. 5** Approximation error of $u$ and $v$

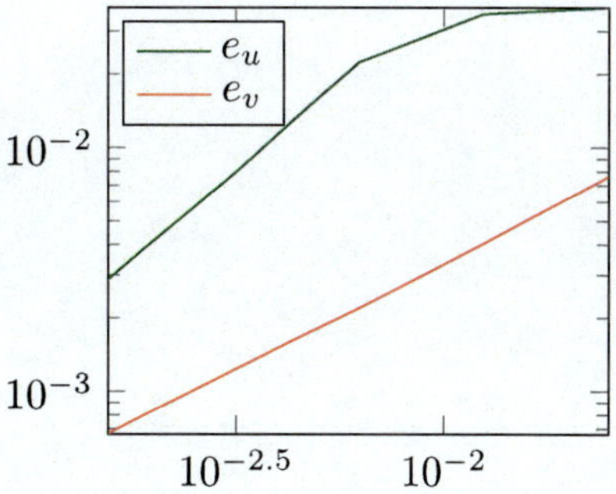


**Acknowledgements** This work originates from the author's master thesis under the direction of Carlotta Donadello and Ulrich Razafison, and is part of an ongoing research project involving Giuseppe M. Coclite, Nicola De Nitti and Carlotta Donadello. The results obtained in collaboration will appear in a forthcoming paper.

## References

1. Adimurthi, S.S.G., Gowda, G.D.V.: Optimal controllability for scalar conservation laws with convex flux. J. Hyperbolic Differ. Equ. **11**(3):477–491 (2014)
2. Ancona, F., Marson, A.: On the attainable set for scalar nonlinear conservation laws with boundary control. SIAM J. Control Optim. **36**(1), 290–312 (1998)
3. Andreianov, B., Donadello, C., Ghoshal, S.S., Razafison, U.: On the attainable set for a class of triangular systems of conservation laws. J. Evol. Equ. **15**(3), 503–532 (2015)
4. Bressan, A.: Hyperbolic systems of conservation laws. The One-dimensional Cauchy Problem, volume 20 of Oxford Lecture Series in Mathematics and its Applications. Oxford University Press, Oxford (2000)
5. Colombo, R.M., Perrollaz, V.: Initial data identification in conservation laws and Hamilton-Jacobi equations. J. Math. Pures Appl. **9**(138), 1–27 (2020)
6. Dafermos, C.M.: Hyperbolic Conservation Laws in Continuum Physics, 3rd ed. Grundlehren der mathematischen Wissenschaften: A Series of Comprehensive Studies in Mathematics. Springer, Dordrecht (2010)
7. Godlewski, E., Raviart, P.-A.: Numerical Approximation of Hyperbolic Systems of Conservation Laws, vol. 118. Springer Science & Business Media (2013)
8. Gosse, L., James, F.: Numerical approximations of one-dimensional linear conservation equations with discontinuous coefficients. Math. Comput. **69**(231), 987–1015 (2000)
9. LeVeque, R.J.: Finite Volume Methods for Hyperbolic Problems. Cambridge Texts in Applied Mathematics. Cambridge University Press (2002)
10. Liard, T., Zuazua, E.: Analysis and numerical solvability of backward-forward conservation laws. HAL-02389808v2 (2020)
11. Liard, T., Zuazua, E.: Initial data identification for the one-dimensional Burgers equation. IEEE Trans. Autom. Control **67**(6), 3098–3104 (2022)

# Two-Layer Exchange Flow with Time-Dependent Barotropic Forcing

S. R. Pudjaprasetya and P. V. Swastika

**Abstract** Exchange flow phenomena were investigated using numerical simulations of the two-layer shallow water equations. Here, we adopt a previously developed momentum-conserving staggered-grid (MCS) scheme, and calculate steady interface in channels with sills, represent maximal and submaximal exchange flow cases. These two results are in good agreement with the analytical solution, indicating the accuracy of our MCS scheme. Next, the effect of time-dependent barotropic flow on the steady interface was investigated; it was evaluated numerically by applying suitable boundary conditions. Here, we successfully model the barotropic effect and obtained the expected results.

**Keywords** Two-layer shallow water equations · Maximal exchange flow · Time-dependent barotropic flow · Momentum conserving staggered-grid scheme

## 1 Introduction

The two-layer exchange flow in a channel with a sill refers to the phenomenon of water flow in a channel where two layers of water with different densities exchange positions due to the presence of a sill barrier. The sill is a horizontal obstruction that can be natural, such as a sandbar or rock outcrop, or artificial, such as a dam or weir. The two-layer exchange flow can have significant environmental implications, as it can affect the mixing and circulation of water, nutrient transport, and the distribution of marine organisms. Understanding the dynamics of this flow is essential to manage estuaries, lagoons, and other coastal environments where it often occurs.

S. R. Pudjaprasetya (✉)
Bandung Institute of Technology, Jalan Ganesha 10, Bandung 40132, Indonesia
e-mail: srpudjap@gmail.com

P. V. Swastika
Faculty of Mathematics and Natural Sciences, Department of Mathematics, Udayana University, Jl. Raya Kampus UNUD, Bukit Jimbaran, Kuta Selatan 803611, Bali, Indonesia

 
E. Franck et al. (eds.), *Finite Volumes for Complex Applications X—Volume 2, Hyperbolic and Related Problems*, Springer Proceedings in Mathematics & Statistics 433,
https://doi.org/10.1007/978-3-031-40860-1_28

One of the key challenges in modeling two-layer exchange flows is to capture the interface between the two layers. This interface can vary in thickness and can be subject to instabilities and mixing. An additional difficulty, the two-layer shallow water equations are a conditionally hyperbolic system. Several numerical schemes have been developed to address these challenges; see, for instance, [1–3, 5].

In this study, we adopt a hydrostatic model for the two-layer fluid flow in a channel with topography, which is solved numerically. In our earlier study [6], we developed a numerical scheme known as the momentum-conserving staggered grid scheme, abbreviated as the MCS scheme; and we successfully used this MCS scheme to simulate several steady interfaces for the two-layer model. This scheme is basically a variant of the finite volume method applied to a staggered grid. Focusing on the exchange flow in a channel with sill, in this contribution, we study the effect of time-dependent barotropic flow on the deformation of steady interface, and examine the flow characteristic at the sill crest.

The presentation of this paper is organized as follows. In the first section, mathematical models are discussed, including steady flow equations. Further, the numerical methods are discussed in Sect. 2, which is followed with implementation for simulating two types of steady solutions, the maximal exchange flow and sub-maximal exchange flow. In Sect. 3, the scheme is modified to include the effect of barotropic force; flow condition under this barotropic force is discussed. Finally, we present our conclusions.

## 2 Governing Equations

Consider a layer of two immiscible fluids, bounded by the free surface $z = \eta_1(x, t)$ and below by a solid bottom $z = d(x)$. Two layers of fluids of constant density $\rho_1$ and $\rho_2$ are separated by the interface $z = \eta_2(x, t)$, as shown in Fig. 1. Under this setting, the thicknesses of the upper and lower layers are $h_1 = \eta_1 - \eta_2$, and $h_2 = \eta_2 - d$, respectively. The velocities of the fluid particles in the upper and lower layers are denoted as $u_1(x, t)$ and $u_2(x, t)$, respectively. Let the cross-sectional area of fluid in each layer$-i$, be denoted by $a_i$; for rectangular channels with constant width $b$, the cross-sectional area is $a_i(x, t) = bh_i(x, t)$, where $i = 1$ and $i = 2$ are for the upper and lower layers, respectively. Furthermore, the mass flow rate for each layer is $q_i(x, t) = a_i(x, t)u_i(x, t)$. The governing equations of the two-layer fluid system written in normalized variables are given by

$$\partial_t a_1 + \partial_x q_1 = 0, \tag{1}$$

$$\partial_t u_1 + u_1 \partial_x u_1 + \frac{1}{\sigma}\partial_x(h_1 + h_2 + d) = 0 \tag{2}$$

$$\partial_t a_2 + \partial_x q_2 = 0, \tag{3}$$

$$\partial_t u_2 + u_2 \partial_x u_2 - \partial_x h_1 + \frac{1}{\sigma}\partial_x(h_1 + h_2 + d) = 0. \tag{4}$$

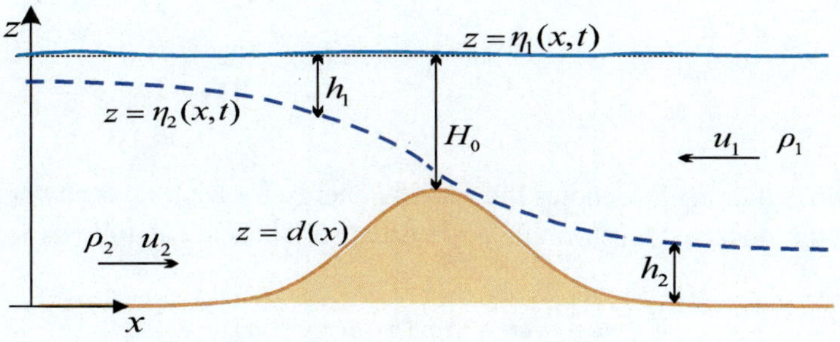


**Fig. 1** Sketch of the problem and notations

The non-dimensional parameter in the above equations is $\sigma = g'/g$, where $g' = g(1-r)$ represents the reduced gravity, with $r = \rho_1/\rho_2$.

The governing equations (1)–(4) are expressed in non-dimensional variables; their relation to physical variables (written with ^) is as follows $d = \frac{\hat{d}}{H_0}$, $h_i = \frac{\hat{h}_i}{H_0}$, $q_i^2 = \frac{\hat{q}_i^2}{g'b^2H_0^3}$, $i = 1, 2$, with $H_0$ the fluid depth reference (see Fig. 1). A similar model like (1)–(4) is used by Helfrich [5] to study two-layer hydrostatic flow in a strait separating two infinite basins; where the weakly hyperbolic condition of this model is also mentioned.

## Steady Equations

When the fluid densities are approximately the same $\rho_2 \approx \rho_1$ (or $r \approx 1$), the composite Froude number for two-layer flows is $G^2 \equiv F_1^2 + F_2^2$, with

$$F_i^2 = \frac{u_i^2}{g'h_i} = \frac{q_i^2}{g'b^2h_i^3}, \quad i = 1, 2, \tag{5}$$

which represents the Froude number for each layer. Two-layer flow is said to be critical at locations for which

$$G^2 = 1, \text{ or } F_1^2 + F_2^2 = 1 \tag{6}$$

The locations with $G^2 < 1$ mean that the flow is subcritical, while if $G^2 > 1$ the flow is supercritical, while if $G^2 = 1$ the flow is called critical and its location is called a control.

As discussed in [4], the steady interface can be determined from two equations; those are the Bernoulli equation and the rigid-lid equation, as follows.

$$\frac{\hat{E}_1-\hat{E}_2}{g'\rho_2 H_0}=\frac{1}{2}\frac{\rho_1}{\rho_2}h_1F_1^2-\frac{1}{2}h_2F_2^2-(h_2+d), \tag{7}$$
$$h_1+h_2+d=1. \tag{8}$$

In the above equation, $\hat{E}_i$ denotes the specific energy for each layer; by neglecting pressure at the surface, their formula in physical variables are as follows

$$\hat{E}_1=\frac{1}{2}\rho_1\hat{u}_1^2+g\rho_1(\hat{h}_1+\hat{h}_2+\hat{d}) \tag{9}$$
$$\hat{E}_2=\frac{1}{2}\rho_2\hat{u}_2^2+g\rho_2(\hat{h}_2+\hat{d})+g\rho_1\hat{h}_1 \tag{10}$$

Furthermore, for $\rho_1\approx\rho_2$, by using (5) (or $h_i=\left(\frac{q_i}{b}\right)^{\frac{2}{3}}F_i^{-\frac{2}{3}}$) and denoting $\Delta E\equiv 1+\frac{\hat{E}_1-\hat{E}_2}{g'\rho_2 H_0}$, we can express (7) and (8) in terms of $F_1^2$ and $F_2^2$, are as follows.

$$F_2^2=q_r\big(2F_1^{-2/3}+F_1^{4/3}-2\Delta E b^{2/3}q_1^{-2/3}\big)^{3/2}, \tag{11}$$
$$q_r^{2/3}F_1^{-2/3}+F_2^{-2/3}=\Big(\frac{q_2}{(1-d)^{3/2}b}\Big)^{-2/3}. \tag{12}$$

Note that both equations (11), (12) are written as non-dimensional variables, with $q_r=\frac{q_1}{q_2}$. For a given bathymetry $d(x)$, we can construct the steady interface of the two fluid layers using these equations, and this will be discussed in Sect. 3.

## 3 Numerical Scheme

To solve the two-layer fluid systems (1)–(4) we implement a numerical method called the momentum-conserving staggered-grid (MCS) scheme. In our earlier article [6], we have successfully developed the MCS scheme and used it for various simulations of the two-layer fluid system, and here we provide a brief review.

Consider a computational domain $\Omega:=[-X,X]$, which is divided into $2N$ number of cells of length $\Delta x/2$ to obtain a staggered grid partition $x_{1/2}=-X,\ldots,x_{j-1/2}$, $x_j, x_{j+1/2},\ldots,x_{N+1/2}=X$, with $N=(2X)/\Delta x$. On this staggered grid, the unknowns are evaluated on different grid points, those such as $u_i$ are evaluated at half grid points $u_i(x_{j+1/2},t^n)\equiv u^n_{i,j+1/2}$, whereas $a_i$ are evaluated on the full grid points $a_i(x_j,t^n)\equiv a^n_{i,j}$. The semi-discrete scheme for a two-layer system (1)–(4) is as follows.

$$\frac{da_{1,j}}{dt} + \frac{q^n_{1,j+1/2} - q^n_{1,j-1/2}}{\Delta x} = 0, \tag{13}$$

$$\frac{du^{n+1}_{1,j+1/2}}{dt} + (u_1\partial_x u_1)^n_{j+1/2} + \frac{1}{\sigma}\frac{\eta^n_{1,j+1} - \eta^n_{1,j}}{\Delta x} = \nu_1 \frac{u^n_{1,j+3/2} - 2u^n_{1,j+1/2} + u^n_{1,j-1/2}}{\Delta x^2}, \tag{14}$$

$$\frac{da_{2,j}}{dt} + \frac{q^n_{2,j+1/2} - q^n_{2,j-1/2}}{\Delta x} = 0, \tag{15}$$

$$\frac{du^{n+1}_{2,j+1/2}}{dt} + (u_2\partial_x u_2)^n_{j+1/2} - \frac{h^n_{1,j+1} - h^n_{1,j}}{\Delta x} + \frac{1}{\sigma}\frac{\eta^n_{1,j+1} - \eta^n_{1,j}}{\Delta x} = \cdots \nu_2 \frac{u^n_{2,j+3/2} - 2u^n_{2,j+1/2} + u^n_{2,j-1/2}}{\Delta x^2} \tag{16}$$

with $\eta^n_{1,j} = h^n_{1,j} + h^n_{2,j} + d_j$, $q^n_{i,j+1/2} = {}^*a^n_{i,j+1/2} u^n_{j+1/2}$, and

$${}^*a^n_{i,j+1/2} = \begin{cases} a^n_{i,j}, & \text{if } u^n_{i,j+1/2} \geq 0, \\ a^n_{i,j+1}, & \text{if } u^n_{i,j+1/2} < 0. \end{cases}$$

In the formula above, the advection terms $u_i \partial_x u_i$, $i = 1, 2$ are calculated using the momentum-conserving approximation, see [6] for details. We resume here, the MCS scheme used to simulate a two-layer shallow flow is (13)–(16). For time integration, we apply the forward time for (13), (15), in combination with the backward time for (14), (16).

According to [1], the two-layer system (1)–(4) is conditionally hyperbolic, that is because the internal eigenvalues may become complex depending on the shear parameter $S \equiv \frac{(u_2-u_1)^2}{h_1+h_2}$, that is when $S > 1$. In the next subsection, the MCS scheme is adopted to simulate steady solutions; while the shear parameter curve $S$ is also plotted. Unlike other numerical schemes that fail when $S > 1$ (see [3]), our MCS scheme continues to work when $S > 1$, that is because our scheme adopts the numerical damping terms (those with coefficients $\nu_1$ and $\nu_2$ in (15) and (16), with typical values $\nu_i = 0.005$ to 0.01 [5]).

## Steady Maximal Exchange Flow

By using the MCS numerical scheme described above, we simulate the formation of a steady interface in a two-layer fluid flow. Consider a channel with spatial domain $[-3, 3]$ where the bottom topography is defined as

$$d(x) = \begin{cases} 0.5 - 0.125\left(\cos\frac{\pi x}{2} + 1\right)^2, & |x| \leq 2, \\ 0.5, & \text{otherwise.} \end{cases} \tag{17}$$

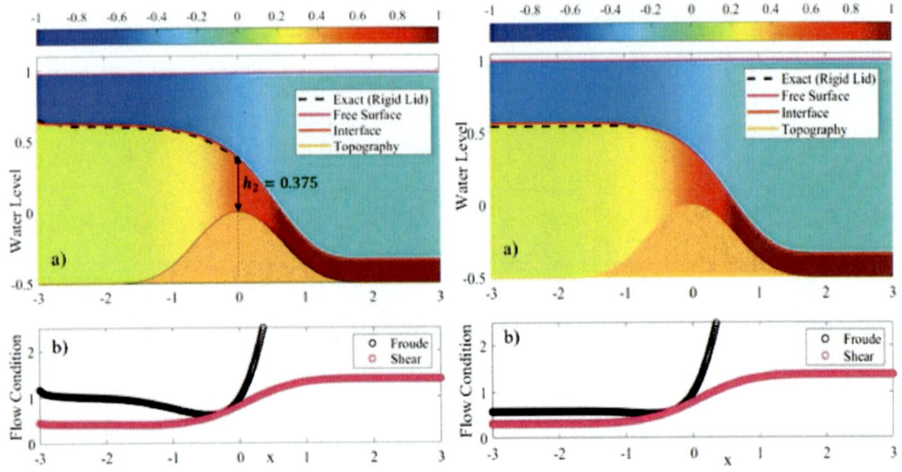


**Fig. 2** (Left) The maximal exchange flow for channel with sill; (right) the sub-maximal exchange flow for channel with sill

The following parameters are used in this simulation; density ratio $r = 0.9894$, the spatial and temporal step size of $\Delta x = 0.01$ and $\Delta t = 0.0005$, respectively. For the initial conditions, we use initial heights $h_1(x, 0) = 1$, $h_2(x, 0) = 0.5$, with zero initial velocities $u_1(x, 0) = 0$, $u_2(x, 0) = 0$.

To simulate maximal exchange flow, here we set the flux ratio $q_r = 1$ and assume that the fluids in the upper and lower layers are moving in opposite directions. As described in [4], the exchange flow will reach a maximum when $q_2 = 0.208$; so, the boundary conditions for the simulations are set as follows.

$$h_1(-3, t) = 0.2370\ ,\ q_1(3, t) = -0.2081 \tag{18}$$

$$q_2(-3, t) = 0.2081\ ,\ h_2(3, t) = 0.2147 \tag{19}$$

Under the influence of the boundary conditions above, the interface starts to deform soon after the simulation begins, and shortly after that the interface reaches a steady state. Figure 2 (left a) shows the interface and the free surface in steady condition, whereas the color scales represent horizontal velocities $u_1$, $u_2$ for each layer. As shown in the figure, the numerical steady interface confirms the exact solution of the rigid-lid model by [4] and therefore indicates the precision of our MCS scheme. Figure 2 (left b) presents the curve of the composite Froude number $G^2$ and the shear parameter $S$ in steady state. From the curve $G^2$ we observe that there were two controls, at the sill crest $x = 0$ and in the upstream part of the channel near $x = -3$. Moreover, the shear parameter curve $S$ illustrates the hyperbolic state of the flow.

The steady solution is called in the maximum-exchange case, where there were two controls that separate the subcritical flow region from the left and right reservoirs. According to the theory of maximum exchange flow by [4], the height of the interface in the sill is $h_2 = 0.375$, and this value can be obtained from the Bernoulli equation

(11) and the criticality condition (6). Our simulation results in Fig. 2 (Left a) show that the interface height on the sill is $h_2 = 0.376$, which shows a good agreement (RMSE = 0.050122), also suggesting the accuracy of our numerical results.

### *Steady Sub Maximal Exchange Flow*

In this subsection, we simulate the sub-maximal exchange flow. For that purpose, we use the same setting like the previous simulation, but the density ratio is now adjusted to $r = 0.9804$. Moreover, we adopt the boundary conditions.

$$h_1(-3, t) = 0.4311 \ , \ q_1(3, t) = -0.2076 \tag{20}$$

$$q_2(-3, t) = 0.2076 \ , \ h_2(3, t) = 0.1617 \tag{21}$$

The simulation result is shown in Fig. 2 (right). In this submaximal exchange flow, it appears that the numerical interface height exhibits good agreement (RMSE= 0.013486) with that found from the exact solution of the rigid lid model by [4]. This is indeed the sub-maximal case, which is characterized by having exactly one control achieved at the sill crest, see Fig. 2 (right).

## 4 Barotropic Forcing

In oceanic situations, two-layer hydraulic flows through straits are affected by tidal flows. Here, we analyze how the maximal exchange flow evolves under the influence of the external barotropic (tidal) flow. Following the approach used by [5], the barotropic flow is considered to be periodically time-dependent in the form of the following harmonic function.

$$q_b(t) = q_{b0} \sin\left(\frac{2\pi t}{\gamma}.\right). \tag{22}$$

In the above formula $q_{b0} = u_{b0}$, $u_{b0}$ denotes the normalized barotropic velocity at the sill crest where the cross-sectional area is one. Here, we study the effect of sinusoidal barotropic effect that depends on two parameters, amplitude $q_{b0}$ and period $\gamma$, on the two-layer exchange flow in a channel with sill.

For the simulation, the maximal exchange flow from the previous calculation is used for the initial condition. On the computational domain $-3 \le x \le 3$, the barotropic forcing (22) is applied to the left and right boundaries. In the simulation, parameters $q_{b0} = 0.5$ and $\gamma = 8$ are taken. Under the influence of barotropic force and after some adjustment period, the interface starts to develop a regular periodic motion. Snapshots of the interface and flow conditions at four points in the barotropic cycle

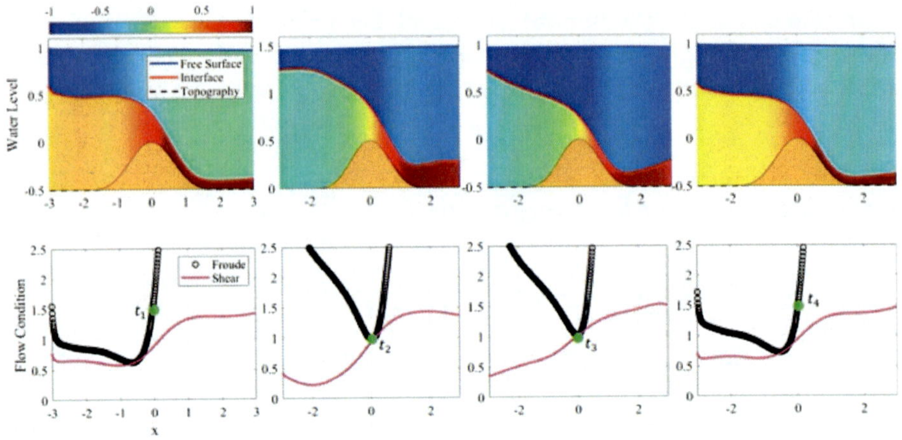


**Fig. 3** Snapshots of interface and flow conditions at four moments of time $t_1, \ldots, t_4$ (green dots indicate the position of controls); a simulation using barotropic forcing (22) with $q_{b0} = 0.5$, $\gamma = 8$

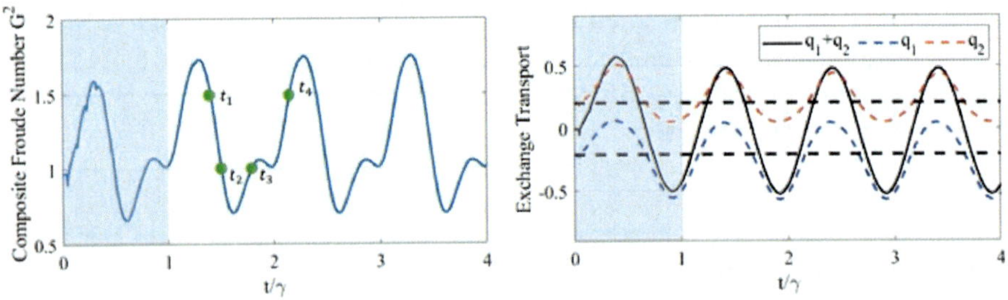


**Fig. 4** Flow condition at the sill crest $x = 0$; (left) the composite Froude Number $G^2$, with green dots showing the position of control at $t_1, \ldots, t_4$; (right) dynamics of momentum $q_1$, $q_2$, $q_1 + q_2$ at sill $x = 0$. Shaded area indicates adjustment period

$0 < t < \gamma$ are plotted in Fig. 3. These results indicate that the external barotropic (tidal) forcing was successfully applied, and the effect was successfully propagated throughout the channel. Under the influence of barotropic forcing (22), Fig. 3 (below) illustrates how the control position swings to the left and right of the sill. The more precise graph can be seen in Fig. 4 (left), in which the Froude number $G^2$ in the sill is plotted as a function of normalized time $t/\gamma$. We make further observations by recording $q_1$ and $q_2$, as well as their sum $(q_1 + q_2)$ on the crest of the sill $x = 0$, and the results are plotted in Fig. 4 (right). It appears that after one period, the barotropic force exerted on the boundary has entered the domain; no damping is observed and the amplitude remains $q_{b0} = 0.5$. The recorded results of $q_1(0, t)$ and $q_2(0, t)$ at the sill crest are also included, and as shown in Fig. 4 (right) the two curves move in the same phase as $(q_1 + q_2)(0, t)$. Moreover, our results show good agreement with the numerical results from [5].

# 5 Conclusions

The momentum conserving staggered-grid (MCS) scheme has been successfully applied to solve the two-layer shallow-water equations. Focusing on a channel with sill, the steady interface has been simulated for the case of sub-maximal as well as the maximal exchange flow. In these two cases, the control locations of the two steady solutions agree with analytical theory and steady interface profiles. The effect of time-dependent barotropic force on the steady interface was simulated by the use of suitable boundary conditions. The simulation results show good results.

**Acknowledgements** The financial support of the Institut Teknologi Bandung and Program Unggulan Udayana (PUU) 2023 [B/1.612/UN14.4.A/PT.01.03/2023]

# References

1. Abgrall, R., Karni, S.: Two-layer shallow water system: a relaxation approach. SIAM J. Sci. Comput. **31**(3), 1603–27 (2009). https://doi.org/10.1137/06067167X
2. Brandt, P., Alpers, W., Backhaus, J.O.: Study of the generation and propagation of internal waves in the Strait of Gibraltar using a numerical model and synthetic aperture radar images of the European ERS 1 satellite. JGR Oceans **101**(C6), 14237–14252 (1996). https://doi.org/10.1029/96JC00540
3. Castro, M.J., Garcıa-Rodrıguez, J.A., González-Vida, J.M., Macıas, J., Parés, C., Vázquez-Cendón, M.E.: Numerical simulation of two-layer shallow water flows through channels with irregular geometry. J. Comput. Phys. **195**(1), 202–235 (2004). https://doi.org/10.1016/j.jcp.2003.08.035
4. Farmer, D.M., Armi, L.: Maximal two-layer exchange over a sill and through the combination of a sill and contraction with barotropic flow. J. Fluid Mech. **164**, 53–76 (1986). https://doi.org/10.1017/S002211208600246X
5. Helfrich, Karl R.: Time-dependent two-layer hydraulic exchange flows. J. Phys. Oceanogr. **25**(3), 359–373 (1995). https://doi.org/10.1175/1520-0485(1995)025<0359:TDTLHE>2.0.CO;2
6. Swastika, P.V., Pudjaprasetya, S.R.: The momentum conserving scheme for two-layer shallow flows. Fluids **6**, 346 (2021). https://doi.org/10.3390/fluids6100346

# Split Form Discontinuous Galerkin Methods for Conservation Laws

**Gero Schnücke**

**Abstract** The goal of this paper is to give a review of the split form nodal discontinuous Galerkin spectral element methods (DGSEM). The split form discontinuous Galerkin (DG) framework provides a tool to construct the numerical approximation in a way that aliasing issues caused by the discretization of DG volume integrals are avoided. In particular the split form DG framework can be used to construct entropy conservative (EC) or kinetic energy preserving (KEP) DGSEM to solve the Euler equations. Various split form DGSEM to solve the three dimensional compressible Euler equations are investigated numerically to show the capabilities of these DG methods.

**Keywords** Split form discontinuous galerkin methods · De-aliasing · Entropy conservative methods · Kinetic energy preserving methods

## 1 Introduction

The construction of high order numerical methods to solve conservation laws and related equations includes the approximation of non-linear flux terms in the volume integrals. These terms can lead to aliasing and stability issues in the DG approximation. Aliasing error arises when the flux terms are composed of products of polynomials, and are then interpolated by a polynomial basis of a lower order than the product of the polynomials. This interpolation essentially transforms higher order modes into lower order modes in the interpolated flux. These errors are generated in the volume integrals and cannot be sufficiently controlled with the dissipation of the numerical surface fluxes alone.

Standard de-aliasing techniques in the DG framework are based on either projection of the non-linear flux functions, e.g. [2, 20], or exact evaluations of the integrals

G. Schnücke (✉)
Faculty of Mathematics and Computer Science, Friedrich Schiller University Jena, Ernst-Abbe-Platz 2, 07743 Jena, Germany
e-mail: gero.schnuecke@uni-jena.de



E. Franck et al. (eds.), *Finite Volumes for Complex Applications X—Volume 2, Hyperbolic and Related Problems*, Springer Proceedings in Mathematics & Statistics 433,
https://doi.org/10.1007/978-3-031-40860-1_29

in the variational formulation, sometimes termed "overintegration" or "consistent integration", e.g., [8, 10, 15]. Another approach is based on the construction of kinetic energy preserving (KEP), kinetic energy dissipating (KED), entropy conserving (EC) or kinetic energy dissipating (KED) DG methods, e.g. [5, 6, 12, 18].

A framework to construct high order entropy conserving (EC) schemes in periodic domains has been given by LeFloch et al. [14]. Fisher and Carpenter [3] combined this approach with SBP (summation-by-parts) operators and proved that two-point EC fluxes can be used to construct high order schemes when the derivative approximations in space are SBP operators. Gassner et al. [5, 6, 12, 18] showed that split form DGSEM for the Euler equations can be discretely recovered when specific numerical volume fluxes in the flux form volume integral of Fisher and Carpenter are chosen. In particular, the work from Gassner et al. provides a framework to construct general split form and kinetic energy preserving (KEP) DGSEM for the Euler equations. Moreover, it is formally possible to construct ES and kinetic energy dissipating (KED) DGSEM, when a two-point ES or KED flux function is used as surface flux in the DG method, the derivative matrix satisfies the SBP property and a two-point EC flux or KEP flux function is used as volume flux in the flux form volume integral of Fisher and Carpenter. Thus, the split framework provides a powerful tool to construct the numerical approximation in a way that aliasing issues are avoided.

In this work, the construction of split form DGSEM are briefly reviewed on the example of one dimensional scalar conservation laws. Afterwards the capability of various split form DGSEM to solve the three dimensional compressible Euler equations is shown.

## 2 The Split Form DGSEM

In this section, the split form DGSEM discretization of the initial value problem

$$u_t + (f(u))_x = 0, \qquad u(x,0) = u_0(x), \tag{1}$$

with suitable boundary conditions on a bounded domain $\Omega \subset \mathbb{R}$ is presented. The domain $\Omega$ is divided in $K$ non overlapping elements $I_j = \left]x_{j-1}, x_j\right[$, with length $\Delta x_j = x_{j+1} - x_j$, such that the closure of the $\bigcup_{j=1}^{K} I_j$ becomes the whole domain $\Omega$. Then for each element $I_j$ the initial value problem (1) can be mapped on the reference element $[-1, 1]$. On the reference element (1) becomes

$$u_t + \frac{2}{\Delta x_j}(f(u))_\xi = 0, \qquad u(\xi,0) = u_0(\xi), \qquad \xi \in [-1,1]. \tag{2}$$

## 2.1 Building Blocks

The discretization is based on a nodal approach with Lagrange basis functions $\left\{\ell_j\left(\cdot\right)\right\}_{j=0}^{N}$ constructed from Legendre-Gauss-Lobatto (LGL) points $\{\xi_i\}_{i=0}^{N}$. The interpolation operator for a function $g:[-1,1]\to\mathbb{R}$ is given by

$$\mathbb{I}_N\left(g\right)\left(\xi\right)=\sum_{j=0}^{N}g_j\ell_j\left(\xi\right)\qquad\text{with}\qquad g_j=g\left(\xi_j\right). \tag{3}$$

Integrals are approximated by a $2N-1$ accurate LGL quadrature formula. Hence, interpolation and quadrature nodes are collocated. For two functions $g,h:[-1,1]\to\mathbb{R}$ the following bracket notation

$$\langle f,h\rangle_N=\sum_{j=0}^{N}\omega_j g\left(\xi_j\right)h\left(\xi_j\right) \tag{4}$$

is used, where $\omega_i$, $i=0,\ldots,N$, are the LGL quadrature weights and $\xi_i$, $i=0,\ldots,N$, are LGL points. The following matrix elements are defined

$$\mathbf{M}_{ij}=\omega_i\delta_{ij},\qquad \mathbf{D}_{ij}=\ell'_j\left(\xi_i\right),\qquad i,j=0,\ldots,N. \tag{5}$$

Gassner [4] showed that a matrix $\mathbf{M}$ and the derivative matrix $\mathbf{D}$ given by (5) satisfy the SBP property

$$\mathbf{MD}+\left(\mathbf{MD}\right)^T=\mathbf{B},\quad \mathbf{B}=\operatorname{diag}\left(-1,0,\ldots,0,1\right). \tag{6}$$

## 2.2 Standard DGSEM

On the reference element $[-1,1]$ the solution and fluxes of the initial value problem (2) are approximated by polynomials of degree $N$, e.g.,

$$u\left(\xi,t\right)\approx U\left(\xi,t\right)=\sum_{j=0}^{N}U_j\left(t\right)\ell_j\left(\xi\right), \tag{7}$$

$$f\left(u\left(\xi,t\right)\right)\approx F\left(\xi,t\right)=\sum_{j=0}^{N}f\left(U_j\left(t\right)\right)\ell_j\left(\xi\right). \tag{8}$$

Derivatives are approximated by exact differentiation of the polynomial interpolants, e.g.

$$f\left(u\left(\xi,t\right)\right)_{\xi}\approx\frac{\partial}{\partial\xi}\mathbb{I}_{N}\left(F\left(U\left(\xi,t\right)\right)\right)=\sum_{j=0}^{N}f\left(U_{j}\left(t\right)\right)\ell_{j}'\left(\xi\right). \tag{9}$$

Then the Eq. (2) is multiplied with test functions $\varphi\in\mathbb{P}^{N}\left([-1,1]\right)$. The resulting equations are integrated and integration-by-parts is used. The integrals in the variational form are approximated with the LGL quadrature. At the element interfaces, a two point numerical surface flux $F^{*}$ is inserted. This flux is depending on solution states from the neighboring element, too. Next, the SBP property (6) is used. Finally, the standard DGSEM strong form is given by: For given initial data $U_0=\mathbb{I}_{N}\left(u_0\right)$ and all $\varphi\in\mathbb{P}^{N}\left([-1,1]\right)$ apply

$$\langle U_t,\varphi\rangle_N+\left\langle\frac{\partial}{\partial\xi}\mathbb{I}_{N}\left(F\left(U\right)\right),\varphi\right\rangle_N+\left(F^{*}-F\left(U\right)\right)\varphi\big|_{-1}^{1}=0. \tag{10}$$

### 2.3 Split Form DGSEM

Differentiation and interpolation only commute if there are no interpolation errors [9]. Thus, the term $\frac{\partial}{\partial\xi}\mathbb{I}_{N}\left(F\left(U\right)\right)$ causes aliasing errors in the standard strong form (10). In [5, 6, 12, 18] the split form DGSEM approach was introduced. In this approach a special derivative operator is introduced to replace the terms that cause aliasing errors. The derivative operator is given by

$$\mathbb{D}_{N}\left(F\left(U\right)\right)_i=\sum_{m=0}^{N}2\,\mathbf{D}_{im}F^{\#}\left(U_i,U_m\right),\qquad i=0,\ldots,N. \tag{11}$$

The coefficients $\mathbf{D}_{im}$ are given by (5) and the function $F^{\#}$ is a two point numerical volume flux. It is important that the volume flux is consistent and symmetric such that for all $i,m=0,\ldots,N$ holds

$$F^{\#}\left(U_i,U_i\right)=f\left(U_i\right),\qquad F^{\#}\left(U_i,U_m\right)=F^{\#}\left(U_m,U_i\right). \tag{12}$$

Then, for each element the semi-discrete split form DGSEM i.e. given by: For given initial data $U_0=\mathbb{I}_{N}\left(u_0\right)$ and all $\varphi\in\mathbb{P}^{N}\left([-1,1]\right)$ apply

$$\langle U_t,\varphi\rangle_N+\langle\mathbb{D}_{N}\left(F\left(U\right)\right),\varphi\rangle_N+\left(F^{*}-F\left(U\right)\right)\varphi\big|_{-1}^{1}=0. \tag{13}$$

## 3 Numerical Results

In this section the capability of the split form DGSEM discretization for the three dimensional compressible Euler equations is presented. The Euler equations are given by

$$\rho_t \quad +\nabla \cdot \rho\vec{u} = 0, \tag{14}$$

$$(\rho\vec{u})_t +\nabla \cdot \left(\rho\vec{u}\vec{u}^T + pI\right) = 0, \tag{15}$$

$$E_t \quad +\nabla \cdot (E + p)\, u = 0, \tag{16}$$

where $\rho$ is the mass density, $\vec{u} = [u_1, u_2, u_3]^T$ the velocity, $p$ is the pressure and the total energy is $E$. Pressure and total energy are connected by the following equation of state

$$E = \frac{1}{2}\rho\,|\vec{u}|^2 + \frac{p}{\gamma - 1}, \tag{17}$$

with gas constant $\gamma > 1$.

In [5, 6, 12, 18] it has been proven that the split form DGSEM discretization with curved moving elements to solve the compressible Euler and Navier-Stokes equations can be a high order KEP or entropy conserving EC schemes by a reasonable choice of the volume and surface flux function in the DGSEM discretization. In the following, the KEP and EC property of the different split forms (choice of the two point volume flux) are investigated numerically. The following two point flux functions are considered: Chandrashekar (CH) [1], Kennedy and Gruber (KG) [7], Kuya, Totani and Kawai (KTK), Pirozzoli (PI) [16], Ranocha (RA) [17]. The implementation of the split form DGSEM used for the upcoming numerical experiments has be done in the open source code FLEXI [11].

### 3.1 Experimental Convergence Rates

In this section, the convergence behavior of the split form DGSEM is assessed for different flux functions using the method of manufactured solutions. The results are used to verify that the proposed methodology retains its high order accuracy. For the following simulations, we assume a solution of the form

$$\rho\,(\vec{x}, t) = \; 2 + 0.1 \sin\left(\pi(x_1 + x_2 + x_3 - 0.6t)\right), \tag{18}$$

$$\rho u_\iota\,(\vec{x}, t) = \; 2 + 0.1 \sin\left(\pi(x_1 + x_2 + x_3 - 0.6t)\right), \quad \iota = 1, 2, 3, \tag{19}$$

$$E\,(\vec{x}, t) = \left[2 + 0.1 \sin\left(\pi(x_1 + x_2 + x_3 - 0.6t)\right)\right]^2, \tag{20}$$

with total energy $E = e + \frac{1}{2}\rho\,|\vec{u}|^2$ on the domain $[-1, 1]^3$ and compute the residual when above states are inserted into the Euler equations. The resulting terms are then used as sources for the simulations, and are discretized as a solution independent part of the computation. The simulations are performed on initially Cartesian grids with an increasing number of elements

$$K = 2^3, 4^3, 8^3, 16^3, 32^3. \tag{21}$$

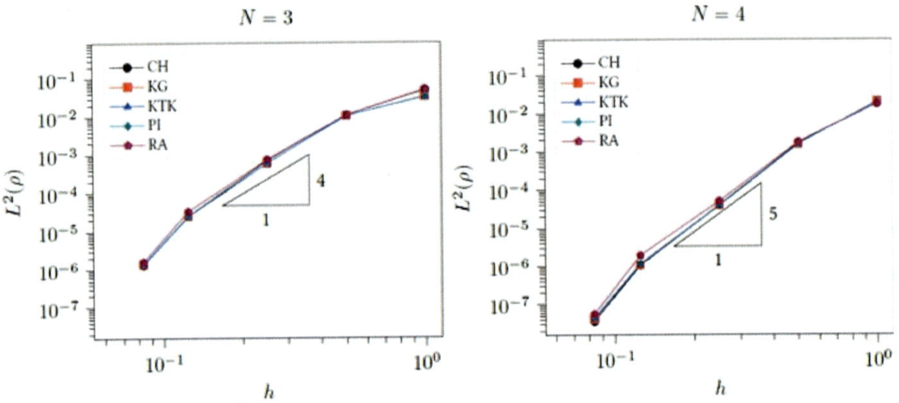


**Fig. 1** Experimental convergence rates for different flux functions. Manufactured solution test case with standing wave mesh movement, for both polynomial degree $N = 3$ and $N = 4$. Shown are the $L^2$ error norms of the density $\rho$ over the mesh size $h$

The simulation is advanced until final time $T = 5$. The surface fluxes are used with Roe-type dissipation terms, which can be found in [18, Appendix C.3]. Figure 1 shows the results obtained for the manufactured solution test cases. The quantity used for comparison is the $L^2$ norm of the error in the density $\rho$.

### 3.2 Numerical Validation of the Entropy and Kinetic Energy Analysis

The Taylor–Green–Vortex (TGV) problem [19] is considered to investigate the KEP and EC property of the different split forms. This is an important problem for laminar-turbulent transition and turbulent flows. The flow is defined on the domain $\Omega = [0, 2\pi]^3$ with periodic boundary conditions on all sides, and its initial conditions are given by

$$\rho(\vec{x}, 0) = 1, \tag{22}$$

$$\vec{u}(\vec{x}, 0) = [\sin(x_1)\cos(x_2)\cos(x_3), -\cos(x_1)\sin(x_2)\cos(x_3), 0]^T, \tag{23}$$

$$p(\vec{x}, 0) = p_0 + \frac{1}{16}(\cos(2x_1) + \cos(2x_2))(\cos(2x_3) + 2), \tag{24}$$

where for all following simulations, $p_0$ is chosen such that the Mach number becomes $\mathrm{Ma} = \frac{1}{\sqrt{\gamma p_0}} = 0.1$, and the influence of compressibility is relatively small. The flow is evolved until final time $T = 13$, which is past the critical point for stability reached at $t \approx 9$, where the maximum of turbulent dissipation occurs. The TGV test problem is chosen, since (i) it has simple periodic boundary conditions, removing the influence of e.g. walls, (ii) is challenging for the stability of the scheme, especially if the

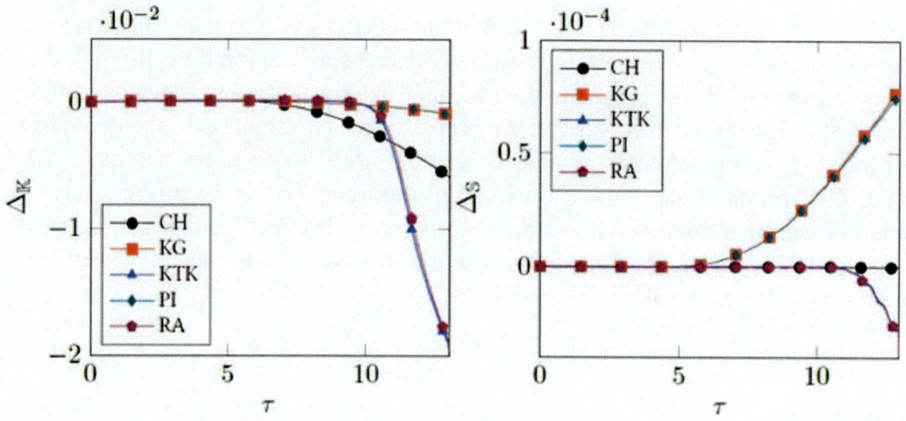


**Fig. 2** The temporal evolution of the error in kinetic energy $\Delta_{\mathbb{K}}$ and entropy $\Delta_{\mathbb{S}}$ during the TGV test case for various split form configurations with polynomial degree $N = 3$

inviscid case is investigated, and (iii) it is analytically isentropic if viscous effects are neglected, allowing to confirm the EC or KEP property of the methods. The mesh is initially Cartesian with $K = 16^3$ elements. The behavior of the discrete error for the temporal evolution of the global entropy $\Delta\mathbb{S}$ and global kinetic energy $\Delta\mathbb{K}$ are investigated. It is important to note that no additional dissipation terms are added to the surface fluxes, all dissipative effects must be attributed to the numerical scheme. This allows to judge the conservation and preservation for all considered flux formulations. We note that the kinetic energy is not a conserved quantity for compressible flows, but as the Mach number is very small, the pressure contribution in the momentum equation decreases and the kinetic energy is almost conserved. The numerical results are plotted in Fig. 2.

## 4 Conclusion

The construction of split form DGSEM has been briefly reviewed on the example of one dimensional scalar conservation laws. Then various split form DGSEM for the Euler equations were analyzed by the TGV problem. Here the key element in the approximation is a two-point EC or KEP two point numerical flux in the derivative operator (11). Likewise a two-point EC or KEP two point numerical flux is used in the discrete surface integrals. The results of the TGV problem (see Fig. 2) show that the split form DGSEM constructed by the two point flux function from Chandrashekar (CH) [1] is an EC method, but the discrete kinetic energy $\Delta_{\mathbb{K}}$ is decreasing around a certain time point. The DGSEM constructed by the two point flux functions from Kennedy and Gruber (KG) as well as Pirozzoli (PI) [16] are KEP methods, but the discrete entropy $\Delta_{\mathbb{S}}$ is increasing for both methods. The methods constructed by the two point flux functions from Kuya, Totani and Kawai (KTK) as well as Ranocha

(RA) [17] show a similar behavior. The methods preserve almost the discrete kinetic energy $\Delta_{\mathbb{K}}$ and the discrete entropy is slowly decreasing around time $T = 10$. It is worth to mention that the two point fluxes from Chandrashekar and Ranocha include terms with logarithmic averages. These terms make the evaluation of the derivative operator (11) computationally expensive. Furthermore, it should be mentioned that in the TGV problem the numerical two point flux functions in the discrete surface integrals were investigated without dissipation terms. If Roe-type dissipation terms are added to these methods, the split form DGSEM become kinetic energy dissipative or entropy stable methods [5, 6, 12, 18].

## References


1. Chandrashekar, P.: Kinetic energy preserving and entropy stable finite volume schemes for compressible Euler and Navier–Stokes equations. Commun. Comput. Phys. **14** (2013)
2. Flad, D., Gassner, G.J.: On the use of kinetic energy preserving DG-schemes for large eddy simulation. J. Comput. Phys. **350** (2017)
3. Fisher, T.C., Carpenter, M.H.: High-order entropy stable finite difference schemes for nonlinear conservation laws: finite domains. J. Comput. Phys. **252** (2013)
4. Gassner, G.J.: A skew-symmetric discontinuous Galerkin spectral element discretization and its relation to SBP-SAT finite difference methods. SIAM J. Comput. **35** (2013)
5. Gassner, G.J., Winters, A.R., Hindenlang, F.J., Kopriva, D.A.: The BR1 scheme is stable for the compressible Navier–Stokes equations. J. Sci. Comput. **77** (2018)
6. Gassner, G.J., Winters, A.R., Kopriva, D.A.: Split form nodal discontinuous Galerkin schemes with summation-by-parts property for the compressible Euler equations. J. Comput. Phys. **327** (2016)
7. Kennedy, C.A., Gruber, A.: Reduced aliasing formulations of the convective terms within the Navier–Stokes equations for a compressible fluid. J. Comput. Phys. **227** (2008)
8. Kirby, R.M., Karniadakis, G.E.: De-aliasing on non-uniform grids: algorithms and applications. J. Comput. Phys. **191** (2003)
9. Kopriva, D.A.: Implementing Spectral Methods for Partial Differential Equations: Algorithms for Scientists and Engineers. Springer Science & Business Media (2009)
10. Kopriva, D.A.: Stability of overintegration methods for nodal discontinuous Galerkin spectral element methods. J. Sci. Comput. **76** (2018)
11. Krais, N., Beck, A., Bolemann, T., Frank, H., Flad, D., Gassner, G., Hindenlang, F., Hoffmann, M., Kuhn, T., Sonntag, M., Munz, C.-D.: FLEXI: A high order discontinuous Galerkin framework for hyperbolic-parabolic conservation laws. Comput. Math. Appl. **81** (2021)
12. Krais, N., Schnücke. G., Bolemann, T., Gassner, G. J.: Split form ALE discontinuous Galerkin methods with applications to under-resolved turbulent low-Mach number flows. Physics **421** (2020)
13. Kuya, Y., Totani, K., Kawai, S.: Kinetic energy and entropy preserving schemes for compressible flows by split convective forms. J. Comput. Phys. **375** (2018)
14. Lefloch, P.G., Mercier, J.-M., Rohde, C.: Fully discrete, entropy conservative schemes of arbitrary order. SIAM J. Numer. Anal. **40** (2002)
15. Mengaldo, G., De Grazia, D., Moxey, D., Vincent, P.E., Sherwin, S.J.: Dealiasing techniques for high-order spectral element methods on regular and irregular grids. J. Comput. Phys. **375** (2018). https://doi.org/10.1016/j.jcp.2015.06.032
16. Pirozzoli, S.: Generalized conservative approximations of split convective derivative operators. J. Comput. Phys. **229** (2010)
17. Ranocha, H.: Generalised summation-by-parts operators and entropy stability of numerical methods for hyperbolic balance laws. Ph.D. thesis. Cuvillier Verlag (2018)

18. Schnücke. G., Krais, N., Bolemann, T., Gassner, G.J.: Entropy stable discontinuous galerkin schemes on moving meshes for hyperbolic conservation laws. J. Sci. Comput. **82** (2020)
19. Shu, C.-W., Don, W.-S., Gottlieb, D., Schilling, O., Jameson, L.: Numerical convergence study of nearly incompressible, inviscid Taylor–Green vortex flow. J. Sci. Comput. **24** (2005)
20. Winters, A.R., Moura, R.C., Mengaldo, G., Gassner, G.J., Walch, S., Peiro, J., Sherwin, S.J.: A comparative study on polynomial dealiasing and split form discontinuous Galerkin schemes for under-resolved turbulence computations. J. Comput. Phys. **372** (2018)

# An Optimally Stable Approximation of Reactive Transport Using Discrete Test and Infinite Trial Spaces

**Lukas Renelt, Christian Engwer, and Mario Ohlberger**

**Abstract** In this contribution we propose an optimally stable ultraweak Petrov-Galerkin variational formulation and subsequent discretization for stationary reactive transport problems. The discretization is exclusively based on the choice of discrete approximate test spaces, while the trial space is a priori infinite dimensional. The solution in the trial space or even only functional evaluations of the solution are obtained in a post-processing step. We detail the theoretical framework and demonstrate its usage in a numerical experiment that is motivated from modeling of catalytic filters.

**Keywords** Optimal stability · Reactive transport · Ultraweak formulation

## 1 Introduction

In this contribution we are considered with stationary reactive transport equations of the form

$$\begin{aligned} A_\circ u := \nabla \cdot (\vec{b}u) + cu &= f_\circ \quad \text{in } \Omega, \\ u &= g_D \quad \text{on } \Gamma_- . \end{aligned} \tag{1}$$

---

L. Renelt (✉) · C. Engwer · M. Ohlberger
Institute for Analysis and Numerics, University of Münster, Einsteinstr. 62, 48149 Münster, Germany
e-mail: lukas.renelt@uni-muenster.de
URL: https://www.uni-muenster.de/AMM

C. Engwer
e-mail: christian.engwer@uni-muenster.de

M. Ohlberger
e-mail: mario.ohlberger@uni-muenster.de



E. Franck et al. (eds.), *Finite Volumes for Complex Applications X—Volume 2, Hyperbolic and Related Problems*, Springer Proceedings in Mathematics & Statistics 433,
https://doi.org/10.1007/978-3-031-40860-1_30

Here, $\Omega \subset \mathbb{R}^n$ denotes an open, bounded polyhedral domain with Lipschitz-boundary $\Gamma := \partial\Omega$, $\vec{b} \in H^1(div, \Omega)$ a given divergence-free[1] transport field and $c \in L^\infty(\Omega)$ a reaction coefficient. Moreover, $f_\circ \in L^2(\Omega)$ denotes the source term and $g_D \in L^2(\Gamma_-)$ the boundary values at the inflow boundary, where the in- and outflow boundary parts are defined as $\Gamma_\pm := \{z \in \Gamma \mid \vec{b}(z) \cdot \nu(z) \gtrless 0\}$ and $\nu(\cdot)$ denotes the outer unit normal. In a weak sense, we interpret $A_\circ : U \to V'$ as an operator that maps from a trial space $U$ to the dual of a test space $V$, where $U, V$ need to be chosen appropriately.

A known challenge in the numerical treatment of transport problems is the stability of the chosen variational formulation and discretization. A common approach for stabilization is the enrichment of the test space by supremizers leading to a Petrov-Galerkin scheme. Depending on the norms on trial and test space, determining the supremizer for a given trial function often requires solving an additional equation with complexity of the full problem. In this contribution, we use an approach using 'optimal' test spaces, i.e. spaces that a-priori include all supremizers and thus are unconditionally stable (see e.g. [4, 5]). However, unlike the discontinuous Petrov-Galerkin method (DPG) and the method defined in [4] we discretize by first choosing the discrete test space and subsequently determine a corresponding trial space as it has been used in [2, 6]. The resulting approach is computationally more efficient and leads to an optimally stable formulation and discretization. We show that it is actually not needed to construct a discrete trial space at all since all computations can be performed using a related normal equation that is defined solely on the test space. Interpreting the solution to the normal equation as a dual variable, this is similar to a (FOS)LL* approach (see [3] and references therein). We provide the theoretical framework in Sect. 2 and the resulting approximation scheme in Sect. 3. Numerical experiments for reactive transport in catalytic filters are given in Sect. 4. The experiments show advantages and challenges of the proposed approach. Resulting perspectives for model order reduction based on the proposed framework are discussed in the conclusion.

## 2 Ultraweak Petrov-Galerkin Variational Formulation and Related Normal Equation

Following the ideas developed in [2], we derive an ultraweak variational Petrov-Galerkin formulation for the reactive transport problem (1) as well as a corresponding normal equation that only depends on the test space. We start with the definition of the formal adjoint operator $A_\circ^* : V \to U'$ given through

[1] We restrict to divergence free velocities. However, non-divergence free fields may be considered as well.

$$v \mapsto A_\circ^*[v], \qquad (A_\circ^*[v])(u) := (u, -b\nabla v + cv)_{L^2(\Omega)} + (u|_{\Gamma_+}, v|_{\Gamma_+})_{L^2(\Gamma_+, |\vec{b}\vec{\nu}|)} \tag{2}$$

using the weighted inner product $(u, v)_{L^2(\Gamma_\pm, |\vec{b}\vec{\nu}|)} := \int_{\Gamma_\pm} u\, v\, |\vec{b}\vec{\nu}|\, ds$.

## 2.1 Choice of Appropriate Function Spaces

Let us start with regular spaces $U = C^\infty_{\Gamma_-}(\Omega)$, the space of $C^\infty$-functions vanishing on $\Gamma_-$ and $V = C^\infty(\Omega)$. For given $v \in V$ the operator $A_\circ^*[v]$ is continuous, and hence in $U'$, if we choose

$$||u||_U := \sqrt{||u||^2_{L^2(\Omega)} + ||u|_{\Gamma_+}||^2_{L^2(\Gamma_+, |\vec{b}\vec{\nu}|)}} \tag{3}$$

as the norm on $U$. Via closure we obtain the ultraweak trial space $\mathcal{X} := \mathrm{clos}_{||\cdot||_U}(U)$ equipped with the continuous extension of the norm $||\cdot||_U$. For this trial space we have the following characterization (without proof).

**Proposition 1** *There exists a linear and continuous trace operator*

$$\gamma : \mathcal{X} \to L^2(\Gamma_+, |\vec{b}\vec{\nu}|)$$

*as well as a linear and continuous projection operator*

$$\mathrm{pr}_{L^2} : \mathcal{X} \to L^2(\Omega) \tag{4}$$

*fulfilling* $\gamma(u) = u|_{\Gamma_+}$ *and* $\mathrm{pr}_{L^2}(u) = u$ *for all* $u \in U \subset \mathcal{X}$.

**Proposition 2** *The trial space* $\mathcal{X}$ *is isometrically isomorphic to the Sobolev-space* $\mathcal{X}_{L^2}$ *defined as*

$$\mathcal{X}_{L^2} := L^2(\Omega) \times L^2(\Gamma_+, |\vec{b}\vec{\nu}|)$$

*equipped with the canonical norm* $||(u, \hat{u})||^2_{\mathcal{X}_{L^2}} := ||u||^2_{L^2(\Omega)} + ||\hat{u}||^2_{L^2(\Gamma_+, |\vec{b}\vec{\nu}|)}$. *The isometry* $\Phi : \mathcal{X} \to \mathcal{X}_{L^2}$ *is given as*

$$\Phi(u) := (pr_{L^2}(u), \gamma(u)).$$

**Corollary 1** *It holds that*

$$\begin{aligned} ||x||^2_{\mathcal{X}} = &\ ||\mathrm{pr}_{L^2}(x)||^2_{L^2(\Omega)} + ||\gamma(x)||^2_{L^2(\Gamma_+, |\vec{b}\vec{\nu}|)} = (x, x)_{\mathcal{X}} \\ \text{with } (x, x')_{\mathcal{X}} := &\ (\mathrm{pr}_{L^2}(x), \mathrm{pr}_{L^2}(x'))_{L^2(\Omega)} + (\gamma(x), \gamma(x'))_{L^2(\Gamma_+, |\vec{b}\vec{\nu}|)}. \end{aligned}$$

*In particular,* $\mathcal{X}$ *is a Hilbert space and there exists the Riesz-map* $R_{\mathcal{X}} : \mathcal{X}' \to \mathcal{X}$.

Hence, by continuous extension we can interpret $A_\circ^*[v]$ as an operator acting on $\mathcal{X}$, where in the following we will write $(u, \hat{u}) \in \mathcal{X}$ in the sense of Proposition 2, i.e.

$$A_\circ^*[v](u, \hat{u}) := (u, -b\nabla v + cv)_{L^2(\Omega)} + (\hat{u}, v|_{\Gamma_+})_{L^2(\Gamma_+, |\vec{b}\vec{\nu}|)}. \tag{5}$$

Provided that $A_\circ^*$ is injective on $V$ (which holds under mild assumptions on the data functions, see [2, Proposition 2.2]) we are now in the position to define a norm on $V$ as follows

$$||v||_V := ||A_\circ^*[v]||_{\mathcal{X}'} = ||R_{\mathcal{X}}(A_\circ^*[v])||_{\mathcal{X}}. \tag{6}$$

By a simple variational argument one sees that for a given $v \in V$ the Riesz-representative $r_v := R_{\mathcal{X}}(A_\circ^*[v]) \in \mathcal{X}$ fulfills $\Phi(r_v) = (-b\nabla v + cv, v|_{\Gamma_+})$. We thus obtain

$$||v||_V^2 = ||r_v||_{\mathcal{X}}^2 = ||\Phi(r_v)||_{\mathcal{X}_{L2}}^2 = ||-b\nabla v + cv||_{L^2(\Omega)}^2 + ||v|_{\Gamma_+}||_{L^2(\Gamma_+, |\vec{b}\vec{\nu}|)}^2.$$

By closure we finally obtain the test space $\mathcal{Y} := \mathrm{clos}_{||\cdot||_V}(V)$ of our ultraweak formulation equipped with the norm

$$||y||_{\mathcal{Y}} := ||A^*[y]||_{\mathcal{X}'} \tag{7}$$

where $A^* : \mathcal{Y} \to \mathcal{X}'$ is the continuous extension of $A_\circ^*$ to $\mathcal{Y}$.

The following proposition gives a characterization of the test space $\mathcal{Y}$:

**Proposition 3** *The test space $\mathcal{Y}$ is isomorphic to the Sobolev-space*

$$H^1(\vec{b}, \Omega) := \{v \in L^2(\Omega) \mid \vec{b}\nabla v \in L^2(\Omega)\}. \tag{8}$$

## 2.2 *Optimally Stable Ultraweak Formulation and Normal Equation*

With the definitions of the trial space $\mathcal{X}$, the test space $\mathcal{Y}$ and the adjoint operator $A^*$ we are now prepared to give an ultraweak variational formulation for reactive transport as follows.

**Definition 1** (*Ultraweak variational formulation of reactive transport*) $u \in \mathcal{X}$ is called a solution of the ultraweak variational formulation, if it satisfies

$$(u, A^*[v])_{\mathcal{X}\times\mathcal{X}'} = f(v) \qquad \forall v \in \mathcal{Y}. \tag{9}$$

with right-hand side $f(v) := (f_\circ, v)_{L^2(\Omega)} - (g_D, v)_{L^2(\Gamma_+, |\vec{b}\vec{\nu}|)}$.

**Proposition 4** *The mappings* $A : \mathcal{X} \to \mathcal{Y}'$ *and* $A^* : \mathcal{Y} \to \mathcal{X}'$ *are isometries, i.e.*

$$\mathcal{Y} = A^{-*}\mathcal{X}', \quad \mathcal{X} = A^{-1}\mathcal{Y}'$$

*and*

$$||A||_{\mathcal{L}(\mathcal{X},\mathcal{Y}')} = ||A^*||_{\mathcal{L}(\mathcal{Y},\mathcal{X}')} = ||A^{-1}||_{\mathcal{L}(\mathcal{Y}',\mathcal{X})} = ||A^{-*}||_{\mathcal{L}(\mathcal{X}',\mathcal{Y})} = 1.$$

***Proof*** The proof follows the argumentation in [4, Proposition 2.1]. □

As a consequence we obtain the following corollary, which shows optimal stability of the ultraweak variational formulation.

**Corollary 2** (Optimal stability) *The variational formulation* (9) *is well-posed and has optimal condition number* $\kappa_{\mathcal{X},\mathcal{Y}}(A) := ||A||\,||A^{-1}|| = 1$.

Since by Proposition 4 every $u \in \mathcal{X}$ has a representation $u = R_{\mathcal{X}} A^*[w]$ for some $w \in \mathcal{Y}$ we can substitute $u$ in the ultraweak formulation (9) to obtain the equivalent (continuous) normal equation.

**Definition 2** (*Normal equation of the ultraweak formulation*) $w \in \mathcal{Y}$ is called a solution of the normal equation of the ultraweak formulation, if it satisfies

$$(A^*[w], A^*[v])_{\mathcal{X}'} \;=\; f(v) \qquad \forall v \in \mathcal{Y} \tag{10}$$

or equivalently

$$(-b\nabla w + cw, -b\nabla v + cv)_{L^2(\Omega)} + (\gamma(w), \gamma(v))_{L^2(\Gamma_+,|\vec{b}\vec{\nu}|)} \;=\; f(v). \tag{11}$$

This is essentially a LL*-method (see e.g. [3]) applied to the minimization of the residual energy $||A[u] - f||_{\mathcal{Y}'}$.

**Remark 1** Let $w$ denote a solution of the normal Eq. (10). If $w$ is regular enough (e.g. $w \in C^2(\Omega)$), then $w$ solves the degenerated Poisson problem

$$\begin{cases} -\nabla \cdot (D\nabla w) & = f_\circ \quad \text{in } \Omega \\ -b\nabla w & = g_D \quad \text{on } \Gamma_- \\ (D\nabla w)\vec{\nu} + w & = 0 \quad \text{on } \Gamma_+ \end{cases} \tag{12}$$

with a rank-1 diffusion tensor $D := \vec{b} \otimes \vec{b}$.

The equivalent formulation of the normal equation will serve as the starting point for the definition of an optimal stable approximation method in the following section.

## 3 A Test-Space only Discretization

As indicated in the previous section, we propose to use the normal Eq. (10) to define an optimally stable approximation scheme. It is thus obvious, that a discretization can be fully based on a discrete approximate test space $\mathcal{Y}^\delta$.

### *3.1 Discrete Normal Equation and Functional Reconstruction*

Let $\mathcal{Y}^\delta \subseteq \mathcal{Y}$ be a conforming discretization of the optimal test space (i.e. using a standard Lagrange finite element space). Based on (10) we then define the *discrete* normal equation using Galerkin-projection.

$$\text{Find } w^\delta \in \mathcal{Y}^\delta : \quad (A^*[w^\delta], A^*[v^\delta])_{L^2(\Omega)} = f(v^\delta) \qquad \forall v^\delta \in \mathcal{Y}^\delta. \tag{13}$$

Note that this is still an optimally conditioned problem. Given the discrete solution $w^\delta$ we may reconstruct the discrete solution $u^\delta = A^*[w^\delta]$. Technically, this solution lies in the finite-dimensional subspace $\mathcal{X}^\delta := A^*[\mathcal{Y}^\delta] \subseteq \mathcal{X}$, however, due to its non-accessible structure this space is of no practical use.

Previous work often used knowledge of the structure of $A^*$ to determine a larger, more traditional (DG-)space $\mathcal{Z}^\delta \supsetneq \mathcal{X}^\delta$ and then assembled the matrix $\underline{A}$ representing the operator $A^* : \mathcal{X}^\delta \to \mathcal{Z}^\delta$ in the respective standard FE-bases. In this case, one can determine the system matrix $\underline{A}^{NE}$ of the normal equation as $\underline{A}^{NE} = \underline{A}^T \underline{M}_\mathcal{Z} \underline{A}$ (where $\underline{M}_\mathcal{Z}$ denotes the inner-product matrix in $\mathcal{Z}^\delta$), solve the linear system

$$\underline{A}^{NE} \underline{w} = \underline{f} \tag{14}$$

and compute the coefficients $\underline{u}$ of $u^\delta \in \mathcal{X}^\delta \subset \mathcal{Z}^\delta$ in the basis of $\mathcal{Z}^\delta$ by simply computing $\underline{u} = \underline{A}\, \underline{w}$.

However, this is suboptimal as the construction of a discrete larger space $\mathcal{Z}^\delta$ is only feasible or even possible with suitable additional assumptions on the data, e.g. (elementwise) constant data functions. For non-constant reaction or velocities one has to resort to a nonconforming choice $\mathcal{Z}^\delta \not\supset \mathcal{X}^\delta$ introducing an additional projection error which might be difficult to estimate or control.

Here, we propose an approach that avoids ever computing a matrix $\underline{A}$ representing the operator $A^*$. The system matrix of the normal equation $\underline{A}^{NE}$ can also be directly assembled in a basis of $\mathcal{Y}^\delta$ which means basically assembling a normal equation using the full infinite dimensional trial space $\mathcal{X}$. The reconstruction $u^\delta := A^*[w^\delta]$ is now seen as an element of $\mathcal{X}$ (we technically know that it lies in the finite dimensional subspace $\mathcal{X}^\delta \subset \mathcal{X}$ but this does not give us any useful information). The crucial insight is that in almost all applications only functional evaluations of $u^\delta$ are needed. Examples include point-evaluations for the visualization of $u^\delta$ or the computation of quantities of interest via numerical quadrature (i.e. $||u^\delta||$). Therefore, we replace the reconstruction by functional evaluations and e.g. do a pointwise reconstruction. Note that in this way we do not introduce any additional projection error.

### *3.2 Conditioning of the System Matrix and Solving the Linear System*

Solving the linear equation system (14) is actually quite a challenging task - a problem that has to our knowledge not been discussed so far. Although Problem (13) is optimally stable in theory, the condition of the system matrix $\underline{A}^{NE}$ still scales quadratically in the inverse grid width $h^{-1}$ and is thus a significant challenge even for moderately large problems. To better understand these seemingly conflicting statements consider the non-symmetric formulation of (13):

$$\text{Find } u^\delta \in \mathcal{X}^\delta : \quad (u^\delta, A^*[v^\delta]) = f(v^\delta) \qquad \forall v^\delta \in \mathcal{Y}^\delta. \tag{15}$$

Let $\{\psi_i\}_{i=1}^N$ be a basis of $\mathcal{Y}^\delta$ (e.g. a finite element basis). Then, the set $\{\varphi_i\}_{i=1}^N$, $\varphi_i := A^*[\psi_i]$ forms a basis of $\mathcal{X}^\delta$ and the matrix $\underline{A}$ representing $A^*$ in these bases is the identity matrix. The condition of the system matrix $\underline{A}^{NE}$ is still of order $\mathcal{O}(h^{-2})$ since the trial functions $\varphi_i$ have, contrary to classic finite elements, in this case a magnitude of $\mathcal{O}(h^{-1})$.

As mentioned in Remark 1, the normal equation can also be seen as the weak form of a specific Poisson-problem with rank-deficient diffusion tensor $D$. In the following numerical experiments we thus employed an algebraic multigrid for preconditioning and a conjugate gradient (CG) solver—methods that are known to perform well for this type of problems. For more complex problems (e.g. for velocity fields with (locally) small magnitude) the efficient preconditioning and solving of the linear equation system (14) still needs further investigation.

## 4 Numerical Experiments

The discretization scheme and the following experiments were implemented using the DUNE-framework [1] and the DUNE-PDELab discretization toolbox[2]. We consider the reactive transport of a pollutant inside a catalytic filter (Fig. 1). Let $\Omega$ be the unit square with boundary $\Gamma := \partial\Omega$. We assume that the velocity $\vec{b}$ is given as the solution to the Darcy-equation

$$\nabla \cdot \vec{b} = 0, \qquad \vec{b} = -k\nabla p$$

[2] https://www.dune-project.org/modules/dune-pdelab.

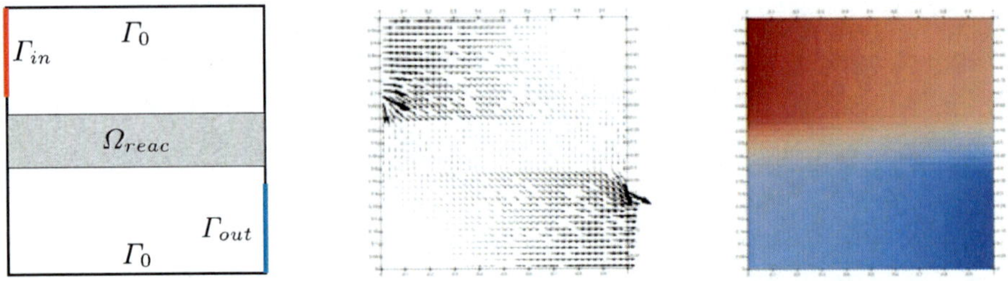


(a) Problem setup (b) Velocity field $\vec{b}$ (c) Pressure field $p$

**Fig. 1** Problem setup and Darcy velocity field $\vec{b}$

**Table 1** Chosen parameters for the catalytic filter problem

| $\Omega$ | $\Omega_{reac}$ | $\Gamma_{in}$ | $\Gamma_{out}$ | $g_D(z)$ | $f_\circ(x)$ | $k_{min}$ | $c_0$ |
|---|---|---|---|---|---|---|---|
| $[0,1]^2$ | $[0,1] \times [0.4, 0.6]$ | $\{0\} \times (\frac{2}{3}, 1)$ | $\{1\} \times (0, \frac{1}{3})$ | $\sin(3\pi z)^2$ | 0 | $10^{-1}$ | 0.5 |

subject to the boundary conditions

$$p = 1 \quad \text{on } \Gamma_{in}, \qquad p = 0 \quad \text{on } \Gamma_{out}, \qquad \vec{b} = 0 \quad \text{on } \Gamma_0 := \Gamma \setminus (\Gamma_{in} \cup \Gamma_{out}).$$

Here, $k \in L^\infty(\Omega)$ denotes the permeability field. We introduce the reactive domain (washcoat) $\Omega_{reac} \subset \Omega$ and assume that the permeability in the reactive domain is significantly smaller, i.e. $k = k_{min} \cdot \mathbb{1}_{\Omega_{reac}} + \mathbb{1}_{\Omega \setminus \Omega_{reac}}, k_{min} \in \mathbb{R}^+$. Similarly, the reaction function is given by $c = c_0 \cdot \mathbb{1}_{\Omega_{reac}}$, $c_0 \in \mathbb{R}^+$. All chosen parameters of the problem are summarized in Table 1.

In order to inspect the reconstruction $u^\delta$ we perform a voxel-wise evaluation on a refined mesh (i.e. a projection into $\mathbb{P}^0$). In Fig. 2 this is depicted alongside a solution obtained by a first order SIPG-method. In particular one notices nonaligned gradients when using a first order discretization of $\mathcal{Y}$ (Fig. 2b). However, since $u^\delta$ is only a $L^2$-like best approximation we do not expect $u^\delta$ to provide meaningful information w.r.t. the gradient (although the magnitude of the derivation still needs to converge in order to achieve $L^2$-convergence). Additionally, $u^\delta$ is the $\mathcal{X}_{L^2}$-best approximation from a non-standard space $\mathcal{X}^\delta$ where no information about its approximation qualities are given.

Finally, we investigate the convergence of the method under $h$-refinement. In [7] it was shown that the rates depend solely on the regularity of the test space solution. In [2] the dependence of the convergence rate on the regularity of the inflow condition (which in turn determines the regularity of the test space solution) has also been numerically evaluated for a linear transport problem. In our chosen catalytic filter problem we observe a convergence order of about 1.2 for linear and 2.3 for quadratic test functions (Fig. 3).

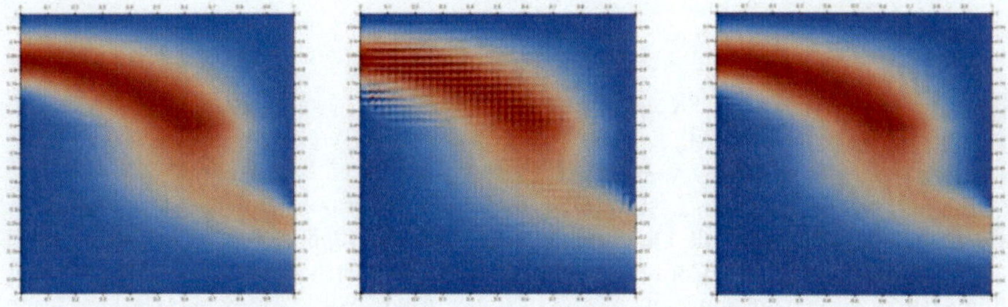

(a) Solution computed with a first-order SIPG-scheme. (b) Reconstructed solution for first order test functions. (c) Reconstructed solution for second order test functions

**Fig. 2** Solutions obtained by different discretization methods. All of them are based on a structured grid with $h^{-1} = 40$

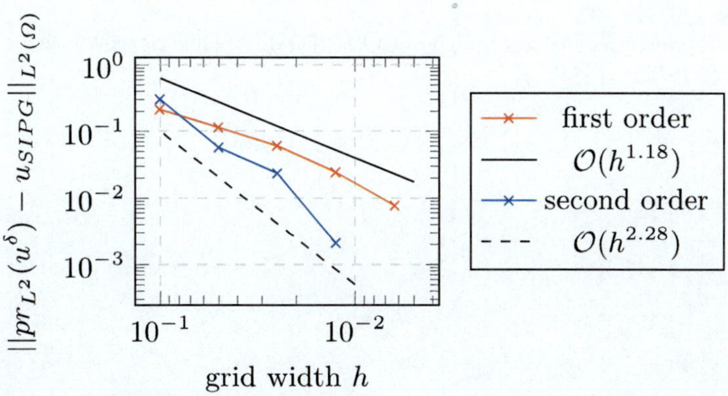


**Fig. 3** Convergence under $h$-refinement

# 5 Conclusion

In this contribution we derived an ultraweak, optimally stable formulation for reactive transport. In contrast to previous work we did not introduce additional boundary conditions on the test space but instead imposed them weakly by including boundary terms in the adjoint operator. We also showed that a standard FE-discretization of the test space is sufficient to solve the normal equations and perform classic functional evaluations of the reconstructed solution without discretizing the trial space explicitly. Future work will concern the robust solving of the normal equations, as well as using model order reduction techniques to efficiently solve parameter-dependent reactive transport problems.

**Acknowledgements** The authors acknowledge funding by the BMBF under contract 05M20PMA and by the Deutsche Forschungsgemeinschaft under Germany's Excellence Strategy EXC 2044 390685587, Mathematics Münster: Dynamics—Geometry—Structure.

## References

1. Bastian, P., Blatt, M., Dedner, A., Dreier, N.-A., Engwer, C., Fritz, R., Grüninger, C., Kempf, D., Klöfkorn, R., Ohlberger, M., Sander, O.: The DUNE framework: basic concepts and recent developments. Comput. Math. Appl. **81**, 75–112 (2021)
2. Brunken, J., Smetana, K., Urban, K.: (Parametrized) first order transport equations: realization of optimally stable Petrov-Galerkin methods. SIAM J. Sci. Comput. **41**(1), A592–A621 (2019)
3. Cai, Z., Manteuffel, T.A., McCormick, S.F., Ruge, J.: First-order system $\mathcal{L}\mathcal{L}^*$ (FOSLL*): scalar elliptic partial differential equations. SIAM J. Numer. Anal. **39**(4), 1418–1445 (2001)
4. Dahmen, W., Huang, C., Schwab, C., Welpers, G.: Adaptive Petrov-Galerkin methods for first order transport equations. SIAM J. Numer. Anal. **50**(5), 2420–2445 (2012)
5. Demkowicz, L., Gopalakrishnan, J.: A class of discontinuous Petrov–Galerkin methods. Part I: the transport equation. Comput. Methods Appl. Mech. Eng. **199**(23–24), 1558–1572 (2010)
6. Henning, J., Palitta, D., Simoncini, V., Urban, K.: An ultraweak space-time variational formulation for the wave equation: Analysis and efficient numerical solution. ESAIM: M2AN **56**(4), 1173–1198 (2022)
7. Keith, B.: A priori error analysis of high-order LL*(FOSLL*) finite element methods. Comput. Math. Appl. **103**, 12–18 (2021)

# An Hybrid Solver to Compute a Turbulent Compressible Model

**Sergey Gavrilyuk, Jean-Marc Hérard, Olivier Hurisse, and Ali Toufaili**

**Abstract** In this paper, we propose a numerical strategy to compute a turbulence model for compressible flows, including a dynamic estimation of the jump in turbulent kinetic energy across shock waves. The model is taken from [4] and the Finite Volume method applies to a specific Riemann solver interface. The entire procedure requires the detection of shock waves in unsteady flows.

**Keywords** Riemann solver · Compressible turbulence · Shock detection

## 1 Introduction

Several compressible turbulence models have been proposed in the literature during the last thirty years, most of them relying on Favre's formalism. One classical difficulty in that framework concerns the correct definition of shock waves, due to the possible occurrence on first-order non-conservative terms in the PDE formulation. Another one is linked with the correct prediction of the turbulent kinetic energy when shock patterns occur in the flow. For that purpose, a specific turbulence model has been proposed in 2006 [4], that aims at predicting the turbulent entropy production through the discontinuity in a dynamical way, still using the standard Euler set of equations grounded on mass balance, momentum and energy balances.

S. Gavrilyuk
IUSTI, Technopôle Château-Gombert, Marseille, France
e-mail: sergey.gavrilyuk@univ-amu.fr

J.-M. Hérard · O. Hurisse · A. Toufaili (✉)
EDF R & D, 6 Quai Watier, 78400 Chatou, France
e-mail: ali.toufaili@edf.fr

J.-M. Hérard
e-mail: jean-marc.herard@edf.fr

O. Hurisse
e-mail: olivier.hurisse@edf.fr

 
E. Franck et al. (eds.), *Finite Volumes for Complex Applications X—Volume 2, Hyperbolic and Related Problems*, Springer Proceedings in Mathematics & Statistics 433,
https://doi.org/10.1007/978-3-031-40860-1_31

Nonetheless, this model is indeed more complex than the simple one proposed in [2], and has never been implemented in CFD codes up to the knowledge of authors. We thus propose in this note a first numerical strategy based on Riemann solvers and the Finite Volume method, in order to handle approximate solutions of this turbulent compressible model. We will first briefly present the model and its main properties. Next we will turn to the numerical procedure. The latter requires defining a suitable shock detection, and it then makes use of a specific interface approximate Riemann solver when some shock wave occurs. Some numerical results will eventually be presented. More details can be found in [3].

## 2 Governing Equations

The system of PDEs governing the time-space evolution of the variables is:

$$\begin{cases} \partial_t(\rho) + \partial_x(\rho u) = 0 \\ \partial_t(\rho u) + \partial_x\left(\rho u^2 + P + \dfrac{2K}{3}\right) = 0 \\ \partial_t(\rho E) + \partial_x\left(u\left(\rho E + P + \dfrac{2K}{3}\right)\right) = 0 \\ \partial_t(\rho\xi) + \partial_x(\rho u\xi) = -M(x,t)\delta_{(x-\sigma t=0)} \end{cases} \tag{1}$$

The quantities $\rho$, u, P, K, E and $\xi$ respectively represent the mean density, the mean velocity, the mean pressure, the turbulent kinetic energy, the mean total energy and the turbulent entropy. We define the mean total energy, the modified pressure and the turbulent entropy, as follows:

$$\rho E = \rho e(P,\rho) + \frac{\rho u^2}{2} + K,\ \ P^* = P + \frac{2}{3}K \text{ and } \xi = K\rho^{-5/3}.$$

The turbulent speed of density waves $\tilde{c}$ is given by:

$$\tilde{c}^2 = c^2 + \frac{10K}{9\rho} = c^2 + \frac{10}{9}\xi\rho^{2/3},\ \text{where } c^2 = \left(\frac{\partial e}{\partial P|_\rho}\right)^{-1}\left\{\frac{P}{\rho^2} - \frac{\partial e}{\partial \rho|_P}\right\}.$$

The Dirac mass is $\delta_{(x-\sigma t=0)}$ located in $x = \sigma t$, $M(x,t)$ is the source term located at the shock position given by formula introduced in [4] and detailed in [3], and $\sigma$ is the velocity of the shock wave. The properties of system (1) in a general framework with respect to the **EOS** are as follows:

**Proposition 1** *The system is hyperbolic. It admits four real eigenvalues:*

$$\lambda_1(W) = u - \tilde{c},\qquad \lambda_{2,3}(W) = u,\qquad \lambda_4(W) = u + \tilde{c}. \tag{2}$$

**Proposition 2** *We introduce the entropy-entropy flux pair* $(\eta,\ f_\eta)$ *with:*

$$\eta(w) = -\rho s, \quad \textit{and} \quad f_\eta(w) = -\rho u s. \tag{3}$$

*Then the following inequality holds for smooth solutions:*

$$\partial_t \eta + \partial_x f_\eta \leq 0. \tag{4}$$

**Proposition 3** *The Rankine-Hugoniot relations associated with model* (1) *are:*

$$\begin{cases} -\sigma[\rho] + [\rho u] = 0, \\ -\sigma[\rho u] + \left[\rho u^2 + P + \dfrac{2K}{3}\right] = 0, \\ -\sigma[\rho E] + \left[u\left(\rho E + P + \dfrac{2K}{3}\right)\right] = 0 \\ -\sigma[\rho \xi] + [\rho u \xi] = -M \end{cases} \tag{5}$$

*By noting the jump:*

$$[f] = f_R - f_L,$$

*between the left and right states on each side of a discontinuity travelling at speed* $\sigma$.

## 3 Numerical Method

The method requires the definition of an interface hybrid solver to calculate approximate solutions for the problem. This hybrid solver is made up of two solvers: a classical approximate Riemann solver when no shock wave is detected, and a specific interface solver dedicated to the interface where a shock wave has been detected. We will refer to this as the Modified Approximate Riemann Solver "MARS". For this we use a classical Finite Volume formulation and we focus on an one-dimensional domain.

In the sequel, we define a practical technique to detect shock waves, then the modified solver "MARS" (the classical approximate Riemann solver is detailed in [1]) and finally the numerical cell scheme.

**a- Shock detector**: For an interface '$i + \frac{1}{2}$' separating the two cells $i$ and $i + 1$, we define the quantity $g_{i,i+1}$ as follows:

$$g^n_{i,i+1} = -\sigma_{i+1/2}[\eta^n]_i^{i+1} + [f^n_\eta]_i^{i+1}, \tag{6}$$

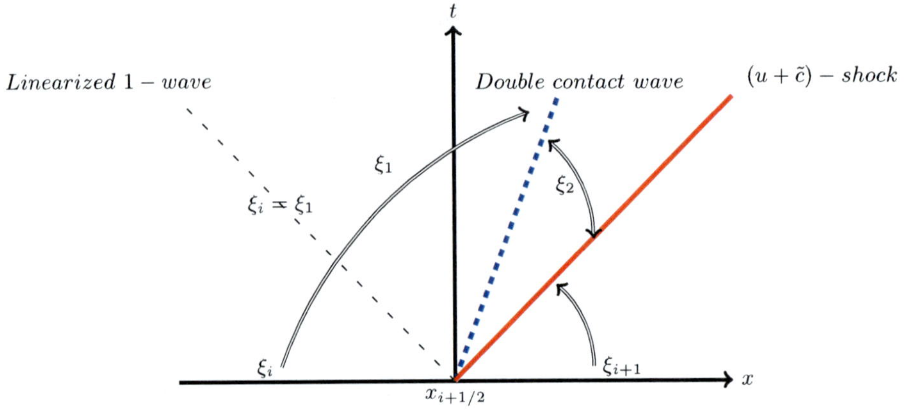


**Fig. 1** Approximate solution of the Riemann problem in terms of $\xi$, when a 4-shock occurs. Intermediate states are noted '1', '2'

where $(\eta, f_\eta)$ denotes the entropy-entropy flux pair, with $\eta = -\rho s$ and $f_\eta = u\eta$. The estimation of the shock velocity is made on the basis of the mass balance:

$$\sigma_{i+1/2} = [(\rho u)^n]_i^{i+1} / [\rho^n]_i^{i+1}. \tag{7}$$

For a discontinuity between cells 'i' and 'i+1' travelling at speed $\sigma_{i+\frac{1}{2}}$, we have the following inequality: $g^n_{i,i+1} < 0$. Though redundant, we also enforce the test: $[u^n]_i^{i+1} < 0$. Using the Lax inequalities (see [5]), we can deduce if it is a 1-shock or a 3-shock.

**b- Modified Approximate Riemann Solver when a 4-shock occurs**: The wave configuration retained for the hybrid solver at the interface is depicted in Fig. 1, when a 4-shock occurs.

We calculate the modified solver in three steps which are:

- use of the jump relation through 4-wave,
- exact connection throught the double contact wave $\lambda_{2,3} = u$,
- connection between state 'i' and '1' through linearized 1-wave.

**Step 1**: The jump relations associated with the 4-wave are as follows:

$$\begin{cases} \rho_{i+1}(u_{i+1} - \sigma) = \rho_2(u_2 - \sigma) = m \\ m^2[\tau]_2^{i+1} + [P^*]_2^{i+1} = 0 \\ [e + K\tau]_2^{i+1} + \bar{P}^*_{i+1,2}[\tau]_2^{i+1} = 0 \\ m[\xi]_2^{i+1} + M_0 = 0 \end{cases} \tag{8}$$

The shock speed $\sigma_{i+1/2}$ is known numerically on the interface thanks to the detection stage (7). Using relation (8a), we have $m = \rho_{i+1}(u_{i+1} - \sigma_{i+1/2})$, with $M_0$ given by [3] and (8d), we find $\xi_2 = \xi_{i+1} + \frac{M_0}{m}$. We solve (8b)–(8c) to get $\tau_2$, and thus $P_2^* = P_{i+1}^* + m^2[\tau]_2^{i+1}$ and $u_2 = \frac{m}{\rho_2} + \sigma_{i+1/2}$.

**Step 2**: Knowing that the Riemann invariants associated with the contact wave are $(u, P^*)$, we set: $u_1 = u_2$ and $P_1^* = P_2^*$.

**Step 3**: $u_1$ and $P_1^*$ are now known, thus using VFRoe-ncv we obtain: $\rho_1 = \rho_i + \frac{[P^*]_i^{\bar{\bar{1}}}}{2c_{i,i+1}^2} - \bar{\rho}\frac{[u]_i^{\bar{\bar{1}}}}{2c_{i,i+1}}$.

**Numerical cell scheme**: The numerical scheme reads:

$$\Delta x_i (Z_i^{n+1} - Z_i^n) + \Delta t(\mathcal{F}_{i+\frac{1}{2}}^n - \mathcal{F}_{i-\frac{1}{2}}^n) + \Delta t B_i^n = 0, \tag{9}$$

with $Z = (\rho, \rho u, \rho E, \rho\xi)$, $B_i^n = (0, 0, 0, M_i^n)$, and:

$$\mathcal{F}_{i+\frac{1}{2}}^n = \begin{cases} \mathcal{F}_{i+\frac{1}{2}}^{lnd} = F(Z(W_{lnd}^*(W_i^n, W_{i+1}^n)) & \text{when no shock wave has been detected,} \\ \mathcal{F}_{i+\frac{1}{2}}^{MARS} = F(Z(W_{MARS}^*(W_i^n, W_{i+1}^n)) & \text{if a shock wave has been detected,} \end{cases} \tag{10}$$

where:

$$F(Z) = (\rho u, \rho u^2 + P^*, u(\rho E + P^*), \rho u \xi). \tag{11}$$

The solution $W_{lnd}^*(W_L, W_R)$ of the approximate Riemann problem, is found with the linearized solver VFRoe-ncv (see [1, 2]) and the solution $W_{MARS}^*(W_L, W_R)$ of the approximate Riemann problem, is found with the modified interface solver 'MARS'. $M_i^n$ is a contribution in cell i to the global jump of turbulent entropy $M_0$. The calculation of $M_i$ is done as follows:

We define: $M_i^n = 0$, except:

$$\begin{cases} M_{i_0}^n = \frac{1}{2} M_0, \\ M_{i_0+1}^n = \frac{1}{2} M_0. \end{cases} \tag{12}$$

where $(i_0 + 1/2)$ is the interface where the shock wave has been detected at time $t^n$.

## 4 Numerical Results

We consider a single 4-shock wave. We assume that $M_0 \neq 0$ is given by the user using [4]. We consider a perfect gas EOS: $P = (\gamma - 1)\rho e$ where $\gamma = 7/5$. The exact solution is shown in Fig. 2. The initial condition of the Riemann problem is given by:

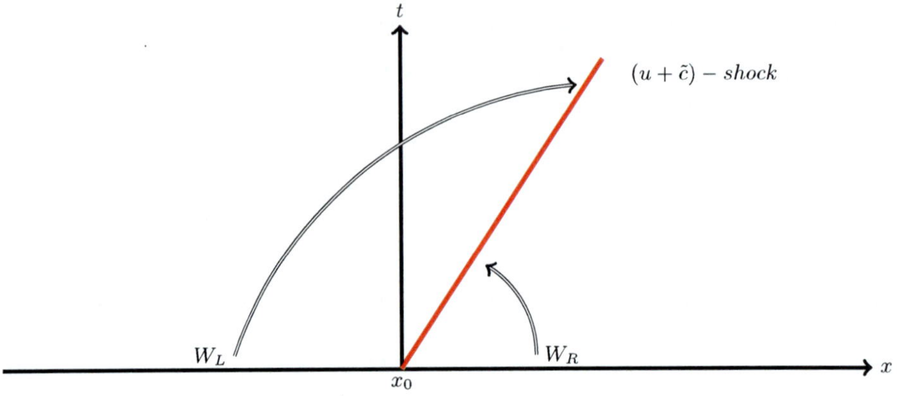


**Fig. 2** The exact solution of the Riemann problem for the 4-shock wave

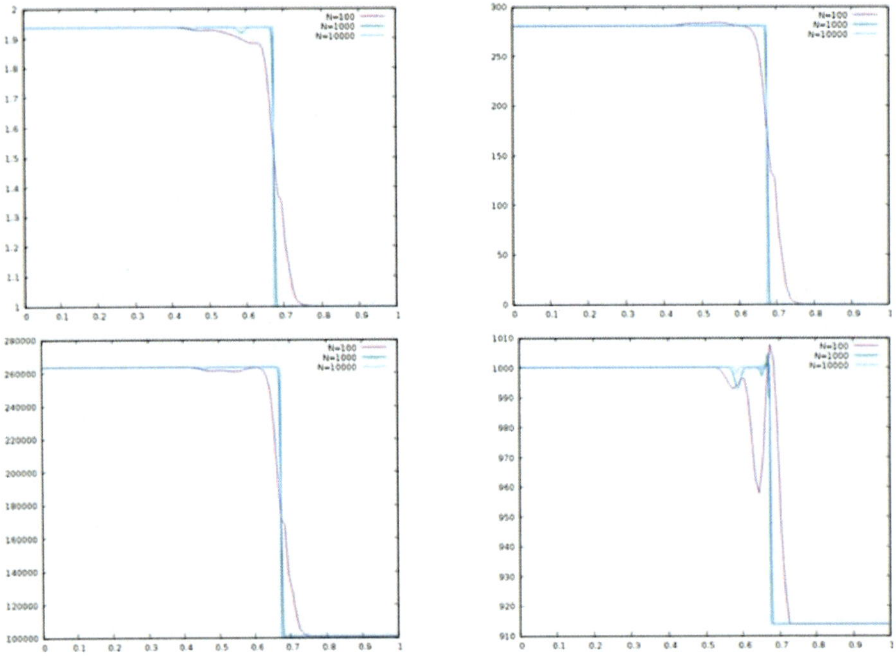


**Fig. 3** Density (top left), velocity (top right), $P^*$ (bottom left) and $\xi$ (bottom right). Profile of the approximate solution for different meshes =100 cells, 1000 cells, 10000 cells at $t = 3 \times 10^{-3}$ s, $CFL = 0.5$

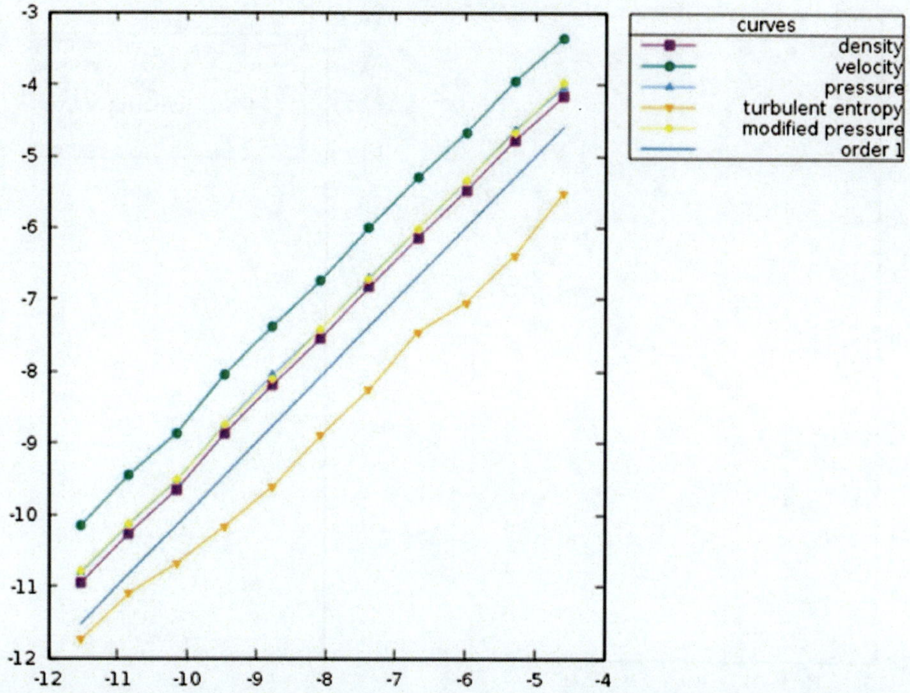


**Fig. 4** Simple 4-shock wave test case 1, with modified scheme. Convergence curves: logarithm of the relative L1-error versus the logarithm of the mesh size with uniform meshes containing from 200 to 200000 cells. The error is plotted for variables, $\rho$, $u$, P, $P^*$ and $\xi$

$$\sigma = 580, \text{ and } M_0 = -50000,$$

$$(\rho_L, u_L, P_L, \xi_L) = (1.936, 280.451, 261265.856, 1000)$$

$$(\rho_R, u_R, P_R, \xi_R) = (1, 650, 10^5, 913.793)$$

where $W_R$, $\sigma$, $M_0$ are given, and $W_L$ calculated with (8). Figure 3 shows the behavior of the density $\rho$, velocity u, modified pressure $P^*$ and turbulent entropy $\xi$ at a given time $T_f = 3 \times 10^{-3}$, on different meshes with 100 cells, 1000 cells and 10000 cells. Figure 4 shows the convergence curve, for the set of variables $\{\rho$, u, $P^*$,$\xi\}$. The error varies as $\approx h^1$ for all variables ($\rho$, u, $P^*$, $\xi$), hence the convergence of all the variables.

Other schemes such as those detailed in [3] may lead to some consistency error (see Fig. 5), in connection with the definition of the mass integration, which is different from (12).

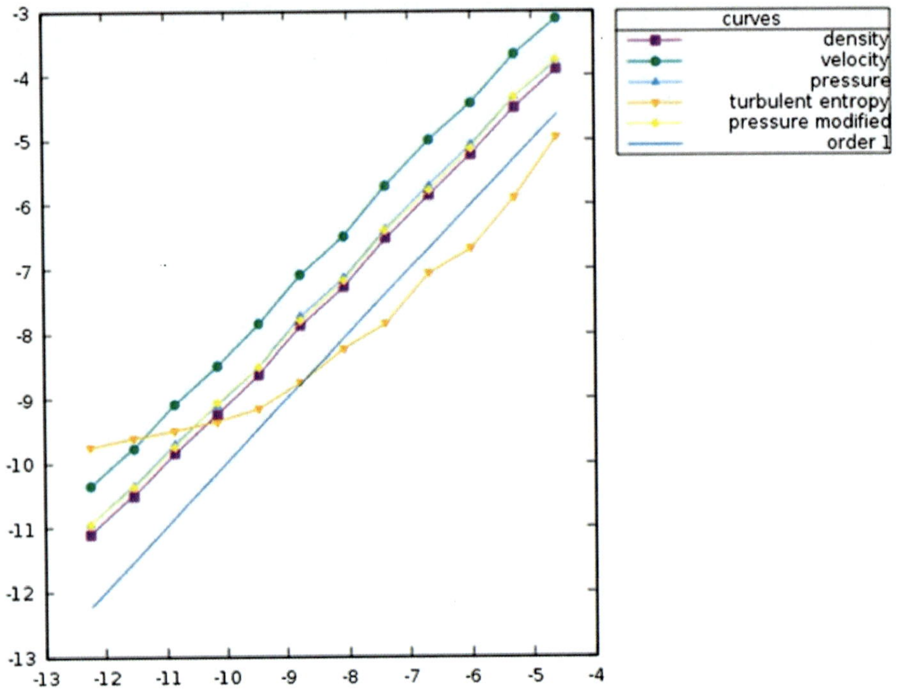


**Fig. 5** Simple 4-shock wave test case 1, with another scheme (see [3])

## References

1. Buffard, T., Gallouët, T., Hérard, J.M.: A sequel to a rough Godunov scheme: application to real gases. Comput. Fluids (2000)
2. Gavrilyuk, S., Hérard, J.M., Hurisse, O., Toufaili, A.: Theoretical and numerical analysis of a simple model derived from compressible turbulence. J. Math. Fluid Mech. (2022)
3. Gavrilyuk, S., Hérard, J.M., Hurisse, O., Toufaili, A.: An hybrid solver to compute a compressible model with dynamic estimation of the turbulent kinetic energy across shock waves. hal-03941246 (2023)
4. Gavrilyuk, S., Saurel, R.: Estimation of the turbulent energy production across a shock wave. J. Fluid Mech. (2006)
5. Smoller, J.: Shock Waves and Reaction—Diffusion Equations. Springer Science & Business Media (1983)

# Author Index





E. Franck et al. (eds.), *Finite Volumes for Complex Applications X—Volume 2, Hyperbolic and Related Problems*, Springer Proceedings in Mathematics & Statistics 433,
https://doi.org/10.1007/978-3-031-40860-1